Earth and Environmental Sciences Library

Earth and Environmental Sciences Library (EESL) is a multidisciplinary book series focusing on innovative approaches and solid reviews to strengthen the role of the Earth and Environmental Sciences communities, while also providing sound guidance for stakeholders, decision-makers, policymakers, international organizations, and NGOs.

Topics of interest include oceanography, the marine environment, atmospheric sciences, hydrology and soil sciences, geophysics and geology, agriculture, environmental pollution, remote sensing, climate change, water resources, and natural resources management. In pursuit of these topics, the Earth Sciences and Environmental Sciences communities are invited to share their knowledge and expertise in the form of edited books, monographs, and conference proceedings.

Shalini Yadav · Makarand M. Ghangrekar · Ram Narayan Yadava
Editors

Emerging Technologies in Biological and Hybrid Wastewater Treatment: Lessons from Developed to Enhancing Practices in Developing Countries

Springer

Editors
Shalini Yadav
Department of Civil Engineering
Rabindranath Tagore University
Raisen, Madhya Pradesh, India

Makarand M. Ghangrekar
Department of Civil Engineering
Indian Institute of Technology Kharagpur
Kharagpur, West Bengal, India

Ram Narayan Yadava
Research and International Affairs Division
Madhyanchal Professional University
Bhopal, Madhya Pradesh, India

ISSN 2730-6674 ISSN 2730-6682 (electronic)
Earth and Environmental Sciences Library
ISBN 978-3-031-74514-0 ISBN 978-3-031-74515-7 (eBook)
https://doi.org/10.1007/978-3-031-74515-7

This Springer imprint is published by the registered company Springer Nature Switzerland AG
The registered company address is: Gewerbestrasse 11, 6330 Cham, Switzerland

Contents

Introduction

An Overview of the Advances in Biological and Hybrid Wastewater Treatment Technologies in Developing Countries with Experience from Developed Countries

Manikanta M. Doki, Anil Dhanda, Shalini Yadav, Makarand M. Ghangrekar, and Ram Narayan Yadava

Abstract Refractory organic contaminants in significant amounts have recently entered wastewater treatment facilities from different sources. These contaminants are usually found in modest concentrations and have been dubbed "emerging contaminants" as they are recalcitrant to conventional biological treatments. These contaminants originate from different sources, including industrial chemicals, pharmaceuticals, personal care products, and agricultural runoff, and necessitate advanced treatments for their removal from the waste streams. Hence, hybrid treatment technologies work in conjunction with contemporary traditional treatment technologies that might serve the intended purpose of removing these contaminants of emerging concern. The present chapter dives deeply into understanding the various sources and types of emerging contaminants that enter wastewater treatment plants together with wastewater, and emphasis is given to the implications of these trace organic contaminants on the operation of conventional treatment systems. Special emphasis is given to the necessity of implementing alternative and sustainable hybrid technologies and present a brief overview of the latest state-of-the-art advancements in this field. This chapter addresses the current state of wastewater treatment, provides background information and an overview of the topics covered in this book. Furthermore, this chapter emphasises the necessity of a biological-hybrid treatment system for the efficient treatment of wastewater generated from various sources to preserve the water quality in the receiving water bodies and safeguard aquatic life.

M. M. Doki · A. Dhanda · M. M. Ghangrekar (✉)
Department of Civil Engineering, Indian Institute of Technology Kharagpur, Kharagpur, West Bengal 721302, India
e-mail: ghangrekar@civil.iitkgp.ac.in

S. Yadav
Rabindranath Tagore University, Bhopal, Madhya Pradesh, India

R. N. Yadava
Madhyanchal Professional University, Bhopal, Madhya Pradesh, India

S. Yadav et al. (eds.), *Emerging Technologies in Biological and Hybrid Wastewater Treatment: Lessons from Developed to Enhancing Practices in Developing Countries*, Earth and Environmental Sciences Library,
https://doi.org/10.1007/978-3-031-74515-7_1

Keywords Conventional techniques · Emerging contaminants · Hybrid wastewater treatment systems · State-of-the-art technologies

1 Introduction

Wastewater treatment is currently a topic of great concern in society because of its direct impact on the environment and human health. With the ongoing urbanisation and industrialisation, there has been a substantial rise in the volume of wastewater generated, which presents a formidable obstacle for conventional wastewater treatment facilities. On the worldwide level, approximately 380,000 million cubic meters of wastewater is generated each year. However, only about 24% of the wastewater produced from the industrial and residential areas is treated before being discharged to the rivers or used for agricultural activities [14, 23]. The primary pollutants in untreated wastewater that cause substantial ecological toxicity problems include heavy metals (such as Cd, Cr, Cu, Ni, Zn, Fe, Pb, Hg, As, and Mn), pathogenic microbes, and emerging contaminants. These contaminants are of considerable concern because of the possible health risks they bring to consumers when introduced into the food chain.

Therefore, developing and scaling techniques that can efficiently mitigate these notorious pollutants economically is the need of the hour. This chapter attempts to rekindle the existing know-how on contemporary biological wastewater treatment technologies and different hybridisation schemes that have been adopted to augment the operational efficiency of these systems. The fundamentals of the well-established biological processes have been briefly discussed, and their mechanism and operational limitations have been underlined. Moving forward, the hybridisation strategies for countering the limitations of traditional approaches have been illustrated, emphasising the most successful coupled systems that have performed exceedingly well in practical settings.

This chapter also explains various classes of hybrid systems with the aim of enabling the identification of the combined technologies based on a particular wastewater composition and sought-after levels of treatment. Essentially, this chapter identifies the drawbacks of conventional technologies for wastewater treatment and calls for adoption of hybrid treatment techniques to handle the problems caused by complex pollutants. Eventually, readers will be enlightened as to the hybrid biological wastewater treatment that is at the forefront and its important function in preserving the fragile aquatic ecosystem from harmful man-made pollution.

2 Sources of Wastewater

The wastewater can either result from natural or anthropogenic activities. The natural phenomenon includes acid deposits in the atmosphere, weathering of rocks, and volcanic eruptions that pollute the indigenous water sources present. However, anthropogenic activities such as household and agricultural activities contaminated with pernicious substances contribute to wastewater generation [23]. Further, the wastewater can be categorised into industrial, domestic, and stormwater runoff [4]. Industrial wastewater is generated from industries like textile, electroplating, pulp and paper, mining, food, dairy, dyeing, etc. The composition of the wastewater generated from a specific sector depends on the particular operation related to the industry. For instance, the paper and pulp industry produces wastewater with high chemical oxygen demand (COD), suspended solids (SS), and colour. If discharged untreated, this wastewater can cause mutagenic impacts and physiological impairment in the aquatic organisms [1]. Also, the nutrients in effluent can promote aquatic vegetation growth, resulting in problems such as eutrophication [8].

The textile industry, which processes manufactured and natural threads into fabrics, also significantly contributes to wastewater generation. Moist chemical processes are required during manufacturing to properly purify, prepare, colour, and finish the majority of textiles. This results in formation of effluent, the contamination burden of which is caused not only by removing impurities from the primary materials but also by the remaining chemical reagents used in processing. The main constituents found in textile wastewaters are elevated SS, COD, temperature, colour, acidity, and several other soluble components [26].

Toxic contaminants stemming from industrial discharges have resulted in the contamination of extensive agricultural land and aquatic environments. Globally, various wastewater treatment methods, such as the activated sludge process (ASP), percolating filters, fluidised aerobic bio-reactor, membrane separation, etc., are currently in use. However, at larger scales, the requirement of high capital costs and operation and maintenance costs make the application of these technologies infeasible. The industrial sector generates nearly 13,500 million litres per day (MLD) of wastewater in India, of which only 60% of the wastewater is treated partially or totally, and the remainder is discharged as waste [23]. Hence, novel hybrid technologies are required to treat the wastewater to reach effluent discharge standards.

Another primary source of wastewater is the wastewater produced from domestic activities like kitchens, bathing, toiletries, and washing. Around 80% of the total water supplied (using natural water bodies or pumped from groundwater) for household activities is converted into wastewater [12]. The wastewater thus generated contains contaminants like salts, inorganic and organic matter in suspended and soluble form, and microbial pathogens like fungi, helminths, viruses, parasites, and bacteria that can cause severe issues. Thus, a dual challenge of depleting potable water sources and increased waste generation arises if wastewater is not treated to discharge standards. According to the Central Pollution Control Board, domestic sewage generated in India from urban areas is about 72,368 MLD. In contrast, the installed operational treatment capacity is only 37% of the wastewater generated, that is 26,869 MLD, leaving 63% of the wastewater generated without proper treatment and thus discharged [10]. Therefore, increasing the reuse of treated wastewater, treatment capacity, and recycling efficiency is critical for successful water resource management.

3 Conventional Wastewater Treatment Technologies

The traditional wastewater treatment technologies are categorised based on how they collect wastewater, i.e., centralised or decentralised [20]. These facilities treat wastewater in three stages: primary, secondary, and tertiary or advanced treatment. The primary treatment eliminates larger solids like paper, wood, plastics, grit, and debris. The larger solids can disrupt subsequent stages of the treatment process, hence, they are removed first. The treatment methods employed for this purpose include screens, grit chambers, comminution, equalisation and sedimentation. The secondary treatment utilises indigenous microorganisms in an engineered environment for the biological degradation of organic matter and, to some extent, nutrients. The tertiary treatment stage removes excess nutrients and emerging pollutants.

The pollutants present in wastewater can be categorised based on their physical features (soluble or insoluble) and chemical composition (organic or inorganic). Further, they can also be classified based on their vulnerability to microbial degradation, i.e., biodegradable or non-biodegradable, source of inception (natural or anthropogenic), and toxicity [18]. The specific degradation methods required for the removal of these contaminants can be classified into three categories, i.e., physical, chemical, and biological. The physical treatment methods use screening, sedimentation, flocculation, skimming, and flotation to remove the large-sized particles and suspended solids. In contrast, chemical treatment requires contacting the added chemicals with contaminants that results in processes of precipitation, coagulation, chemical oxidation, neutralisation, stabilisation, and disinfection. The physical processes combined with chemical processes are known as physico-chemical processes. Biological processes are facilitated by naturally occurring microorganisms, which metabolise the contaminants to fulfil their energy needs and support their growth. These processes can be further categorised as oxic, anoxic, or anaerobic [13].

These biological treatment methods have shown promising results for wastewater treatment and nutrient removal. These treatment approaches have been popularly utilised in wastewater treatment facilities. A detailed discussion of these treatment technologies is provided in the subsequent chapters. However, below is a short discussion on their working, limitations, and the need to hybridise them with other technology.

4 Existing Biological Wastewater Treatment Technologies

4.1 Ponds

The ponds offer nature-based solutions for treating wastewater. They can be classified as aeration ponds, facultative ponds, and anaerobic ponds. Tropical and temperate climate regions are ideal for these ponds as the reaction kinetics inside the ponds majorly depend on the sunlight and temperature [30]. Additionally, these ponds require periodic maintenance, including preventing inlet and outlet blockage, grit removal, deterring mosquitoes, clearing macrophytes and periodic settled sludge removal. The types of ponds are discussed below.

4.1.1 Oxidation Ponds

The oxidation ponds are aerobic ponds with a depth of around 1.0–1.15 m. These ponds are designed to maintain dissolved oxygen throughout their depth. The wastewater treatment occurs due to symbiotic relation among bacteria and algae utilising sunlight. The process of oxidation ponds involves algae growth, photosynthesis, bacterial activity, sludge accumulation, and algae removal. There is also a subclass of oxidation ponds known as maturation ponds. These ponds are used at the last and their main purpose is to remove pathogens, nitrogen, and phosphorus. The depth plays a pivotal role in the ponds and it is kept at less than 0.5 m for effective passage of sunlight, which influences pathogen removal and temperature inside the pond [34].

4.1.2 Anaerobic Ponds

Anaerobic ponds are normally constructed 2–5 m deep. These ponds focus on solids sedimentation and anaerobic digestion and are highly influenced by temperature and pH. The influent to these ponds should be neutralised and sulphate concentrations should be regulated to manage odour problems [34], and maintaining pH (7.0–7.5) during the acclimatisation phase is crucial for developing methanogenic archaeal populations in the pond [13].

4.1.3 Facultative Ponds

These ponds can be primary or secondary. The depth of these ponds is between 1.5 and 3 m, and they can hold water for longer than maturation ponds. The algae plays a vital role in these ponds by providing oxygen that bacteria need to grow and lowering biological oxygen demand.

The abundance and change in algal population serve as critical indicators for the pond operation and are greatly influenced by the intensity of light, pH, organic loading rate (OLR) and temperature of the influent. The syntrophic relation of aerobic microbes and algae also plays a pivotal role in the aerobic zone for the stabilisation of organic matter. Part of the organic matter present in the wastewater to be treated in this facultative pond undergoes aerobic oxidation. The remaining is removed by facultative microbes present at the intermediate layer, between the aerobic and anaerobic zones, and by anaerobic degradation of organic solids and microbial cells settled at the bottom of the pond.

4.2 Activated Sludge Process

The ASP is a well-established technology for treating municipal and industrial wastewater. The ASP is an aerobic process (aerated mechanically) where microbial biomass is cultivated and stabilises organic matter. This process demonstrates exceptional performance in removing organic matter and withstanding shock loads, even in changing climates. The process consists of two components, i.e., the aeration tank and the settling tank. The aeration tank supplies oxygen to support microbial acclimatisation. The settling tank is provided for the separation of solids and liquid fractions. Further, the separated solids (microbes) are recycled in the aeration tank to maintain the desired microbial population [13]. Flocculation of these microbes assists in enhancing their density and capture in the clarifier, thereby enriching the microbial populations.

4.3 Up-Flow Anaerobic Sludge Blanket Reactor

Up-flow anaerobic sludge blanket (UASB) reactor is an anaerobic process and utilises a cylindrical or rectangular reactor provided with a gas-liquid-solid separator. The treatment process involves upward movement of influent through a thick sludge blanket provided at the bottom. The higher microbial population in the sludge blanket allows higher COD loading rates and dense granular sludge formation (1–3 mm granules with sludge volume index < 20 mL/g) [13]. Due to these granulated sludge flocs, the UASB reactor has varying solids concentrations (40–100 g/L at the bottom and 5–40 g/L at the top). Biological reactions occur at the bottom of the active anaerobic sludge bed, where soluble organic matter is converted to methane and carbon dioxide. Gas bubbles are trapped, separated, and prevented from allowing the washing out of essential microorganisms [16]. The UASB reactor requires minimal investment, forms dense granular sludge without support, and offers higher biomass retention, good settling, and tolerance to organic loading and toxic compounds. It operates with less energy and minimal sludge production.

The UASB reactor poses significant advantages over the other technologies, such as no need for supportive media due to granular sludge formation, ability to treat high-strength wastewater with higher COD removal rates, low hydraulic retention time

(HRT) and higher solid retention time (SRT). The reactor allows automatic mixing inside the reactor without external energy due to the formation of gas bubbles. Further, it also allows effective treatment even at high OLRs and low temperatures. Moreover, simplicity in construction, maintenance, and operation with low capital and energy costs reduces the overall cost of the technology. Beside the sludge produced in UASB reactor is stabilised and in minimal quantities, which can directly utilised as a manure.

4.4 Constructed Wetlands

Constructed wetlands (CWs) are an engineered version of conventional wetlands with similar functioning, better operational control, and higher performance [28]. The CWs are resilient and adaptable systems that tolerate different loading rates without sacrificing treatment efficacy. Additionally, these facilities are characterised by affordability, requiring minimum installation, operation, and maintenance expenses. Moreover, the CWs use simple treatment methods that may be efficiently managed and maintained by those with minimum expertise or training. The CWs can be categorised based on plant presence and direction of wastewater flow. Plants in the treatment bed distinguish CWs into floating macrophytes, submerged macrophytes, or rooted emergent systems. They can be divided into free water surface (FWS) and subsurface flow (SSF) systems. The SSF-CW can be horizontal or vertical. The FWS maintains a water column above the bed, exposing it to ultraviolet (UV) sterilisation. In contrast, SSF-CW has a hydraulic line below the bed, enhancing wastewater-microbe interaction and treatment efficiency.

The CWs use a variety of natural processes to remove pollutants. These include physical processes like sedimentation and filtration [32] and biological processes like microbial metabolism [31]. The prominent mechanism takes place through microbial degradation for pollutant removal in CWs. The CW bed is a niche for diverse microorganisms and ensures removal of a wide range of pollutants. The presence of vegetation creates a redox gradient by forming aerobic and anaerobic zones inside the bed and favours different microbial processes [9]. This can also be further supported by physicochemical processes like phosphate fixation through plant uptake, which contributes to pollutant removal in CWs. Broadly, CWs are versatile efficient systems for municipal wastewater treatment in terms of removing wide-range of pollutants.

4.5 Sequencing Batch Reactor

The aerobic sequencing batch reactors (SBRs) operates similar to ASP, utilising fill-and-draw cycles in either single or multiple tanks. These SBRs can perform simultaneous equalisation, aeration and clarification within a single batch reactor. The process consists of five phases which includes fill, react, settle, decant and idle as illustrated in Fig. 1 [11]. The fill phase introduces raw wastewater into the reactor.

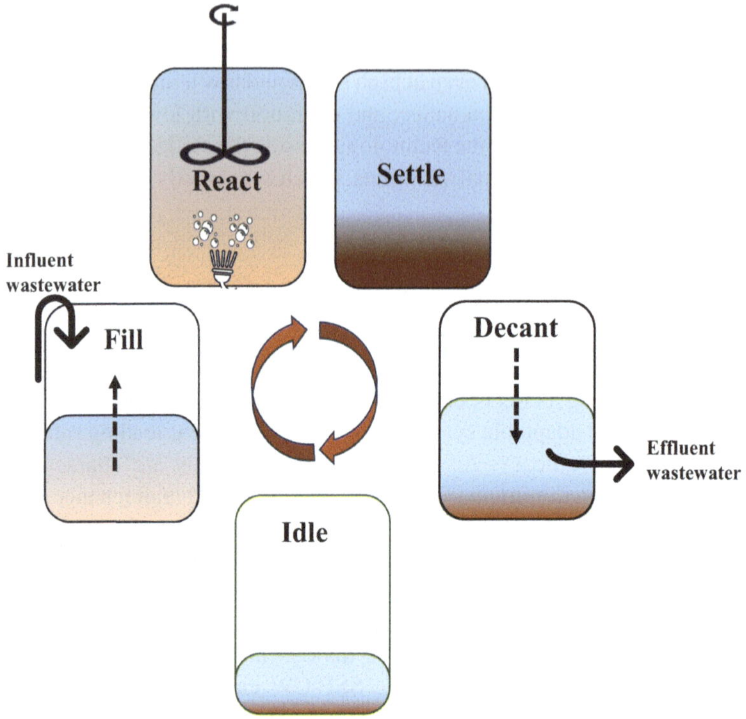

Fig. 1 Operation cycles of sequencing batch reactor

Depending on the influent characteristics and desired target removal of contaminants, the wastewater can be filled with different variations, such as static, mixed, or aerated fill. The react phase focuses on biological reactions and nutrient removal, and it is controlled by air supply to achieve aerobic, anoxic, or anaerobic conditions. In the settle phase, the reactor performs the operation of a clarifier, whereas, during the decant phase, it utilises a decanter to collect the treated supernatant. Sparsely populated areas usually employ single-tank operation; however, multiple tanks offer high-yield efficiency [17]. The SBRs combine aeration, mixing, reaction, and settling processes, eliminating the requirement for a separate clarifier. Besides, the SBRs provide operational versatility, internal equalisation, and ability to control the kinetics of microbes by regulating aeration. Specifically, they are highly efficient in removing nitrogen during the anoxic stage of the react phase. Similar to ASP, the SBR is also prone to sludge bulking due to operational adversities. To counter this, SBR is designed with a bio-selector unit, which can promote floc-forming bacteria over filamentous bacteria by varying food-to-microorganism ratios and supplying low oxygen.

5 Advanced Biological Technologies

These technologies are modified versions of conventional biological technologies and combine the benefits of suspended or attached growth processes.

5.1 Moving Bed Biofilm Reactor

Moving bed biofilm reactors (MBBRs) are aerobic wastewater treatment technology that uses suspended biofilm carriers to remove pollutants from wastewater. This treatment technology alone can provide the advantages of ASP and biofilm processes. Aeration or mechanical mixing is usually employed to keep the biofilm carriers in constant motion (suspended form), which allows contact between the microorganisms and pollutants in the wastewater for enhanced removal efficiencies. The MBBR has several advantages over other biofilm-based reactors, including uniform biofilm distribution, less volume, and resilience to hydraulic and mechanical failures. First developed in the late 1980s, they became a widely accepted wastewater treatment technology [5]. Several factors, including mechanical mixing energy, airflow, and carrier surface properties, govern the thickness of biofilm on the carriers. The commercially available carriers have a surface area that ranges from 400 to 1200 m^2/m^3 [5]. The versatility of MBBRs makes them effective in eliminating majorly available pollutants such as organic matter, nutrients and pathogens. The operational flexibility and ease in maintenance makes MBBRs, a probable solution for the treatment of municipal and industrial wastewaters.

5.2 Membrane Bioreactor

As shown in Fig. 2, membrane bioreactor (MBR) technology implements a two-stage process for the elimination of contaminants present in wastewater. This can be made possible through the commingle action of microbial degradation and low-pressure membrane filtration, either by microfiltration or ultrafiltration techniques, as indicated by Hansen et al. [15]. High-quality treated water from wastewater is mainly achieved by the polymeric membrane materials, usually, of polyethylene and polyacrylonitrile, with the latter having better fouling resistance [37]. The MBRs can either be provided submerged or as external, with submerged MBR being preferred for its higher permeability and minimal fouling, along with lower energy expenditure. The activated biomass transforms organic matter into flocs, which are subsequently segregated utilising membranes. The microbial biofilms on the membranes enables the systems to produce superior effluent quality by facilitating a sieving effect and enhancing the removal of contaminants.

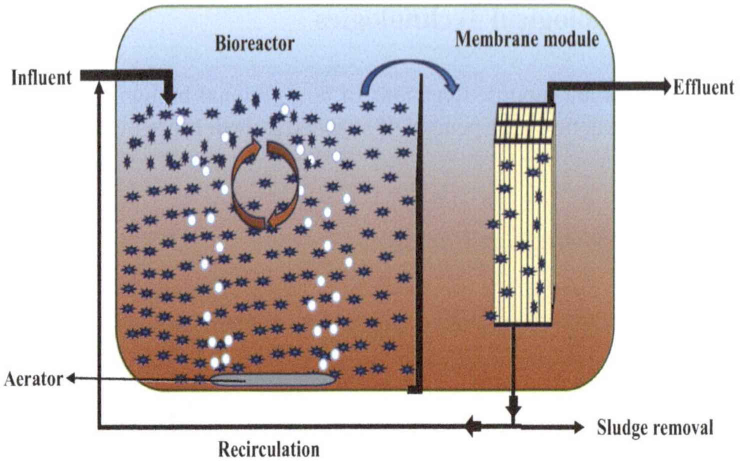

Fig. 2 Schematics of membrane bioreactor

Moreover, different mechanisms, such as physical retention by the membrane, sorption, biodegradation, and air stripping, contribute to contaminant removal in MBR [21]. The biodegradation assists in the removal of hydrophilic compounds, whereas biosorption onto activated sludge assists in handling hydrophobic compounds. Retention on membranes increases the biomass concentration, which improves nutrient removal through nitrification and denitrification. Wastewater with low alkalinity can affect the performance of nitrification and denitrification. Hence, addition of sodium bicarbonate and external carbon sources enhances the performance of nitrification and denitrification [29].

6 Drawbacks of Existing Wastewater Treatment Systems

Conventional wastewater treatment processes can effectively remove the organic matter. However, they are not designed to remove emerging contaminants. These contaminants are ubiquitously available in the wastewaters in trace concentrations (ng/L to μg/L) and are refractory, which poses difficulty in effectively degrading by employing conventional treatments. Due to the complexity of emerging contaminants, they require longer retention times, higher chemical doses, larger areas, and high energy requirements during the operation of conventional treatment methods [24]. Besides, the entry of emerging contaminants through wastewater treatment plants into the environment can be fatal to the natural flora and fauna and also lead to endocrine disruption, development of resistant genes in microbial cultures and toxicity to the reproductive system [19]. Emerging contaminants are classified into personal care products, pharmaceutical organic compounds, pesticides, plasticisers, industrial chemicals, surfactants, endocrine-disrupting compounds and flame retardants [2, 25].

Carballa and co-researchers investigated the fate of hormones (17β-estradiol, estrone, 17α-ethinylestradiol), cosmetics (tonalide, galaxolide) and pharmaceuticals (iopromide, roxithromycin, diclofenac, diazepam, carbamazepine, naproxen, sulfamethoxazole, ibuprofen) at all stages of sewage treatment plant, which includes conventional treatment scheme consisting of grit chamber, primary settling tank (PST) and ASP [7]. The inlet wastewater contained tonalide (0.9–1.7 μg/L), galaxolide (2.1–3.4 μg/L), naproxen (1.8–4.6 μg/L), ibuprofen (2.6–5.7 μg/L), estrone (2.4 ng/L), sulfamethoxazole (approximately 0.6 μg/L), iopromide (6–7 μg/L), and the remaining compounds are less than the limit of quantification (LOQ).

Due to the high solid–liquid partition coefficients for tonalide and galaxolide, a 47% removal by adsorbing onto the solids during primary treatment was observed. However, there was no significant reduction in ibuprofen, naproxen, sulfamethoxazole, estrone, and iopromide due to their low solid-liquid partition coefficients after the primary treatment. After the ASP, 45–50% for tonalide, 30–40% for galaxolide, 40–55% for naproxen, 60–70% for ibuprofen, 67% for sulfamethoxazole, and 47% for 17β-estradiol removal efficiencies were observed. However, iopromide was not removed significantly even after secondary treatment due to its high stability. Moreover, negative removal efficiencies were observed after primary treatment and secondary treatment as the 17β-estradiol readily oxidises into the estrone, further showing an increase in estrone concentration after primary and secondary treatments. Thus, it can be noted that standalone treatment technologies are insufficient to remove these contaminants. Therefore, hybrid technologies should be used.

7 Necessity of Hybrid Wastewater Treatment Systems

As discussed in the previous section, the performance of standalone existing wastewater treatment systems for removing trace organic contaminants is insufficient. In this regard, hybrid technologies are necessary to make this wastewater reusable. Hybrid systems play an important role in treating trace organic contaminants to avoid their potential impact on health and the environment. The hybrid system involves implementing two or more technologies to enhance the system's overall performance. The traits of an integrated approach allow them to overcome the drawbacks of the individual units during the treatment of emerging contaminants.

Further, this enables them to treat wide range of pollutants and versatile to modifications in influent wastewater characteristics and operational parameters. These hybrid technologies can be viable alternative low-cost and sustainable technologies to standalone conventional treatment technologies in treating emerging contaminants. In general, hybrid systems are usually available with physical and chemical processes in conjunction with biological systems. However, the selection of the hybrid system depends mainly on the flow rate, influent and effluent wastewater characteristics [3,

22]. These hybrid systems play a crucial role in bringing the effluent wastewater quality below the permissible limits for reuse, such as for irrigation, horticulture, aquifer recharge, and toilet flushing [27, 35].

Tran and co-researchers conducted an extensive study on the presence and removal of 25 targeted emerging contaminants, encompassing various compound classes, in a full-scale water reclamation plant [33], where they found 19 trace organic contaminants, including beta-blockers, anti-itching agents, lipid regulators, nonsteroidal anti-inflammatory drugs, neuro regulating drugs, plasticiser, repellent, stimulant, X-ray contrast media and UV-filters, in the concentrations of atenolol—4.12 μg/L, crotamiton—49.85 ng/L, gemfibrozil—337.2 ng/L, naproxen—6.247 μg/L, indomethacin—359.1 ng/L, ibuprofen—33.5 μg/L, salicylic acid—3.03 μg/L, diclofenac—378.2 ng/L, acetaminophen—92.5 μg/L, fenoprofen—1.2 μg/L, gabapentin—13.15 μg/L, carbamazepine—323.8 ng/L, sulpiride—375.7 ng/L, bisphenol A—5.7 μg/L, N, N-diethyl-m-toluamide—2.3 μg/L, caffeine—37.182 μg/L, iohexol—115.436 μg/L, iopamidol—34.89 μg/L, and oxybenzone—1.734 μg/L. Two distinct treatment trains, Train A and Train B, were employed, where Train A comprised of PST and an ASP (Modified Ludzack–Ettinger (MLE), followed by a secondary settling tank (SST)) and Train B implemented a MBR system (PST, MLE tanks followed by an MF membrane unit with a pore size < 0.45 μm). The mixed liquor suspended solids (MLSS), HRT, and SRT for Train A were from 1428–1994 mg/L, 7.8–8.9 h, and 4.9–6.7 days, and for Train B these values were 6403–15,350 mg/L, 6.1–7.7 h, and 14–28 days, respectively.

The comparative analysis demonstrated that the MBR system outperformed the ASP for compounds like bisphenol A, crotamiton, diclofenac, gabapentin, gemfibrozil, iohexol, naproxen, oxybenzone, and salicylic acid, exhibiting removal efficiencies of 89.7%, 14.4%, 19%, 78.8%, 88.3%, 71.9%, 67.3%, 97.2%, and 75.3%, respectively. Conversely, their removal efficiencies for ASP were 86.2%, 4.3%, 3.6%, 58.9%, 76%, 40.4%, 65.3%, 95.3%, and 68.9%, respectively. However, compounds including caffeine (100%), fenoprofen (> 99%), acetaminophen (98%), ibuprofen (> 97%), iopromide (38.5%), and carbamazepine (4.8%) demonstrated similar performances in both the processes. Moreover, certain compounds exhibited negative removal efficiencies, probably due to cleavage of the conjugated forms, which led to the re-emergence of the parent compounds through microbial action. This signifies the limitations of standalone treatment systems for completely removing these compounds; hence, it emphasises the need for hybrid technologies to achieve enhanced removal across a broader spectrum of contaminants with consistent performance.

8 Classification of Hybrid Wastewater Treatment Systems

The conventional technologies as standalone or combined with another technology (anaerobic/aerobic) are inefficient in removing the emerging contaminants. For example, in one investigation, the prevalence of pharmaceuticals in the discharged effluent and on the sludge of a sewage treatment plant in Brazil was observed [6]. The treatment scheme consisted of screens, a grit chamber, and secondary treatment using a UASB reactor (HRT—5 h) followed by ASP (HRT—7 h) with final disinfection and sludge disposal management. A total of 13 pharmaceuticals were detected in influent wastewaters and on sludge, which include caffeine (32.19 ± 5.46 μg/L, 0.635 ± 0.858 mg/kg), enrofloxacin (0.037 μg/L, non-detected (n.d.)) ciprofloxacin (0.385 ± 0.534 μg/L, 2.217 ± 2.451 mg/kg), ofloxacin (0.281 ± 0.320 μg/L, 1.142 ± 0.365 mg/kg), clindamycin (0.039 ± 0.023 μg/L, n.d.), metronidazole (0.023 μg/L, n.d.), oxytetracycline (0.641 μg/L, n.d.), sulfathiazole (0.049 μg/L, n.d.), sulfadiazine (0.057 ± 0.027 μg/L, n.d.), sulfamethoxazole (0.980 ± 0.466 μg/L, n.d.), trimethoprim (0.042 ± 0.025 μg/L, n.d.), tylosin (< LOQ, n.d.), paracetamol (13.64 ± 2.136 μg/L, n.d.), doxycycline (n.d., < LOQ), fenbendazole (n.d., 0.026 mg/kg), and norfloxacin (n.d., 0.271 ± 0.173 mg/kg). A 50% removal for sulfamethoxazole was observed after treatment in the UASB reactor, which increased to 70% after the combined UASB reactor + ASP system. At the same time, the concentrations of clindamycin, sulfathiazole, oxytetracycline and trimethoprim increased throughout the treatment. The concentration increase was attributed to the conjoining of the parent compounds or their conjugated metabolites during the treatment. Hence, it can be interpreted that these biological systems should be followed by an (electro) chemical system, which can enhance the removal efficiencies of these emerging contaminants.

Hybrid systems often entail the amalgamation of physical and chemical processes with biological systems. The combinations of these integrations results in the formation of physical-biological, (electro)chemical-biological, or physical-(electro)chemical-biological configurations [27]. Furthermore, (electro)chemical-biological hybrid systems utilise biological unit processes such as trickling filter (TF), ASP, MBR, and MBBR in combination with (electro)chemical processes such as photocatalysis, UV treatment, ozonation, UV/H_2O_2 treatment, Fenton oxidation, and photo-Fenton oxidation. Likewise, physical processes like nanofiltration (NF), MF, reverse osmosis (RO), UF, and adsorption are incorporated in physical-biological hybrid systems. Additionally, implementing physical-(electro)chemical-biological hybrid systems proves advantageous due to the involvement of three different types of processes targeting different contaminants at different stages [27].

For instance, Wang et al. [36] studied the feasibility of the two integrated combinations of MBR with RO or NF for removing 27 types of pharmaceutical and personal care products in municipal sewage. The results indicated that the effluent removal rates of the single MBR ranged from 41% to 95%, corresponding to the lowest and highest removal rates of carbamazepine and estriol, respectively. When MBR integrated with NF, 13 compounds were significantly removed with final concentrations below detection limits. At the same time, 20 compounds were completely removed when MBR complimented with RO. This disparity in removal efficiencies might be attributed to the difference in the mechanisms of RO and NF. The RO membrane effectively obstructs most dissolved organic matter due to electrostatic interaction, steric hindrance, and hydrophobic interaction between the compounds and the membrane. Thus, it can be concluded that hybrid treatment technologies

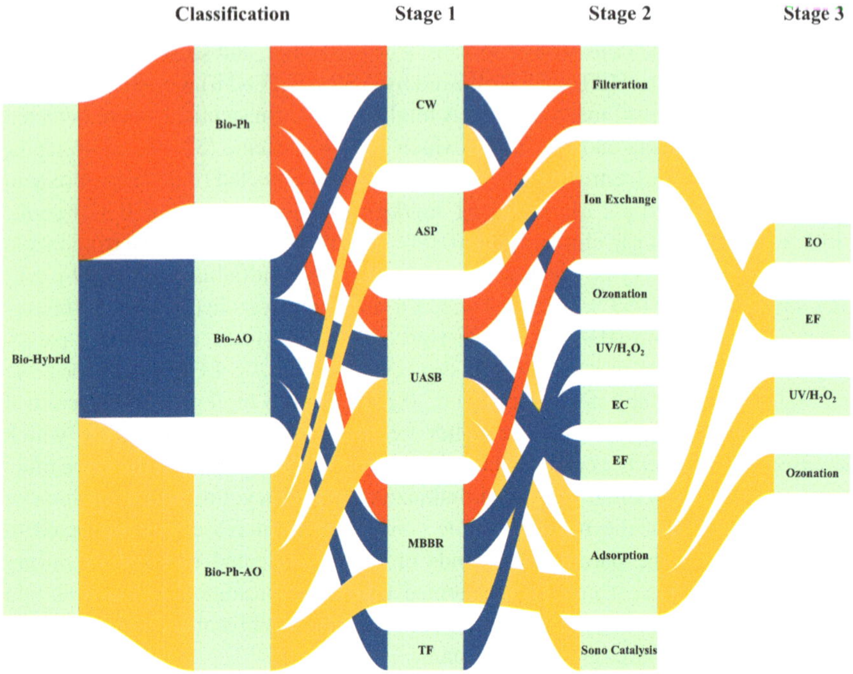

Fig. 3 Different possible configurations for hybrid treatment systems. AO—advanced oxidation (chemical), ASP—activated sludge process, Bio—biological, CW—constructed wetland, EC—electrocoagulation, EF—electro Fenton, EO—electro oxidation, MBBR—moving bed biofilm reactor, Ph—physical, TF—trickling filter, UASB—up-flow anaerobic sludge blanket reactor, UV—ultra violet

are required for the complete treatment of wastewater. The hybrid treatment trains can use different technologies at different stages to target different contaminants. For example, a hybrid biological-physical treatment scheme can consist of a UASB reactor to effectively reduce the organic matter, followed by an electro-Fenton process to remove the emerging contaminants. The other schemes can consist of different configurations, as presented in Fig. 3.

9 Summary

This chapter, in its entirety, exhaustively discuss the treatment of wastewater, which began with its introduction to the jinx and various sources of wastewater. Exploratively, preliminary understanding on the various stand-alone conventional treatment technologies that is, facultative ponds, oxidation ponds, anaerobic ponds, the activated sludge process, up-flow sludge blanket reactors, and sequencing batch reactors have been obtained on the mechanisms put in place to minimize the adverse environmental impact of this wastewater. However, a long-term examination of these traditional approaches has revealed their inherent drawbacks, which paved the way for the implementation of hybrid wastewater treatment systems. The stated limitations, such as spatial demands, limited efficiencies and the necessity for continuous operation, have driven the development and emergence of hybrid technologies.

Furthermore, it is essential to incorporate such novel and hybrid technologies due to the increase in complexity and volume of wastewater. The categorisation of these systems can also offer a systematic pathway to traverse through the dynamic field of wastewater treatment. These hybrid systems can result in sustainable and practical solutions. Primarily, this chapter provides a framework for comprehending the current status of wastewater treatment and emphasises the encouraging trajectory in which the field is heading.

References

1. Abdelbasir SM, Shalan AE (2019) An overview of nanomaterials for industrial wastewater treatment. Korean J Chem Eng 36(8):1209–1225. https://doi.org/10.1007/s11814-019-0306-y
2. Ahmed MB, Zhou JL, Ngo HH, Guo W, Thomaidis NS, Xu J (2017) Progress in the biological and chemical treatment technologies for emerging contaminant removal from wastewater: a critical review. J Hazard Mater 323:274–298. https://doi.org/10.1016/j.jhazmat.2016.04.045
3. Ahmed SF, Mofijur M, Nuzhat S, Chowdhury AT, Rafa N, Uddin MA, Inayat A, Mahlia TMI, Ong HC, Chia WY, Show PL (2021) Recent developments in physical, biological, chemical, and hybrid treatment techniques for removing emerging contaminants from wastewater. J Hazard Mater 416:125912. https://doi.org/10.1016/j.jhazmat.2021.125912
4. Al-Othman ZA, Ali R, Al-Othman AM, Ali J, Habila MA (2016) Assessment of toxic metals in wheat crops grown on selected soils, irrigated by different water sources. Arab J Chem 9:S1555–S1562. https://doi.org/10.1016/j.arabjc.2012.04.006
5. Barwal A, Chaudhary R (2014) To study the performance of biocarriers in moving bed biofilm reactor (MBBR) technology and kinetics of biofilm for retrofitting the existing aerobic treatment systems: a review. Rev Environ Sci Bio/Technol 13(3):285–299. https://doi.org/10.1007/s11157-014-9333-7
6. Bisognin RP, Wolff DB, Carissimi E, Prestes OD, Zanella R (2021) Occurrence and fate of pharmaceuticals in effluent and sludge from a wastewater treatment plant in Brazil. Environ Technol 42(15):2292–2303. https://doi.org/10.1080/09593330.2019.1701561
7. Carballa M, Omil F, Lema JM, Llompart M, García-Jares C, Rodríguez I, Gómez M, Ternes T (2004) Behavior of pharmaceuticals, cosmetics and hormones in a sewage treatment plant. Water Res 38(12):2918–2926. https://doi.org/10.1016/j.watres.2004.03.029
8. Christenson L, Sims R (2011) Production and harvesting of microalgae for wastewater treatment, biofuels, and bioproducts. Biotechnol Adv 29(6):686–702. https://doi.org/10.1016/j.biotechadv.2011.05.015
9. Corbella C, Garfí M, Puigagut J (2014) Vertical redox profiles in treatment wetlands as function of hydraulic regime and macrophytes presence: surveying the optimal scenario for microbial fuel cell implementation. Sci Total Environ 470–471:754–758. https://doi.org/10.1016/j.scitotenv.2013.09.068
10. CPCB (2021) Central pollution control board (CPCB), 4. https://cpcb.nic.in/status-of-stps/
11. Dutta A, Sarkar S (2015) Sequencing batch reactor for wastewater treatment: recent advances. Curr Pollut Rep 1(3):177–190. https://doi.org/10.1007/s40726-015-0016-y

12. Dutta V (2021) Water conflicts and resistance: issues and challenges in South Asia. In: Water conflicts and resistance: issues and challenges in South Asia, pp 1–285. https://doi.org/10.4324/9780429351006
13. Gupta S, Mittal Y, Panja R, Prajapati KB, Yadav AK (2021) Conventional wastewater treatment technologies, chap 3. In: Kumar S, Kumar R, Pandey A (eds). Elsevier, Amsterdam, pp 47–75. https://doi.org/10.1016/B978-0-12-821009-3.00012-9
14. Hamedani EA, Abasalt A, Talebi S (2024) Application of microbial fuel cells in wastewater treatment and green energy production: a comprehensive review of technology fundamentals and challenges. Fuel 370:131855. https://doi.org/10.1016/j.fuel.2024.131855
15. Hansen SH, Nierychlo M, Christensen ML, Nielsen PH, Jørgensen MK (2021) Fouling of membranes in membrane bioreactors for wastewater treatment: planktonic bacteria can have a significant contribution. Water Environ Res 93(2):207–216. https://doi.org/10.1002/WER.1392
16. Hickey RF, Wu WM, Veiga MC, Jones R (1991) Start-up, operation, monitoring and control of high-rate anaerobic treatment systems. Water Sci Technol 24(8):207–255. https://doi.org/10.2166/WST.1991.0226
17. Irvine RL, Busch AW (1979) Sequencing batch biological reactors. An overview. J Water Pollut Control Fed. https://www.jstor.org/stable/25039819
18. Lakatos G (2018) Biological wastewater treatment. In: Wastewater and water contamination: sources, assessment and remediation, pp 105–128. https://doi.org/10.1201/B13775/BIOLOGICAL-WASTEWATER-TREATMENT-LESLIE-GRADY-JR-GLEN-DAIGGER-NANCY-LOVE-CARLOS-FILIPE
19. Mohapatra DP, Kirpalani DM (2019) Advancement in treatment of wastewater: fate of emerging contaminants. Can J Chem Eng 97(10):2621–2631. https://doi.org/10.1002/CJCE.23533
20. Muga HE, Mihelcic JR (2008) Sustainability of wastewater treatment technologies. J Environ Manage 88(3):437–447. https://doi.org/10.1016/j.jenvman.2007.03.008
21. Mutamim NSA, Noor ZZ, Hassan MAA, Olsson G (2012) Application of membrane bioreactor technology in treating high strength industrial wastewater: a performance review. Desalination 305:1–11. https://doi.org/10.1016/j.desal.2012.07.033
22. Parida VK, Saidulu D, Majumder A, Srivastava A, Gupta B, Gupta AK (2021) Emerging contaminants in wastewater: a critical review on occurrence, existing legislations, risk assessment, and sustainable treatment alternatives. J Environ Chem Eng 9(5):105966. https://doi.org/10.1016/j.jece.2021.105966
23. Pratap B, Kumar S, Nand S, Azad I, Bharagava RN, Romanholo Ferreira LF, Dutta V (2023) Wastewater generation and treatment by various eco-friendly technologies: possible health hazards and further reuse for environmental safety. Chemosphere 313:137547. https://doi.org/10.1016/j.chemosphere.2022.137547
24. Priyadarshini M, Ahmad A, Das S, Ghangrekar MM (2022) Application of innovative electrochemical and microbial electrochemical technologies for the efficacious removal of emerging contaminants from wastewater: a review. J Environ Chem Eng 10(5):108230. https://doi.org/10.1016/j.jece.2022.108230
25. Raj R, Das S, Das S, Ghangrekar MM (2023) Environmental aspects of endocrine-disrupting compounds in the aquatic ecosystem and the application of electrochemical technologies for their abatement. Groundwater Sustainable Dev 20:100905. https://doi.org/10.1016/j.gsd.2023.100905
26. Ranade VV, Bhandari VM (2014) Industrial wastewater treatment, recycling, and reuse: an overview, chap 1. In: Ranade VV, Bhandari VM (eds). Butterworth-Heinemann, pp 1–80. https://doi.org/10.1016/B978-0-08-099968-5.00001-5
27. Saidulu D, Gupta B, Gupta AK, Ghosal PS (2021) A review on occurrences, eco-toxic effects, and remediation of emerging contaminants from wastewater: special emphasis on biological treatment based hybrid systems. J Environ Chem Eng 9(4):105282. https://doi.org/10.1016/j.jece.2021.105282
28. Sayadi MH, Kargar R, Doosti MR, Salehi H (2012) Hybrid constructed wetlands for wastewater treatment: a worldwide review. Proc Int Acad Ecol Environ Sci 2(4):204–222

29. Shahid MK, Kashif A, Rout PR, Aslam M, Fuwad A, Choi Y, Banu JR, Park JH, Kumar G (2020) A brief review of anaerobic membrane bioreactors emphasising recent advancements, fouling issues and future perspectives. J Environ Manage 270:110909. https://doi.org/10.1016/j.jenvman.2020.110909
30. Sharafi K, Moradi M, Karami A, Khosravi T (2015) Comparison of the efficiency of extended aeration activated sludge system and stabilisation ponds in real scale in the removal of protozoan cysts and parasite ova from domestic wastewater using Bailenger method: a case study, Kermanshah, Iran. Desalin Water Treat 55(5):1135–1141. https://doi.org/10.1080/19443994.2014.923333
31. Stottmeister U, Wießner A, Kuschk P, Kappelmeyer U, Kästner M, Bederski O, Müller RA, Moormann H (2003) Effects of plants and microorganisms in constructed wetlands for wastewater treatment. Biotechnol Adv 22(1):93–117. https://doi.org/10.1016/j.biotechadv.2003.08.010
32. Sundaravadivel M, Vigneswaran S (2001) Constructed wetlands for wastewater treatment. Crit Rev Environ Sci Technol 31(4):351–409. https://doi.org/10.1080/20016491089253
33. Tran NH, Gin KYH (2017) Occurrence and removal of pharmaceuticals, hormones, personal care products, and endocrine disrupters in a full-scale water reclamation plant. Sci Total Environ 599–600:1503–1516. https://doi.org/10.1016/j.scitotenv.2017.05.097
34. Verbyla ME, von Sperling M, Maiga Y (2019) Waste stabilization ponds. In: Water and sanitation for the 21st century: health and microbiological aspects of excreta and wastewater management (Global Water Pathogen Project). https://doi.org/10.14321/WATERPATHOGENS.65
35. Villarín MC, Merel S (2020) Paradigm shifts and current challenges in wastewater management. J Hazard Mater 390:122139. https://doi.org/10.1016/j.jhazmat.2020.122139
36. Wang Y, Wang X, Li M, Dong J, Sun C, Chen G (2018) Removal of pharmaceutical and personal care products (PPCPs) from municipal waste water with integrated membrane systems, MBR-RO/NF. Int J Environ Res Publ Health 15(2). https://doi.org/10.3390/ijerph15020269
37. Warsinger DM, Chakraborty S, Tow EW, Plumlee MH, Bellona C, Loutatidou S, Karimi L, Mikelonis AM, Achilli A, Ghassemi A, Padhye LP, Snyder SA, Curcio S, Vecitis CD, Arafat HA, Lienhard JH (2018) A review of polymeric membranes and processes for potable water reuse. Prog Polym Sci 81:209–237. https://doi.org/10.1016/j.progpolymsci.2018.01.004

A Global Overview of Traditional Biological Treatment Technologies

Azhan Ahmad, Monali Priyadarshini, Shalini Yadav, Makarand M. Ghangrekar, and Ram Narayan Yadava

Abstract Trace organic contaminants like pharmaceuticals, pesticides, surfactants, and phenolic compounds impose a threat to human health as well as aquatic plants and animals. A significant source of these contaminants entering the aquatic ecosystem is through the discharge of sewage from households, veterinary, agricultural runoff, pharmaceutical industries, and others. Besides, biological treatments are cost-effective and environmentally friendly technologies for wastewater treatment. Thus, it is essential to discuss and compare the traditional biological technologies that effectively treat trace organics from sewage. Different microbes present in the biological setups might break these organics into more environmentally liable and non-toxic compounds. However, further development of efficacious biological technologies is crucial for future full-scale application and commercialization. Therefore, this chapter deals with the technologies, such as activated sludge process, trickling filter, up-flow anaerobic sludge blanket reactor, stabilization ponds, and others used to remove organics present in the wastewater. The factors impacting the removal efficiency, like organic loading rate, and hydraulic retention time have also been elucidated. In addition, the mechanism of removal of organics in traditional biological technologies is comprehensively discussed. Lastly, the need for hybrid treatment technologies for remediation of organic pollutants has been discussed.

A. Ahmad · M. M. Ghangrekar (✉)
Department of Civil Engineering, Indian Institute of Technology Kharagpur, Kharagpur 721302, India
e-mail: ghangrekar@civil.iitkgp.ac.in

M. Priyadarshini · M. M. Ghangrekar
School of Environmental Science and Engineering, Indian Institute of Technology Kharagpur, Kharagpur 721302, India

S. Yadav
Rabindranath Tagore University, Bhopal, India

R. N. Yadava
Madhyanchal Professional University, Bhopal, India

S. Yadav et al. (eds.), *Emerging Technologies in Biological and Hybrid Wastewater Treatment: Lessons from Developed to Enhancing Practices in Developing Countries*, Earth and Environmental Sciences Library,
https://doi.org/10.1007/978-3-031-74515-7_2

Keywords Activated sludge process · Biological treatment technologies · Trace organic contaminants · Up-flow sludge blanket · Wastewater treatment

1 Background

Industrial, domestic, and agricultural practices across the globe are releasing a wider variety of pollutants like pharmaceuticals, nutrients, heavy metals, dyes, and others in the wastewater [3]. Furthermore, the common practice of untreated wastewater disposal from various sources into nearby surface water bodies poses a risk to the aquatic ecosystems. For instance, wastewater with high levels of organic matter when released in the water body typically lowers dissolved oxygen (DO) available. The oxygen-demanding ingredients and other organic matters present in water aid as a food source for aquatic microorganisms. These microorganisms usage the oxygen present in the water for their metabolic development, leading to the reduction of available DO, thereby affecting the higher life forms [48]. Elevated nutrient levels like phosphate and nitrate accelerate the maturation of water bodies by encouraging the development of phytoplankton and algae that can induce eutrophication. Thus, creating dead zones in the water bodies, which can diminish the DO content in the water [4]. Hence, reducing the organics from contaminated water and treating it to a certain level can ensure its safety for reuse or discharge to the environment, which is crucial for environmental preservation and sustainable development.

The treatment of wastewater can be achieved by employing traditional technologies, as centralized or decentralized facilities either solely or in hybridization with another traditional or advanced technology, depending on the characteristics of wastewater and collection pattern [45]. There are different traditional technologies practiced for wastewater treatment, and each has pros and cons (Fig. 1). However, the present chapter mainly focuses on understanding the mechanism and application of technologies, such as moving bed biofilm reactor (MBBR), activated sludge process (ASP), up-flow anaerobic sludge blanket (UASB) reactor, and ponds systems. Moreover, the factors affecting the performance of traditional technologies like pH, temperature, sludge retention time (SRT), and others have been discussed in the present chapter. Lastly, the future perspective of traditional wastewater treatment technologies for organic matter removal has also been elucidated in this chapter.

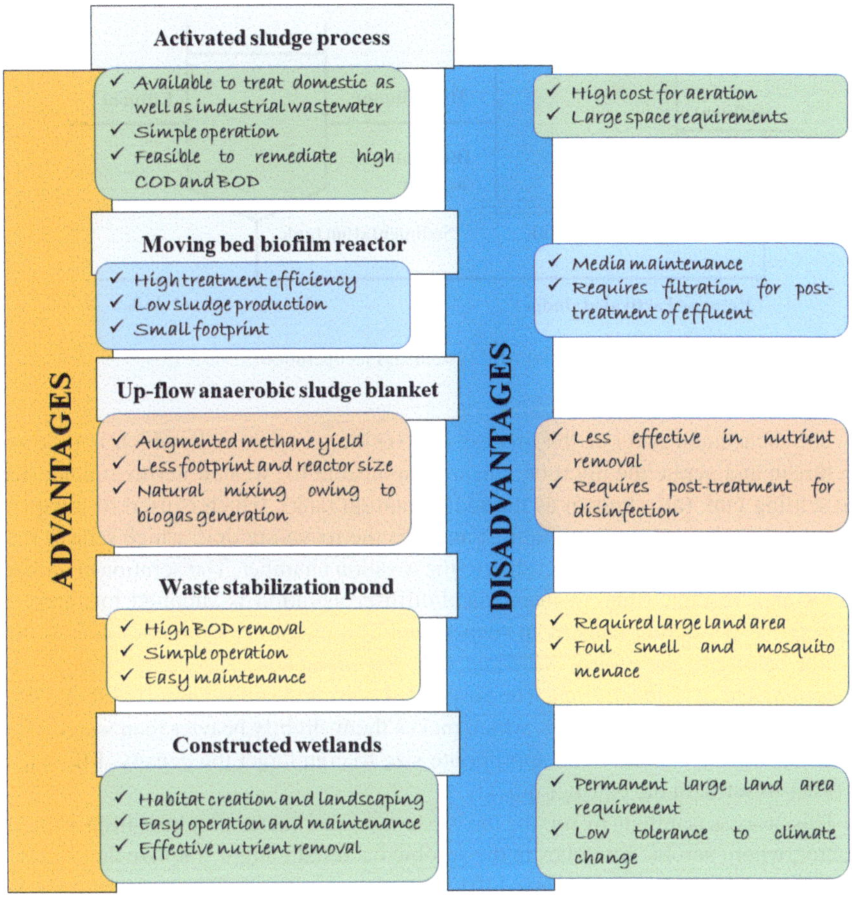

Fig. 1 Advantages and disadvantages of traditional wastewater treatment methods

2 Conventional Wastewater Treatment Technologies

2.1 Activated Sludge Process

The ASP is a widely used conventional biological method in municipal and industrial wastewater treatment plants (WWTPs), utilised to remove nutrients and oxidize carbonaceous and nitrogenous material, primarily focus on ammonium removal (Fig. 2). W. T. Lockett and E. Ardern invented this method in 1914 while serving at Manchester Sewage Works in England. This technique is highly efficient in removing organic matter and offers operational flexibility owing to its ability to tolerate the fluctuation of organic loads.

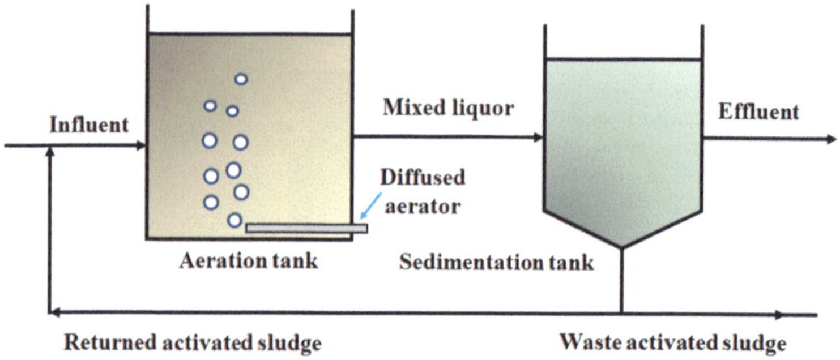

Fig. 2 Schematic representation of an ASP system and its operation

The three main parts of the process are: (i) the aeration tank, which is used to acclimate and aerate the microorganisms responsible for wastewater treatment; (ii) the settling tank (also known as the sedimentation tank), which is used to separate biological sludge solids from liquids; and (iii) the recycling unit, which returns the separated solids containing microbes to the aeration chamber. The aeration chamber serves as a reactor, wherein mechanical/diffused aeration is supplied to improve the activity of microorganisms in suspension. The majority of the microbes in the reactor belong to the groups *Nitrosomonas*, *Pseudomonas*, *Nitrobacter*, *Zoogloea*, *Nocardia*, and *Mycobacterium*. The suspended activated solids comprise about 70–80% of water outside their cells, which makes them slightly heavier than water. The solids need to flocculate to an appropriate size to counteract the density difference in order to settle in the clarifier quickly.

This process is initiate by mixing the wastewater and bacterial culture in an aerated reactor, where aerobic microbes in the aerobic bacterial culture stabilize the organic content present in the wastewater [19]. The retention time of the aeration tank is then adjusted in such a manner to make sure that the wastewater is adequately treated before undergoing a liquid/solid separation in a sedimentation tank [46]. Afterward, the collected settled activated sludge is recirculated into the aeration chamber to preserve a constant concentration of depolluting microbes [19]. Higher concentrations of activated solids in the aeration chamber decrease the liquid holding period required for wastewater treatment. In ASP, organics degrade via three routes: volatilization, which occurs during aeration, microbial mechanisms (metabolic or co-metabolic and biodegradation), and adsorption onto sludge flocs. Among all the processes, adsorption onto activated sludge flocs and subsequent oxidation is a significant pathway for removal of some compounds.

The ASP also treats the wastewater produced by pig farms. According to an investigation, an ASP removed 82 and 92% of chemical oxygen demand (COD) and total Kjeldahl nitrogen (TKN) from pig farms' wastewater [9]. Similarly, coagulation-coupled ASP containing an organic loading of 0.06 gCOD/g TS day eliminated 97% of the COD from the wastewater generated by the meat industry [8]. Moreover, 68, 72, 99, and 98.8% removal efficiencies for enrofloxacin, tetracycline, ibuprofen, and paracetamol present in slaughterhouse effluent were observed during treatment in ASP [10]. The compounds were either absorbed on the surface of solids or degraded by the activated biomass.

2.2 *Moving Bed Biofilm Reactor*

Biofilm-based technologies, including trickling filters, fluidized bed reactors, rotating biological contractors (RBCs), and aerated submerged fixed bed biofilm reactors, have been used to treat wastewater. The common drawbacks of these technologies include uneven biofilm dispersion and mechanical and hydraulic breakdowns. Considering these, a novel biofilm-based method known as the moving bed biofilm reactor (MBBR) was developed in the late 1980s and operated at Lardnal, Norway, in the 1990s [23]. In this system, biomass grows on the surface of plastic carriers (1–4 cm diameter) called "biocarriers," constantly moving inside the reactor to improve the contact between the microbes and contaminants present in wastewater (Fig. 3). A biofilm is collection of microorganisms, which is adhered to the surface of biocarriers, containing a self-produced matrix of extracellular polymeric substances (EPS). The matrix of EPS can shield the microorganisms from UV rays, harsh temperatures, high salinities, extreme pH, antibiotics, and other environmental stresses by serving as "protective clothing".

Furthermore, the support matrix allows the growth of a broad range of microbes on it, which facilitates nitrification and denitrification along with the removal of carbonaceous organic contaminants from wastewater [36]. These carriers are prepared from high-density polyethylene. The thickness of biofilm growth on biocarriers can be controlled by airflow and mechanical mixing. Furthermore, surface characteristics of the carriers, such as surface area and roughness, affect the biomass accumulation. These carriers are commercially available with variable surface areas ranging

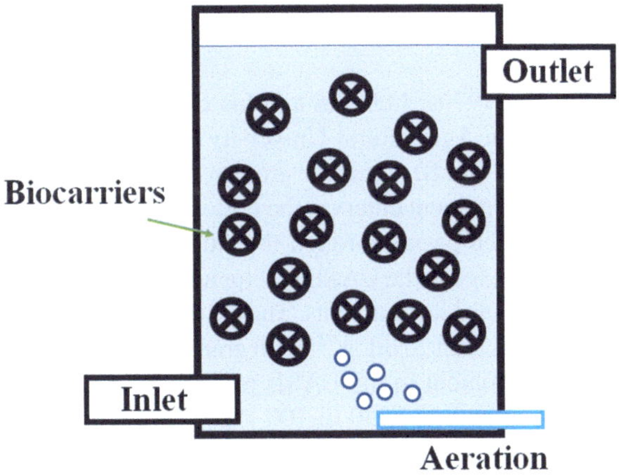

Fig. 3 Schematic view of a MBBR reactor

between 400 and 1200 m^2 m^{-3} [33]. In addition to plastic carriers, polyurethane sponges and other materials can also be utilized as biocarriers. For instance, Ngo and co-authors investigated the efficacy of sponge-based biocarriers in MBBR for eliminating organic material, TN, PO_4-P, and total organic carbon (TOC) [35]. Similarly, Luo and co-researchers evaluated the abatement of five pollutants over a 24-h-batch trial utilizing sponge-supported carriers [29].

The MBBR can be operated in either anaerobic or aerobic conditions. The mobility of biocarriers in an aerobic or oxic reactor is aided by agitation with airflow. As a result, aeration serves a dual purpose, including oxygenating the microbes and keeping the carriers moving in the chamber. Further, the necessity of an aerator for the proper movement of carriers and aeration raises the operating costs of the system [33]. Aeration is often provided by a percolating stainless-steel tube or perforated polymer resins that is positioned at the reactor's bottom. In anaerobic reactors, biocarriers are suspended using vertical or horizontal shaft agitators [33]. Installing a sieve at the tank outlet allows biocarriers to be retained inside the tank. The biocarriers can rotate in three axes once the liquid flows through the passages around the outside or inside the carriers.

The optimal functioning of the MBBR is preserved by appropriately designing the aeration grids and sieve and by fully utilizing the reactor's volume without any dead space or short circuits. Wastewater passes through these biocarriers, facilitating nutrient absorption and decomposition of organic materials by bacterial action. Furthermore, the growth of slow-growing microbes inside the biocarriers makes MBBR perfect for nitrification, which converts ammonium to nitrate. Numerous factors, including the total ammonium nitrogen, temperature, alkalinity, organic load, pH, DO, and the biofilm on the biocarrier, generally affect nitrification rates [41]. A high concentration of organic matter will inhibit the development of nitrifying microorganisms. The MBBR can be implemented in conjunction with other technologies and as a pre-and post-treatment process [47].

2.3 Up-Flow Anaerobic Sludge Blanket Reactor

In the late 1970s, Lettinga and his colleagues first invented the UASB reactor at Wageningen Agricultural University in the Netherlands [28]. A reactor and a gas-liquid-solid (GLS) separator are essential components of UASB reactor. In UASB reactor, the influent enters at the bottom of the reactor travels upward through sludge bed, and then passes through the sludge blanket (Fig. 4). The production of thick granular sludge in the UASB reactor allows the elimination of higher COD compared to other anaerobic treatments. These granulated sludge particles have a sludge volume index of less than 20 mL g^{-1}, with granule sizes ranging from 1 to 3 mm. Furthermore, the solids content in the UASB reactor fluctuates from 5 to 30 g L^{-1} at the top of sludge bed and from 30 to 100 g L^{-1} at the bottom of the sludge bed. Moreover, distinct biological processes take place in the sludge bed (containing active microbes) located at the bottom of the UASB reactor. The soluble organics existing in the

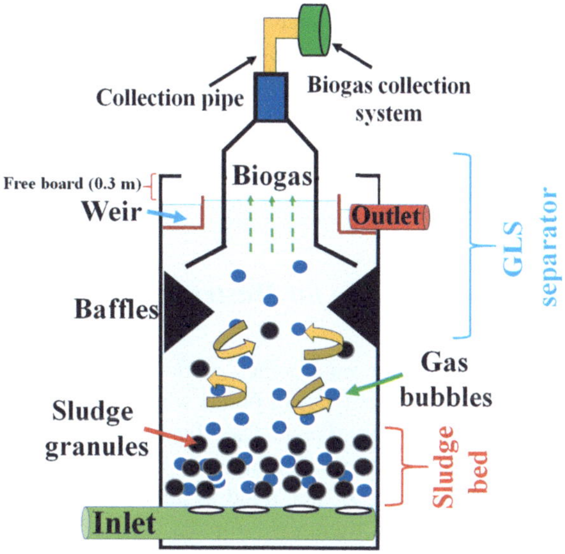

Fig. 4. Typical schematic of a UASB reactor and mechanism involved

polluted water transform into biodegradable intermediates and finally to CH_4 and CO_2, as it passes through the sludge bed zones. The biogas is produced, and sludge captures the biogas and these two phases are separated at the GLS separator with the help of baffles, which prevent the washing out of microbes or granulated sludge from the reaction chamber [24].

The UASB reactor necessitates the least capital investment compared to alternative biological technologies [11]. Compared to the suspended culture medium, this technique is unique because it produces granular sludge without any support medium by utilizing the self-immobilization ability of anaerobes, leading to a dense population of microorganisms, increased biosolids retention, excellent settling, endurance to hazardous chemicals, as well as resilience to higher organic matter loading [20, 21]. Additionally, this technique produces less sludge and uses very little energy in its operation. According to reports, a UASB reactor with a height of about 4 m needs yearly sludge removal [39]. A mean COD elimination of more than 60% has been observed in different types of municipal wastewaters [11], whereas while treating food processing industrial effluents, like dairy, brewery, sugar, etc., COD removal efficiency of more than 80% can be obtained.

Optimizing the organic loading rate for maximum organic matter removal and energy recovery in the form of methane remains a challenge while operating the UASB reactor. Moreover, the process of producing CH_4 takes a longer time when methanogenic biomass is unavailable, owing to the sluggish growth rate of methanogens. Anaerobic conditions present in the reactor might also result in the production of H_2S if sulphate is present in the wastewater. Meeting the post-treatment criteria after the UASB reactor is another difficulty owing to the incomplete removal of nutrients (nitrogen and phosphorous) from wastewater. The accumulation of fatty acid obstructs COD elimination and biogas recovery at higher organic loading rates. Moreover, the build-up of fatty acids in the reactor inhibits the growth of aceto-clastic methanogens. The performance of the reactor is also significantly impacted by operating factors such as pH, organic loading rate, SRT, temperature, food-to-microorganisms (F/M) ratio, and pH buffering capacity of the wastewater being treated.

2.4 Ponds System for Wastewater Treatment

Ponds are shallow waterbodies made up of naturally occurring earthen basins composed of various types of ponds which are connected serially in a specific area. There are three types of ponds: aerobic, facultative, and anaerobic [23]. The organic pollutants and pathogen elimination from wastewater in pond systems have been addressed earlier [34]. For pond systems, temperature and sunlight are essential controlling factors, therefore, tropical climate is suitable for these systems [44]. However, these systems require large land area compared to other conventional treatment systems like ASP.

(i) ***Aerobic Pond***: A microbial community resembling ASP coexists alongside algae in an aerobic pond. The algal population uses the CO_2 released by the aerobic population for their growth. In turn, algae release oxygen, which contributes to the pond's ability to remain aerobic. An aerobic pond with a very shallow depth (0.15–0.45 m) is used to treat wastewater in order to remove nitrogen through the growth of algae. In general, 0.5–1.2 m depth of oxidation pond can be employed for wastewater treatment. To maintain aerobic conditions across the whole pond, photosynthesis has to be supported by solar radiation that reaches the bottom of the pond. When utilised for tertiary wastewater treatment, shallow aerobic oxidation ponds (0.5 m deep) with modest organic loads are referred to as maturation ponds.

(ii) ***Facultative Ponds***: These are another type of pond system, which primarily designed for wastewater treatment. These ponds are separated into three zones: the upper zone, is an aerobic zone, where aerobic biodegradation and algal photosynthesis typically occur. At the bottommost zone, the cells that are produced in the aerobic zone settle down and simultaneously, the settled suspended organic material existing in wastewater undergoes anaerobic breakdown (Fig. 5). Additionally, the algal population in a facultative pond is assumed to indicate the pond's health [37]. Meanwhile, anaerobic and aerobic activities coexist in the intermediate zone [26]. The organic waste decomposes in this zone by facultative bacteria. The existence of the top aerobic zone eliminates the nuisance caused by the anaerobic reaction. Therefore, maintaining an aerobic environment at the top layer is crucial for the facultative stabilization pond to function properly. This condition is dependent on temperature, biochemical oxygen demand (BOD) loading, wastewater properties, and solar

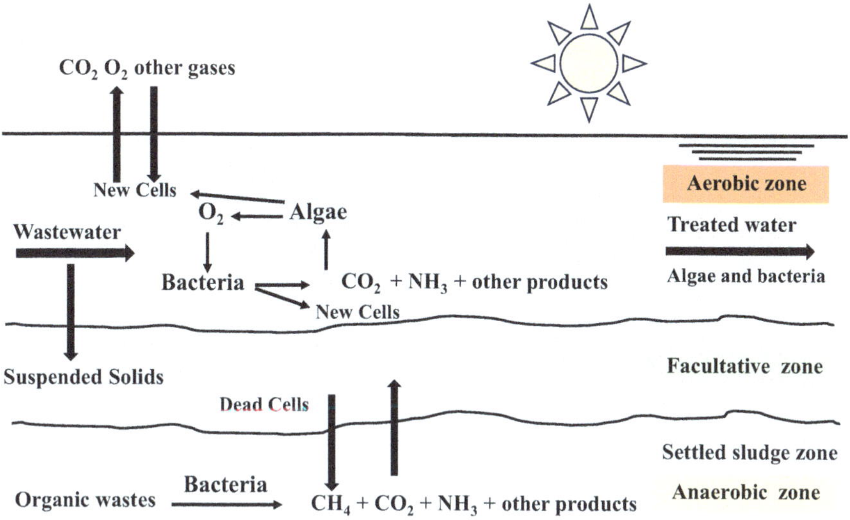

Fig. 5 Schematic of facultative stabilization pond used for wastewater treatment and mechanism involved

radiation. These ponds are highly comparable with other conventional treatment systems.

(iii) ***Anaerobic Pond***: In this type of pond system the entire depth is under anaerobic conditions except for an extremely shallow top layer. For comprehensive treatment, these ponds are typically utilised in succession with facultative or aerobic ponds. These ponds depth ranges from 2.5 to 6 m. They are often used to treat highly concentrated industrial wastewater, while occasionally, they are also utilised to treat municipal wastewater and sludges. Anaerobic ponds may retain wastewater for up to 50 days, depending on its strength. Nowadays, polyethylene sheets are used to cover anaerobic lagoons in order to recover biogas, eliminate odours and reduce greenhouse gas emissions into the environment.

2.5 *Constructed Wetlands*

Constructed wetlands (CW) are artificial green systems designed to remediate wastewater via the use of microbial populations, soils, and wetland plants (Fig. 6). These systems replicate the natural wetlands' inside a planned and regulated setting. For the treatment of different types of wastewater, such as stormwater runoff, industrial effluents, and household sewage, the CW can be a practical and sustainable solution [50]. In general, the primary mechanisms involve in CW are physical filtration biological and chemical reactions [2]. The wetland substrates, such as gravel, sand and soil serves as physical filter, drawing suspended materials out of the wastewater.

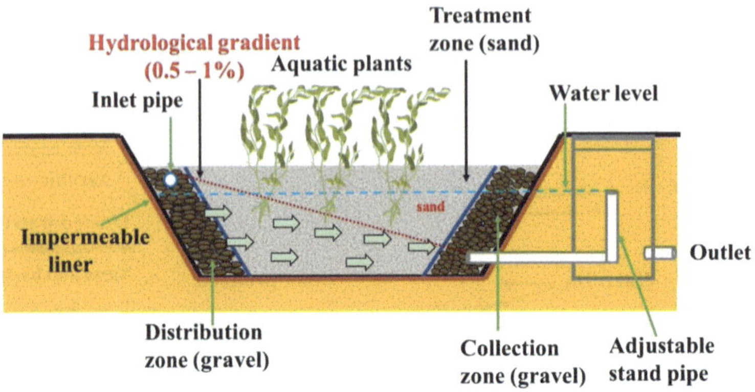

Fig. 6 Typical schematic of CW and mechanism involved

Meanwhile the microorganisms decompose and absorb nitrogen, organic matter, and other pollutants at the roots of plants and within the substrate. Furthermore, CW can aid in the elimination of nutrients by facilitating chemical reactions such as adsorption, precipitation, and ion exchange [1]. There are basically surface and subsurface flow type of CW systems. In surface flow wetland the wastewater flows over the surface of wetland and contacts the roots of wetland plants by flowing horizontally across the layer of the gravel or soil. This kind is appropriate for handling rainfall and household sewage. On the other hand, subsurface flow type CW can be either horizontal or vertical flow types. The contaminated water seeps through the ground below the surface. Municipal wastewater is often handled by horizontal flow systems, although smaller-scale applications are better served by vertical flow systems. Table 1 shows a few instances of different types of wastewater being treated using different conventional treatment methods.

3 Factors Governing Performance of Conventional Treatment Technologies

The operating parameters such as hydraulic retention time (HRT), operating temperature, SRT, pH, DO, and organic loading rate impact the performance efficiency of wastewater treatment technologies in terms of organic pollutant removal. Hence, it is essential to discuss these factors.

Table 1 Treatment of various pollutants containing wastewater in conventional processes

Wastewater type/ Pollutants	Treatment method	HRT (h)	COD removal (%)	Reference
Pharmaceutical	Anaerobic sequencing batch reactor	2856	99.4%	Amin et al. [7]
Pharmaceutical	UASB reactor	168	80–85	Rodríguez-Martinez et al. [40]
Pharmaceutical	UASB reactor	23.5	97	Chen et al. [12]
Antibiotic	ASP	–	99.8	Costanzo et al. [15]
Pharmaceutical	Ponds system	24	92–99	Matamoros et al. [31]
Pharmaceutical	CWs	–	94–99	Shah et al. [43]
Domestic sewage	CWs	–	35–97 (winter) 57–95 (summer)	Dan et al. [16]
Livestock	CWs	7	27–85	Choi et al. [13]
Industrial effluent	MBBR	26.4 ± 2.4	43–76	Mazioti et al. [32]
Estrone (E1)	MBBR	–	65–95	Luo et al. [29]
17β-Estradiol (E2)	MBBR	4–16	95–100	Amin et al. [6]
17α-Ethynylestradiol (EE2)	MBBR	–	90–98	Luo et al. [30]

3.1 *Impact of Sludge Retention Time*

The sludge age or SRT represents the average residence period of microbes in the reactor and it is also an indicator of the growth rate of microbes present in the reactor. Mathematically, it is estimated by the ratio of total mass of the solids present in the reactor (in kg or g) to mass of solids wasted per day from the reactor (kg/day or g/day). Elevated sludge age permits the substantial improvement of the biomass and it is beneficial for the growth of slow-rate enrichment autotrophic bacteria like nitrifiers. These nitrifiers might also release some bio-enzymes that can help in cleaving low-degradable molecules with aromatic structures such as pharmaceuticals like diclofenac. Moreover, the effluent concentration of a few organics like pharmaceuticals ibuprofen and carbamazepine is independent of the influent dosages; however, affected by the operated SRT. Therefore, SRT is a basic and vital factor used to design the reactors for wastewater treatment. For instance, Clara et al. suggest the SRT of 10–15 days for efficient remediation of organic pollutants from wastewater [14]. However, for treating any specific compound, the SRT is to be optimized (critical SRT) to get the desired or permissible effluent concentration. However, in some cases, SRT of more than 30 days may be required to efficiently remove the organic pollutants. For instance, Xia et al. noticed that an SRT of 30 days was required to obtain more than 80% removal of Sulfamethoxazole antibiotic (500 $\mu g\ L^{-1}$) in an aerobic membrane bioreactor [51].

3.2 Impact of Hydraulic Retention Time

The HRT represents the mean residence time of the liquid phase in the chamber. This factor highly affects the F/M ratio. The F/M ratio represents the amount of food available in terms of BOD, COD, and TOC relative to the quantity of microbes present in the chamber. In the case of UASB reactor, HRT directly affects the up-flow velocity, which plays a vital role in entrapping the suspended solids. The organics abatement in terms of COD is highly affected by up-flow velocity because higher velocity reduces the contact period between the organics and the solids; thereby decreasing the COD removal rate [17]. Moreover, at high velocities, elevated sludge washout might also occur, which decreases the efficiency of the UASB reactor. Typically HRT between 6 and 18 h is required for efficient removal of organics from low-strength wastewater (municipal sewage) using the UASB reactor [22]. However, for high-strength wastewater (COD > 10000 mg/L), HRT of more than 24 h is required for about 65–75% of COD removal in the UASB reactor. Similarly, Vieno and co-authors reported that a decrease in HRT resulted in an increase in atenolol pharmaceutical removal [49]. However, there is no straightforward relationship between HRT and the removal of contaminants because in some cases, it was also seen that the decrease in HRT does not affect the removal efficiency much. This difference can be associated to the difference in the reactor design, operating procedures, and range of HRT. Hence, it is essential to assess the critical HRT for the efficient removal of organic pollutants from wastewater.

3.3 Impact of Dissolved Oxygen

Another important factor that can affect the performance of aerobic wastewater treatment systems is the DO content present in wastewater. For example, ammonia-oxidizing microorganisms like the nitrifying bacteria growth is favoured by high DO content, and these microbes also have the potential to degrade the trace emerging contaminants from wastewater [38]. However, in the case of an attached growth process like MBBR, elevating the aeration rate to achieve high DO might result in the generation of high shear force on the surface of the biofilm grown on the media. This can induce mechanical stresses on the biofilm, thereby causing its breakage or detachment from the surface of the media [27]. Therefore, high aeration might result in the removal of biomass from the biofilm, which may deteriorate the biological activity of the biofilm as well as the reduction in performance efficiency. Hence, if the aeration rate is controlled appropriately and sufficient DO is present in the wastewater, it can enhance the redox properties of the biofilm, which can be beneficial in improving the biodegradation of the pollutants. Thus, the presence of critical molecular oxygen can be useful to mitigate the wide range of organic pollutants like pharmaceuticals, surfactants, and phenols from wastewater. Also, in wastewater treatment facilities, aeration frequently consumes the energy, thus, optimizing the DO levels can help to minimise the energy usage and maximise the aeration efficiency.

3.4 Impact of pH

In the case of anaerobic processes like UASB reactor, pH plays a vital role because the growth of methanogens can be efficacious at a pH range between 6.5 and 7.8 [17]. Therefore, for treating domestic sewage in UASB reactor, no external chemical is required to change the pH, as the pH of domestic sewage lies within the range required for the growth of methanogens. Using an anaerobic reactor at pH of 7.0 to treat kitchen wastewater improves both the hydrolysis and acidogenesis rates. As pH of 7.0 creates the ideal condition for anaerobic digestion, which removes more than 80% of the soluble TOC (11.82 g L^{-1}) and COD (53.18 g L^{-1}) [52]. Similarly, the highest trichloroethylene (TCE) removal of 96.4% was obtained at pH of 7.0 in an UASB reactor with initial TCE concentration of 36.5 mg L^{-1}, which was 16% higher than the removal obtained at pH of 6.0 (80.7%). A lower removal at the acidic pH can be associated to the lower diversity of microbes in the reactor that is required for the biodegradation of TCE [53]. Moreover, in another experiment the COD removal efficiency was decreased from 93% to 79% (initial COD of 16 g L^{-1}), when the pH of the wastewater was raised from 7.3 to 8.1, respectively [42]. Increment in pH can lead to change the physical characteristics of granules, such as decreased size, density and volatile solid contents, as result, the microbial metabolism might get decreased and the lower removal of COD was observed. Thus, it is important to optimize the pH to treat the wastewater efficiently.

3.5 Impact of Temperature

Another important parameter is the temperature within the reactor for treating wastewater. Temperature can influence the efficacy of the process, by affecting the microbial activity, oxygen transfer rate, and redox reactions occurring within the system. In the case of ASP, the temperature has a major impact as the elevated temperatures stimulate microbial activity, which in turn accelerates biological responses and enhances organic pollutant removal [25]. Besides, the high temperature might decrease the solubility of oxygen in water, thereby influencing the efficiency of aerators, which require high energy for proper oxygen supply [18]. This can affect the economics of the treatment. In addition, high temperature might also be useful in pathogen destruction during anaerobic digestion, thereby contributing to the safety of the treated sludge. Therefore, it is vital to consider seasonal variations and temperature fluctuation when designing wastewater treatment plants.

4 Need of Hybrid Technologies

Sometimes a single treatment technology might not be effective to remediate the organic pollutants from wastewater efficaciously. Therefore, hybridization of the treatment processes can be carried out, wherein two or more treatment techniques are combined together to enhance the overall performance of the system. Multiple treatment processes may be integrated in hybrid systems to handle a wide range of pollutants such as pharmaceuticals, pesticides, dyes and surfactants from wastewater. Hybrid technologies offer versatility in treating different types of wastewater arising from municipal and industrial sources, with varying influent characteristics. Thus, a combination of two or more technologies is helpful for optimizing an overall treatment system based on the specific characteristics of the wastewater.

In addition, as conventional treatment technologies are not effective in treating recalcitrant compounds like antibiotics, phenols, and pesticides, the removal of these recalcitrant can be improved by the synergy of two or more treatment processes. For example, Alvarino and co-authors reported more than 84% removal of sulfamethoxazole (40 $\mu g\ L^{-1}$) using a UASB reactor coupled aerobic MBR process; however, the removal efficiency was barely 10% using MBR alone [5]. Besides, providing an advanced technology such as electro-Fenton (EF) prior to conventional biological treatment can help in improving the biodegradability of wastewater containing organics like pharmaceuticals. Thus, it is easy for the microbes to readily biodegrade the organics present in the partially treated EF effluent for effective treatment of wastewater. Also, the risks incurred by the failure of a single treatment component can be reduced by hybrid systems. Other components can take over whenever one process fails to operate or is momentarily unavailable, thereby lowering the chance of receiving inadequate treatment. Thus, to facilitate effective treatment of wastewater, hybridization of either conventional technologies or conventional with advanced technologies can be carried out.

5 Conclusion

The traditional wastewater treatment practices like ASP, UASB, MBBR, and facultative stabilization pond systems have been successfully employed to accomplish more than 80% of nutrient and COD removal from wastewater. In addition, constructed wetlands have also shown promising results in eliminating emerging contaminants from wastewater. However, a prolonged retention period is required to effectively remove emerging contaminates. Hence, hybridizing CW with advanced technologies, such as EF-coupled CW, can minimize the retention time as partial degradation can be achieved in EF prior to CW treatment. Besides, owing to the recalcitrant nature of organics like pharmaceuticals, membrane fouling in case of membrane-based technologies, and adsorption on activated sludge flocs remain significant obstacles impeding the biodegradation of emerging organic pollutants using traditional technologies.

In the case of the UASB reactor, efficient removal of COD can be achieved even with low-strength wastewater at lower energy consumption along with minimal sludge production. In addition, even with high organic loading and low mesophilic operating temperature, the performance of the UASB reactor is satisfactory for carbonaceous matter removal. However, the efficiency of the UASB reactor is low for nutrient removal from wastewater. Hence, future investigations can focus on the optimization of advanced biological and electrochemical processes as downstream treatment systems for nutrient removal from the effluent of the UASB reactor. Even though the traditional technologies have shown overall satisfactory performance, there is still a need for future research in terms of economics, the desired quality of treated water for reusability, meeting discharge limits, and removal of xenobiotics from wastewater. Hence, there is still an enormous prospect to enhance the present conventional methods in order to achieve high efficacy while utilizing lesser resources.

6 Recommendations/Future Prospective

There is a wide scope of application for traditional biological technologies for the removal of organic pollutants from wastewater. However, there are still few lacunas, which need to overcome for the commercialization of these technologies. For example, the tradition biological technologies are not effective for the remediation of emerging contaminants like pharmaceuticals, pesticides and surfactants due to the low biodegradability of these contaminants. Hence, hybridizing the conventional technologies with advanced technologies might be beneficial in mitigating the emerging contaminants from wastewater. Even though high-rate reactors have advanced significantly and show great promise as a sewage treatment method, there are still issues that need to be worked out. These include the prolonged start-up time, the reduced removal of organic matter, and the higher concentrations of nutrients, pathogens, and volatile fatty acids in the effluent when compared to conventional aerobic systems, which makes them incompatible with local discharge or reuse regulations. Furthermore, the effectiveness of ASP for eliminating toxic pollutants from wastewater can be improved via. integrating to advanced treatment technologies like UV treatment or ozone, optimizing the bacterial activity through bioaugmentation with specific degrading strains, and utilizing of real-time observing systems to ensure accurate control of operating factors. The performance of CW can be boosted by integrating a diverse range of plant species, which hold high pollutant uptake capabilities and dense root structures that encourage bacteriological activity. Similarly, improving the carrier media to maximize surface area, optimization of the HRT and organic loading rates in MBBR can ensure effectual bacterial colonization, which might enhance its performance.

References

1. Abou-Elela SI, Golinielli G, Abou-Taleb EM, Hellal MS (2013) Municipal wastewater treatment in horizontal and vertical flows constructed wetlands. Ecol Eng 61:460–468. https://doi.org/10.1016/j.ecoleng.2013.10.010
2. Ahmad A, Priyadarshini M, Raj R, Das S, Ghangrekar MM (2023) Appraising efficacy of existing and advanced technologies for the remediation of beta-blockers from wastewater: a review. Environ Sci Pollut Res 30(10):25427–25451. https://doi.org/10.1007/s11356-021-18287-4
3. Ahmad A, Priyadarshini M, Yadav S, Ghangrekar MM, Surampalli RY (2023) Evaluation of waste medicine wrappers as an efficacious low-cost novel electrode material in electrocoagulation for the remediation of Coomassie Brilliant Blue from wastewater. J Environ Chem Eng 11(5):110484. https://doi.org/10.1016/j.jece.2023.110484
4. Ali J, Yang Y, Pan G (2023) Oxygen micro-nanobubbles for mitigating eutrophication induced sediment pollution in freshwater bodies. J Environ Manage 331(January):117281. https://doi.org/10.1016/j.jenvman.2023.117281
5. Alvarino T, Suárez S, Garrido M, Lema JM, Omil F (2016) A UASB reactor coupled to a hybrid aerobic MBR as innovative plant configuration to enhance the removal of organic micropollutants. Chemosphere 144:452–458. https://doi.org/10.1016/j.chemosphere.2015.09.016
6. Amin MM, Bina B, Ebrahim K, Yavari Z, Mohammadi F (2018) Biodegradation of natural and synthetic estrogens in moving bed bioreactor. Chin J Chem Eng 26(2):393–399. https://doi.org/10.1016/j.cjche.2017.06.006
7. Amin MM, Zilles JL, Greiner J, Charbonneau S, Raskin L, Morgenroth E (2006) Influence of the antibiotic erythromycin on anaerobic treatment of a pharmaceutical wastewater. Environ Sci Technol 40(12):3971–3977. https://doi.org/10.1021/es060428j
8. Bohdziewicz J, Sroka E, Lobos E (2002) Application of the system which combines coagulation, activated sludge and reverse osmosis to the treatment of the wastewater produced by the meat industry. Desalination 144(1–3):393–398. https://doi.org/10.1016/S0011-9164(02)00349-1
9. Bortone G, Piccinini S (1991) Nitrification and denitrification in activated-sludge plants for pig slurry and wastewater from cheese dairies. Bioresour Technol 37(3):243–252. https://doi.org/10.1016/0960-8524(91)90191-L
10. Carvalho PN, Pirra A, Basto MCP, Almeida CMR (2013) Activated sludge systems removal efficiency of veterinary pharmaceuticals from slaughterhouse wastewater. Environ Sci Pollut Res 20(12):8790–8800. https://doi.org/10.1007/s11356-013-1867-7
11. Chan YJ, Chong MF, Law CL, Hassell DG (2009) A review on anaerobic-aerobic treatment of industrial and municipal wastewater. Chem Eng J 155(1–2):1–18. https://doi.org/10.1016/j.cej.2009.06.041
12. Chen Z, Wang H, Ren N, Cui M, Nie S, Hu D (2011) Simultaneous removal and evaluation of organic substrates and NH_3–N by a novel combined process in treating chemical synthesis-based pharmaceutical wastewater. J Hazard Mater 197:49–59. https://doi.org/10.1016/j.jhazmat.2011.09.053
13. Choi YJ, Kim LH, Zoh KD (2016) Removal characteristics and mechanism of antibiotics using constructed wetlands. Ecol Eng 91:85–92. https://doi.org/10.1016/j.ecoleng.2016.01.058
14. Clara M, Kreuzinger N, Strenn B, Gans O, Kroiss H (2005) The solids retention time—a suitable design parameter to evaluate the capacity of wastewater treatment plants to remove micropollutants. Water Res 39(1):97–106. https://doi.org/10.1016/j.watres.2004.08.036

15. Costanzo SD, Murby J, Bates J (2005) Ecosystem response to antibiotics entering the aquatic environment. Mar Pollut Bull 51(1–4):218–223. https://doi.org/10.1016/j.marpolbul.2004.10.038
16. Dan A, Yang Y, Dai Y, Chen C-X, Wang SYu, Tao R (2013) Removal and factors influencing removal of sulfonamides and trimethoprim from domestic sewage in constructed wetlands. Bioresour Technol 146:363–370. https://doi.org/10.1016/j.biortech.2013.07.050
17. Daud MK, Rizvi H, Akram MF, Ali S, Rizwan M, Nafees M, Jin ZS (2018) Review of upflow anaerobic sludge blanket reactor technology: effect of different parameters and developments for domestic wastewater treatment. J Chem 2018. https://doi.org/10.1155/2018/1596319
18. Drewnowski J, Remiszewska-Skwarek A, Duda S, Łagód G (2019) Aeration process in bioreactors as the main energy consumer in a wastewater treatment plant. Review of solutions and methods of process optimization. Processes 7(5). https://doi.org/10.3390/pr7050311
19. Eckenfelder WW, Cleary JG (2013) In activated sludge technologies for treating industrial wastewaters. DEStech Publications, Inc.
20. El-Kamah H, Tawfik A, Mahmoud M, Abdel-Halim H (2010) Treatment of high strength wastewater from fruit juice industry using integrated anaerobic/aerobic system. Desalination 253(1–3):158–163. https://doi.org/10.1016/j.desal.2009.11.013
21. Ergüder TH, Demirer GN (2005) Investigation of granulation of a mixture of suspended anaerobic and aerobic cultures under alternating anaerobic/microaerobic/aerobic conditions. Process Biochem 40(12):3732–3741. https://doi.org/10.1016/j.procbio.2005.05.005
22. Ghangrekar MM, Asolekar SR, Ranganathan KR, Joshi SG (1996) Experience with UASB reactor start-up under different operating conditions. Water Sci Technol 34(5-6–6 pt 3):421–428. https://doi.org/10.1016/0273-1223(96)00674-9
23. Gupta S, Mittal Y, Panja R, Prajapati KB, Yadav AK (2021) Conventional wastewater treatment technologies. In: Current developments in biotechnology and bioengineering: strategic perspectives in solid waste and wastewater management, pp 47–75. https://doi.org/10.1016/B978-0-12-821009-3.00012-9
24. Hickey RF, Wu WM, Veiga MC, Jones R (1991) Start-up, operation, monitoring and control of high-rate anaerobic treatment systems. Water Sci Technol 24(8):207–255. https://doi.org/10.2166/wst.1991.0226
25. Hu H, Shi Y, Liao K, Ma H, Xu K, Ren H (2019) Effect of temperature on the characterization of soluble microbial products in activated sludge system with special emphasis on dissolved organic nitrogen. Water Res 162:87–94. https://doi.org/10.1016/j.watres.2019.06.034
26. Kayombo STSA, Mbwette TSA, Katima JHY, Ladegaard N, Jørgensen SE (2005) Waste stabilization ponds and constructed wetlands design manual. UNEP-IETC with the Danish International Development Agency (Danida), pp 1–59
27. Leiknes TO, Ødegaard H (2007) The development of a biofilm membrane bioreactor. Desalination 202(1–3):135–143. https://doi.org/10.1016/j.desal.2005.12.049
28. Lettinga G, van Velsen AFM, Hobma SW, de Zeeuw W, Klapwijk A (1980) Use of the upflow sludge blanket (USB) reactor concept for biological wastewater treatment, especially for anaerobic treatment. Biotechnol Bioeng 22(4):699–734. https://doi.org/10.1002/bit.260220402
29. Luo Y, Guo W, Ngo HH, Nghiem LD, Hai FI, Zhang J, Liang S, Wang XC (2014) A review on the occurrence of micropollutants in the aquatic environment and their fate and removal during wastewater treatment. Sci Total Environ 473–474:619–641. https://doi.org/10.1016/j.scitotenv.2013.12.065
30. Luo Y, Jiang Q, Ngo HH, Nghiem LD, Hai FI, Price WE, Wang J, Guo W (2015) Evaluation of micropollutant removal and fouling reduction in a hybrid moving bed biofilm reactor-membrane bioreactor system. Bioresour Technol 191:355–359. https://doi.org/10.1016/j.biortech.2015.05.073
31. Matamoros V, Rodríguez Y, Albaigés J (2016) A comparative assessment of intensive and extensive wastewater treatment technologies for removing emerging contaminants in small communities. Water Res 88:777–785. https://doi.org/10.1016/j.watres.2015.10.058
32. Mazioti AA, Stasinakis AS, Pantazi Y, Andersen HR (2015) Biodegradation of benzotriazoles and hydroxy-benzothiazole in wastewater by activated sludge and moving bed biofilm reactor systems. Bioresour Technol 192:627–635. https://doi.org/10.1016/j.biortech.2015.06.035

33. McQuarrie JP, Boltz JP (2011) Moving bed biofilm reactor technology: process applications, design, and performance. Water Environ Res 83(6):560–575. https://doi.org/10.2175/106143010x12851009156286
34. Moumouni DA, Andrianisa HA, Konaté Y, Ndiaye A, Maïga AH (2016) Inactivation of *Escherichia coli* in a baffled pond with attached growth: treating anaerobic effluent under the Sahelian climate. Environ Technol (United Kingdom) 37(9):1054–1064. https://doi.org/10.1080/09593330.2015.1098732
35. Ngo HH, Guo W, Xing W (2008) Evaluation of a novel sponge-submerged membrane bioreactor (SSMBR) for sustainable water reclamation. Bioresour Technol 99(7):2429–2435. https://doi.org/10.1016/j.biortech.2007.04.067
36. Ødegaard H, Gisvold B, Strickland J (2000) The influence of carrier size and shape in the moving bed biofilm process. Water Sci Technol 41(4–5):383–391. https://doi.org/10.2166/wst.2000.0470
37. Pearson HW, Mara DD, Bartone CR (1987) Guidelines for the minimum evaluation of the performance of full-scale waste stabilization pond systems. Water Res 21(9):1067–1075. https://doi.org/10.1016/0043-1354(87)90028-5
38. Ren YX, Nakano K, Nomura M, Chiba N, Nishimura O (2007) Effects of bacterial activity on estrogen removal in nitrifying activated sludge. Water Res 41(14):3089–3096. https://doi.org/10.1016/j.watres.2007.04.028
39. Rodríguez-Gómez R (2011) Upflow anaerobic sludge blanket reactor: modelling. In: Licentiate thesis (chemical engineering), Department of Chemical Engineering and Technology, Royal Institute of Technology, Stockholm, Sweden. https://www.diva-portal.org/smash/get/diva2:394741/FULLTEXT02.pdf
40. Rodríguez-Martinez J, Garza-Garcia Y, Aguilera-Carbo A, Martínez-Amador SY, Sosa-Santillan GJ (2005) Influence of nitrate and sulfate on the anaerobic treatment of pharmaceutical wasterwater. Eng Life Sci 5(6):568–573. https://doi.org/10.1002/elsc.200520101
41. Rusten B, Eikebrokk B, Ulgenes Y, Lygren E (2006) Design and operations of the Kaldnes moving bed biofilm reactors. Aquacult Eng 34(3):322–331. https://doi.org/10.1016/j.aquaeng.2005.04.002
42. Sandberg M, Ahring BK (1992) Anaerobic treatment of fish meal process waste-water in a UASB reactor at high pH. Appl Microbiol Biotechnol 36(6):800–804. https://doi.org/10.1007/BF00172198
43. Shah AI, Din Dar MU, Bhat RA, Singh JP, Singh K, Bhat SA (2020) Prospectives and challenges of wastewater treatment technologies to combat contaminants of emerging concerns. Ecol Eng 152(Dec 2019):105882. https://doi.org/10.1016/j.ecoleng.2020.105882
44. Sharafi K, Moradi M, Karami A, Khosravi T (2015) Comparison of the efficiency of extended aeration activated sludge system and stabilization ponds in real scale in the removal of protozoan cysts and parasite ova from domestic wastewater using Bailenger method: a case study, Kermanshah, Iran. Desalin Water Treat 55(5):1135–1141. https://doi.org/10.1080/19443994.2014.923333
45. Sravan JS, Matsakas L, Sarkar O (2024) Advances in biological wastewater treatment processes: focus on low-carbon energy and resource recovery in biorefinery context. Bioengineering 11(3). https://doi.org/10.3390/bioengineering11030281
46. Tchobanoglous G, Burton LF, Stensel DH (2014) Metcalf & Eddy: wastewater engineering: treatment and reuse, vol 7. McGraw Hill Companies, Inc., p 421
47. Theodoridis T, Kraemer J (1998) Post denitrification in a moving bed biofilm reactor process. Water Sci Technol 38(1):31–38. https://doi.org/10.1016/S0273-1223(98)00387-4
48. Vaage B, Myrick C (2022) Growth, metabolism, and dissolved oxygen tolerance of juvenile burbot. Aquaculture 552(Sept 2021):737980. https://doi.org/10.1016/j.aquaculture.2022.737980
49. Vieno N, Tuhkanen T, Kronberg L (2007) Elimination of pharmaceuticals in sewage treatment plants in Finland. Water Res 41(5):1001–1012. https://doi.org/10.1016/j.watres.2006.12.017
50. Wu H, Zhang J, Ngo HH, Guo W, Hu Z, Liang S, Fan J, Liu H (2015) A review on the sustainability of constructed wetlands for wastewater treatment: design and operation. Bioresour Technol 175:594–601. https://doi.org/10.1016/j.biortech.2014.10.068

51. Xia S, Jia R, Feng F, Xie K, Li H, Jing D, Xu X (2012) Effect of solids retention time on antibiotics removal performance and microbial communities in an A/O-MBR process. Bioresour Technol 106:36–43. https://doi.org/10.1016/j.biortech.2011.11.112
52. Zhang B, Zhang LL, Zhang SC, Shi HZ, Cai WM (2005) The influence of pH hydrolysis and acidogenesis of kitchen wastes in two-phase anaerobic digestion. Environ Technol 26(3):329–340. https://doi.org/10.1080/09593332608618563
53. Zhang Y, Hu M, Li P, Wang X, Meng Q (2015) Analysis of trichloroethylene removal and bacterial community function based on pH-adjusted in an upflow anaerobic sludge blanket reactor. Appl Microbiol Biotechnol 99(21):9289–9297. https://doi.org/10.1007/s00253-015-6800-1

Membrane Bioreactors and Emerging Membrane Technologies

Walaa M. Thabet and Abeer A. Moneer

Abstract The growing need for a reliable water supply has outstripped the availability of modern technologies for water treatment. Advanced water treatment methods using membranes have a strong track record of producing clean water. This chapter aims to present the progress of membrane technology and its applications, types; challenges, and the necessity to develop membrane technologies for long-term water treatment. By collecting the literature about membrane technology, the process description and its applications, advantages and disadvantages, investigation of monitoring and control the membrane technologies and comparing with different types of membrane bioreactors and how to improve the efficiency of membrane technology while reducing the adverse impacts. on the environment. While there is a huge attention to green and sustainable membranes, also, analysis of membrane materials and several factors affecting its efficiency such as fouling or toxicity of contaminants so there is a need for further studies to be done and it will be discussed on this chapter. Therefore, many traditional membrane synthesis methods are presented in this chapter. In recent years, various sustainable alternatives have been proposed. In addition, this chapter will examine some of the latest developments in membrane technology for desalination and wastewater treatment in order to improve membrane sustainability. Some of the current trends in membrane science and technology for sustainable water treatment include membrane reuse, the reuse of wasted brine or sludge, the collection of energy from waste, and the application of membrane antifouling techniques to reduce waste. Concluded at last how emerging membrane technology has a promising future for water purification and treatment.

Keywords Membrane bioreactors · Membrane processes · Emerging membrane technology · Water purification · Sustainability

W. M. Thabet (✉) · A. A. Moneer
National Institute of Oceanography and Fisheries (NIOF), Cairo, Egypt
e-mail: thabetniof@yahoo.com

S. Yadav et al. (eds.), *Emerging Technologies in Biological and Hybrid Wastewater Treatment: Lessons from Developed to Enhancing Practices in Developing Countries*, Earth and Environmental Sciences Library,
https://doi.org/10.1007/978-3-031-74515-7_3

1 Introduction

Membrane phenomena firstly appear at the eighteen century and in the nineteen and early twentieth had no commercial or industrial applications [7]. Membrane technology comprises linked engineering and scientific techniques for allowing or rejecting the passage of parts, species, and substances through or via membranes. A number of separation methods are referred to as "membrane technology". Membrane technology defines the mechanical separation processes used to separate streams of gas or liquid [38]. At the past two decades, membrane technology becomes a leading separation technology. When the membrane acts as a permselective barrier between two phases that have a gradient in pressure, partial pressure of vapour, concentration, or electrical potential, the desired species is separated. Most frequently, flux and selectivity are utilized to evaluate membrane performance [30]. The volume, mass, or quantity of moles of a substance that flow across a membrane in a specific length of time and space is known as the flux. Retention or separation factor is related to a membrane's selectivity to a fluid combination [2]. The first classification of separation processes lists as follows microfilteration (MF), ultrafilteration (UF), nanofilteration (NF), reverse osmosis (RO), gas separation, pervaporation, dialysis, electrodeionization and membrane bioreactor (MBR), Each of these categories has a different method of separation, such as size exclusion, solution-diffusion mechanism, Knudsen diffusion, and molecular diffusion, depending on the driving force and solute to be separated [2]. MF is at the upper end of the pressure-driven membrane technology spectrum, while, RO membranes are the pressure-driven membrane processes with the smallest holes and the ability to retain all dissolved constituents of a feed stream, including monovalent ions. Figure 1 depicts some of the key phases in the evolution of membrane technology. This chapter focuses on membrane bioreactor treatment technology and its emerging trends.

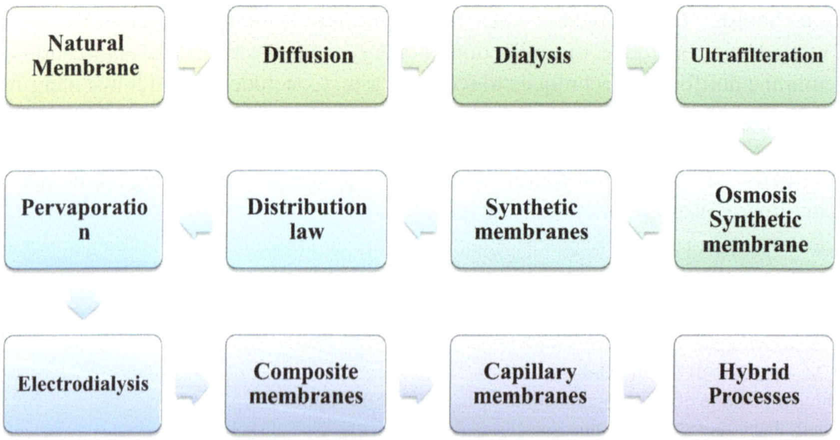

Fig. 1 Major steps of development of membrane technology

2 Membrane Bioreactors: Process Description and Applications

Membrane bioreactors (MBRs) are wastewater treatment systems that combine membrane filtration and biological treatment. In membrane bioreactors, the physical space of a biocatalyst—which may be a bacterium, an enzyme, or a plant or animal cell—can be limited. A catalyst's ongoing usage, or reuse, is made feasible by limiting its physical space, which prevents loss from washing out from the reactor. Given the cost of the biocatalyst and the fact that the outlet stream will be free of biocatalyst, this feature provides a considerable advantage that will make the process of product recovery much simpler [10]. Membrane bioreactors combine membrane filtration with conventional biological treatment (e.g., activated sludge), providing a higher degree of organic and suspended particle removal. When built properly, these systems can also remove nutrients more effectively. In an MBR system, an aerated biological reactor makes use of submerged membranes [38].

By eliminating the traditional sedimentation and filtering steps used in wastewater treatment, only high-quality effluent can pass through the membranes at this level of filtration. This eliminates the need for sedimentation, allowing the biological activity to occur at a much higher mixed liquor concentration. As a result, far less process tankage is required, and updating many current facilities can be done without adding additional tanks [49]. Figure 2 illustrated the schematic membrane bioreactor system. Membrane separation offers a wider variety of environmental and industrial applications since it is a clean technology that saves energy and can take the place of conventional processes including distillation, filtration, chemical treatment systems, and ion exchange. Other benefits include its capacity for high-quality product production and adaptability in system design. Figure 3 summarises the major benefits and drawbacks of membrane technology. Separation can be done continuously in a pleasant environment, using minimal energy and no chemicals. The method can also be combined with other separation techniques to produce hybrid processes., also, produce high quality effluent with low sludge production, it can be reused or recycled immediately. The limitations of this method are low membrane lifetime, concentration polarization, membrane fouling, low selectivity and flow, and high cost [38, 49]. There are several applications of the usage of membrane bioreactors. Figure 4 summarizes of the different uses for membrane processes.

3 Types of Membrane Bioreactors and Configurations

3.1 Types of Membrane Bioreactors (MBRs)

There are two types of MBRs: aerobic and anaerobic. MBR, sometimes known as "aerobic MBR," aims to recover energy and nutrients from wastewater to water reclamation. Furthermore, aeration, a significant electricity consumer to support the

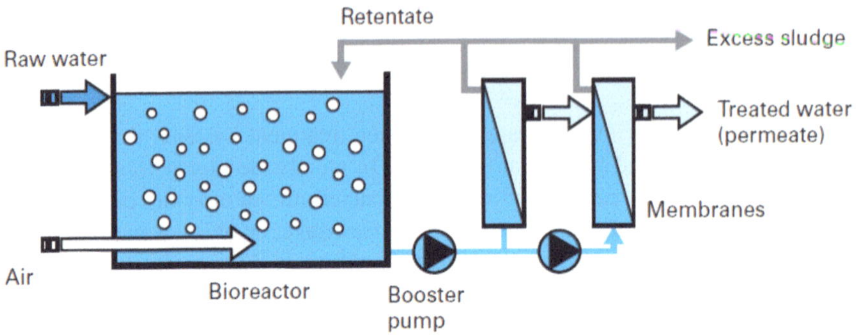

Fig. 2 Schematic membrane bioreactor system

Fig. 3 Lists of advantages and disadvantages of membrane technology

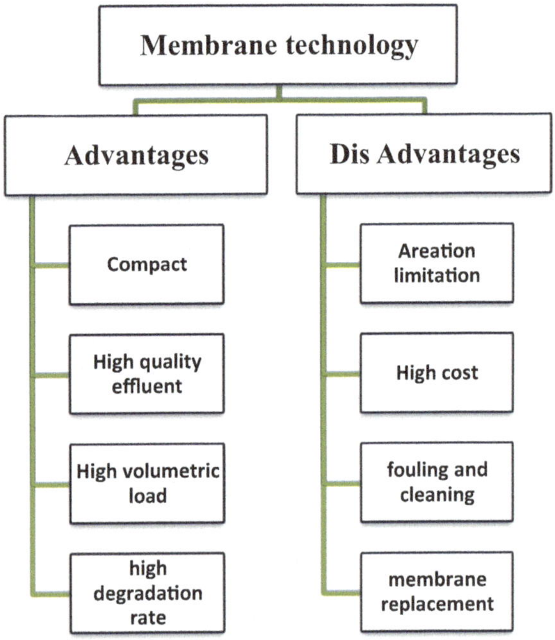

Conventional Activated Sludge (CAS) treatment, makes aerobic MBR an energy-intensive process [22]. Anaerobic MBR processes nutrients (such phosphorus and nitrogen) into more chemically accessible forms to make them easier to recover or utilise directly in agriculture. Anaerobic digestion and membrane separation are combined to maximize resource recovery from wastewater. Anaerobic MBR produces methane-rich biogas by the bacterial conversion of organic compounds in wastewater. AnMBR produces more biogas and produces better effluent than traditional anaerobic digestion. In terms of energy consumption, pollutant removal

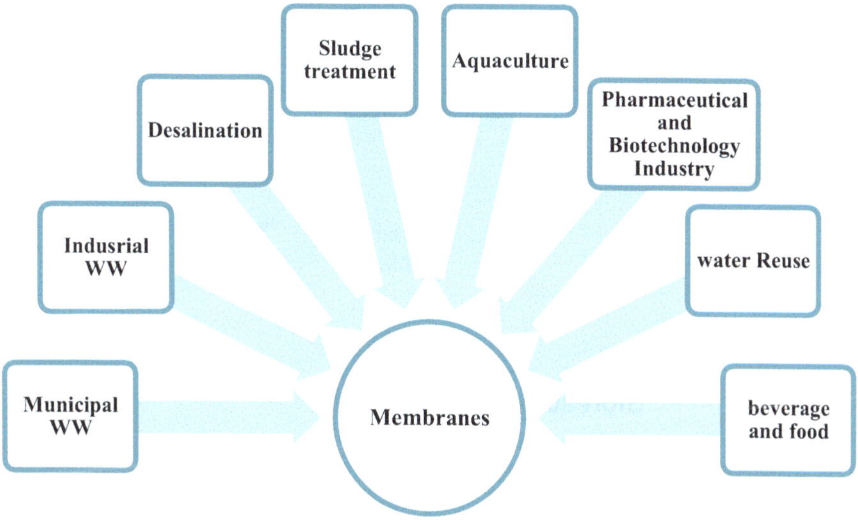

Fig. 4 Schematic illustration of the uses for membrane processes

effectiveness, and volume throughput, AnMBR differs greatly from AeMBR [45] Fig. 5 shows schematic diagrams of AeMBR and AnMBR systems [26].

Also, there are several steps for MBR operating procedure as shown in Fig. 6. Before passing through a grit chamber to remove sand and other heavy particles, the wastewater is filtered to remove larger debris. This is referred to as the pretreatment phase. Next, the water is pushed through a membrane filtration system, which filters out microbes and small particles using a semi-permeable membrane. To eliminate any leftover microbes and lower the risk of illness, the treated water is then disinfected. Monitoring and control are vital phases, to guarantee that the water is being complied with the proper standards, the entire process is continuously monitored, and modifications are made as needed to maximize efficiency. Routine maintenance is crucial for keeping the system working smoothly and replacing or cleaning the

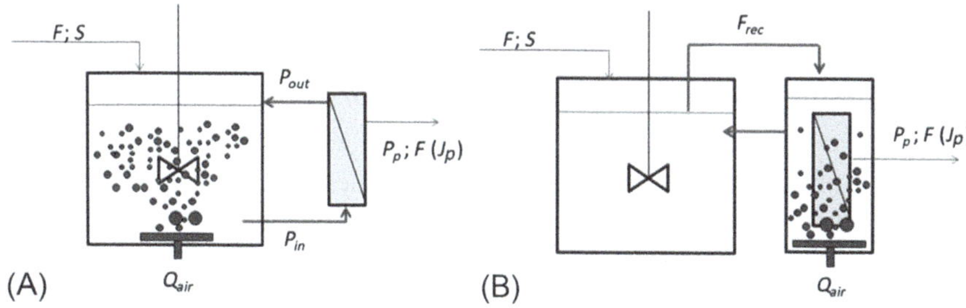

Fig. 5 Schematic diagrams **a** AeMBR and **b** AnMBR systems [26]

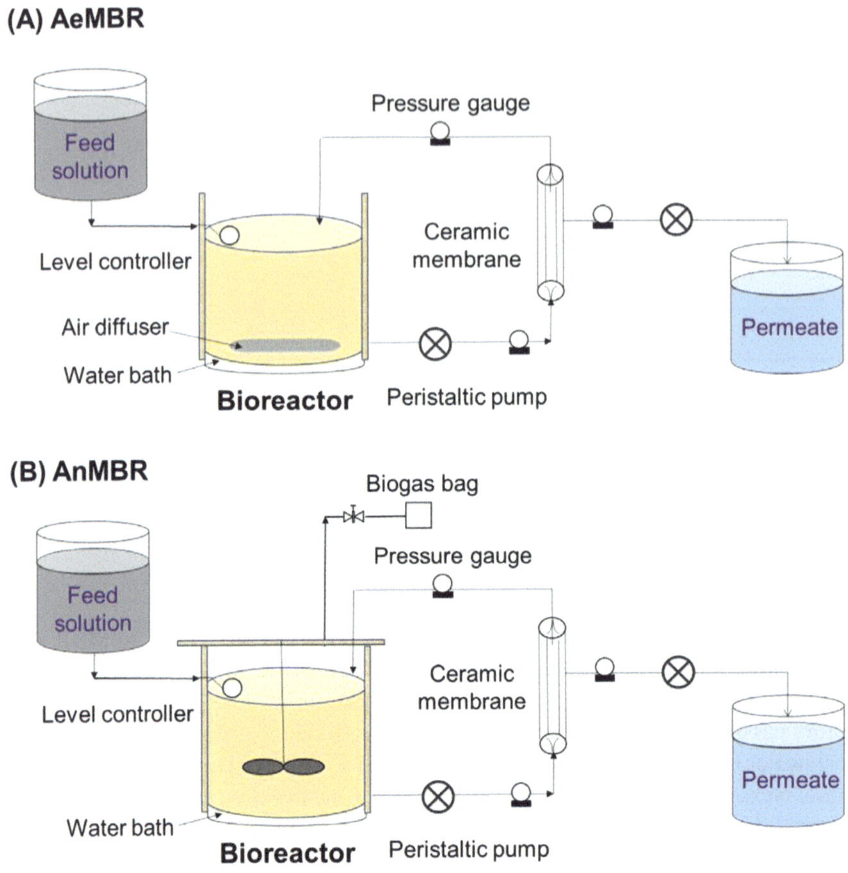

Fig. 6 Membrane bioreactor operating procedure

membranes as necessary. The efficiency of MBR system depends on several factors including wastewater quality and special treatment goals.

3.2 Fundamental of Membrane Bioreactors Configurations

The Design of the system depends on the characteristics of wastewater such as flow statistics, effluent needs and influent characteristics, configuration of the system, biotreatment process and separation process [23]. There are six configurations of MBRs each design with different countenance and advantages with the same purpose to combine membrane filtration with biological wastewater treatment to obtain high levels of treatment effectiveness and water quality, however there are only two primary types: submerged membrane bioreactors (SMBR) and external membrane bioreactors (EMBR). The membrane removes small particles and bacteria, while microorganisms in the biological treatment stage break down pollutants in water to produce high-quality effluent.

a. SMBR is a type of MBR in which water is forced across the membrane while the membrane module is immersed in a biological treatment tank. The most common applications desalination, water reuse, sludge treatment, municipal and industrial wastewater treatment. Figure 7a illustrates SMBR.
b. In an MBR-like system known as an EMBR, the membrane module is situated outside the biological treatment reservoir, and the effluent that has undergone biological treatment is pumped through the membrane. The applications are the same as SMBR, Fig. 7 represents the different configuration of internal and external MBR [33]. Figure 7b illustrates EMBR.
c. Hybrid Membrane Bioreactors: is a class of MBR that gathers the advantages of both SMBR and EMBR. The membrane module in this form of MBR is situated both within and outside the biological treatment tank. This approach has several advantages, including reduced physical space requirements, lower sludge formation rates, longer sludge retention time (SRT), and improved effluent quality. When membrane fouling occurs, the efficacy of filtration is diminished and transmembrane pressure is increased [29].
d. Integrated Membrane Bioreactors (IMBR): is a kind of MBR created to combine membrane filtration and biological treatment in a single, portable device. IMBRs often use less energy and have a smaller environmental impact than conventional MBR systems [28].
e. Air bubbles are used by a particular kind of MBR known as a membrane aerated biofilm reactor to supply oxygen to the bacteria during the biological treatment phase. The membrane gives the bacteria a large surface area to thrive on and

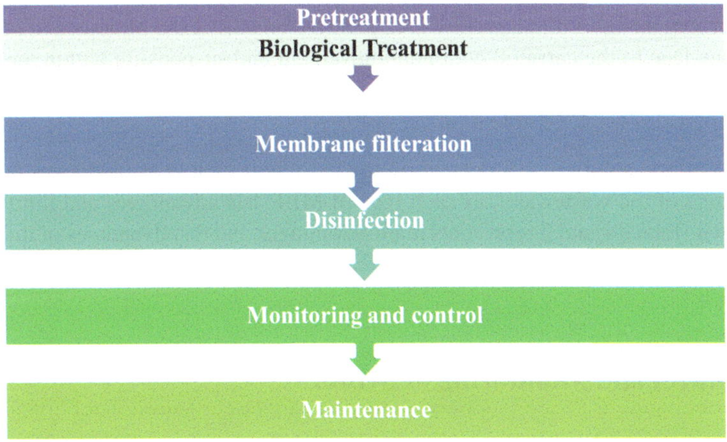

Fig. 7 **a** Internal submerged SMBR **b** external MBRs for wastewater treatment [10]

Table 1 Comparison of advantages various MBR types

<table>
<tr><th>Submerged membrane bioreactor (SMBR)</th><th>External MBR systems</th><th>Hybrid MBR</th><th>Integrated MBR</th><th>Membrane aerated biofilm reactor</th><th>Moving bed bioreactor</th></tr>
<tr><td colspan="6">Common advantages:
High treatment efficiency
High-quality effluent</td></tr>
<tr><td>Compact and space-saving design</td><td colspan="5">Reduced fouling and clogging of the membrane</td></tr>
<tr><td>Low energy consumption</td><td colspan="3">Easy maintenance</td><td>Low energy consumption</td><td>Low maintenance costs</td></tr>
<tr><td></td><td></td><td>Reduced energy consumption compared to SMBRs</td><td>Compact and space-saving design</td><td>Low maintenance costs</td><td>Increased surface area for biofilm growth</td></tr>
<tr><td></td><td></td><td></td><td>Low energy consumption</td><td>Increased surface area for biofilm growth
Compact and space-saving design</td><td>Compact and space-saving design
Low energy consumption</td></tr>
</table>

provides support for the biofilm. The bubbles are injected through the membrane itself [48].

f. Moving bed bioreactor: is a kind of MBR that uses plastic holders to give the microorganisms a broad surface area to thrive on during the biological treatment stage. Due to the carriers' continual movement and suspension within the treatment tank, there is excellent mixing and oxygen transmission, known mainly as biological sewage treatment process [40].

The optimal solution will depend on the particular treatment objectives and the parameters of the wastewater, as each type of MBR has advantages and limitations of its own. Tables 1 and 2 represents main advantages and disadvantages of different membrane bioreactors types.

4 Control and Monitoring of MBRs

There are two main monitoring approaches to minimize costs, decrease fouling and raising the monitoring ability. As a result, several characterization and monitoring approaches created and researched to improve monitoring over the past few years.

Table 2 Comparison of disadvantages various MBR types

Submerged membrane bioreactor (SMBR)	External MBR systems	Hybrid MBR	Integrated MBR	Membrane aerated biofilm reactor	Moving bed bioreactor
Common disadvantage: High maintenance and initial costs and membrane replacement costs Require skilled personnel to operate and maintain					
Membrane fouling	Space-consuming design				
Potential for clogging of the membrane	Higher energy consumption compared to SMBRs		Limited flexibility in terms of scalability	Limited flexibility in terms of scalability Potential for membrane damage due to air bubbles	Limited flexibility in terms of scalability Potential for clogging of the carriers

Due to the close ties between the two processes, each other is impacted by how each one performs. While biological performance influences the membrane, primarily because of fouling formation, and selectivity of the membrane [41].

4.1 *Offline Monitoring*

In wastewater treatment systems, influents and effluents are characterized using a variety of analytical techniques, such as sum parameters chemical oxygen demand (COD), 5-day biochemical oxygen demand (BOD), biosolids, and total organic carbon (TOC), ionic species, and other molecules that are important to the environment [10].

4.2 *Online Monitoring*

The use of real-time monitoring systems, which can instantaneously evaluate various biological and membrane performance parameters, greatly simplifies the operation of MBRs, even though off-line characterization techniques are reliable and offer inclusive data about the exact compounds in an MBR. Due to the development of new commercial sensors, some of the parameters that are routinely monitored in MBR systems can now be measured online (COD, TOC, Nitrogen compounds), including dissolved oxygen and transmembrane pressure (TMP). The employment of particular sensors, however, raises monitoring expenses at the same rate as the number of parameters to be tracked does. Spectroscopic techniques can be used in situ (using optical probes) and are nondestructive, quick, and employable without the use of chemicals. Spectroscopic approaches have so far been successfully associated with parameters like BOD, COD, TOC, extracellular polymeric substances EPS, mixed liquor suspended solids MLSS, nitrate, ammonia, and nitrogen, among others, in biological processes for wastewater treatment [10].

5 Membrane Technology Challenges

Drinking water production, desalination, wastewater reclamation, and reuse, numerous membrane types are used. The creation of new membranes and innovative operational techniques provided promising industry. The main limitations that keep membrane technology from progressing beyond its current use frequently have to do with economic issues: membranes are not cheap, and membrane systems are still energy-intensive, both of which have a negative impact on running expenses. For the production of drinking water, desalination, wastewater reclamation, and reuse, numerous membrane types are used. The creation of new membranes and innovative operational techniques provided industry with exciting potential. Other membranes have taken an alternative approach, making them extremely specialized for uses where there isn't already a market for them. Additional modern issues are related to the policies of the membrane corporations: In addition to the fact that application-specific membranes are not yet being made, membrane designs are usually manufacturer-specific, making it difficult to switch to a competitor's membranes if needed. However, for an industrial usage to be successful and profitable, membranes must maintain their integrity over the duration of their lifetime [17].

5.1 Membrane Fouling and Cleaning

The accumulation of materials (called foulants) at the membrane's surface or in its pores that lowers the permeate flux through the membrane is referred to as membrane fouling [13]. Fouling can generally be split into two categories: internal and external. The buildup of discarded particles or foulants on the membrane's outside surface is known as external membrane fouling. Internal membrane fouling results from minute particles or large molecules sticking to a membrane's internal pore structure. Even in dynamic conditions, such fouling lowers the performance of the membrane [39]. Colloidal particles, salts with limited solubility, dissolved organic solvents, bacteria, protein molecules, and other particulates are examples of sources of fouling (foulants) [19]. Fouling can be classified into four categories based on the material that is deposited on the membrane's surface: colloidal, organic, inorganic, and biofouling [51].

The accumulation of the colloidal particles on the membrane surface causes what is known by colloidal fouling. The organic fouling is the adsorption of organic substances on the membrane surface. As the membrane structure becomes crystallised with solid salts present in the feed and deposit or precipitate as inorganic colloidal particles and particulates, inorganic fouling (scaling) results. Finally, biofouling (microbial fouling) happens as a result of bacteria, viruses, fungus, and other microorganisms adhering to the membrane's surface and growing there [39]. Fouling is a convoluted phenomenon that is prejudiced by a variety of circumstances whether it develops in the pores or on the membrane's surface. According to their source, these influences or contributors to membrane fouling can be classified as; (1) feed water characteristics; (2) foulant characteristics; (3) membrane materials and their properties; and (4) processing and operational conditions [14].

5.2 Osmotic Pressure and Concentration Polarization (CP)

CP is a phenomenon that taking place when concentration of salts near the membrane surface is higher than that in the bulk of the solution. CP happens as a consequence of the retained solutes' increased concentration at the feed or concentrates side. Due to the solvent going through the membrane and the solute being held, the flux is reduced [44]. The amount of concentration polarization is influenced by variables, including the mass transfer coefficient close to the membrane's surface, filtration flux, solute retention, and solute concentration. Figure 8 shows reducing concentration polarization effects.

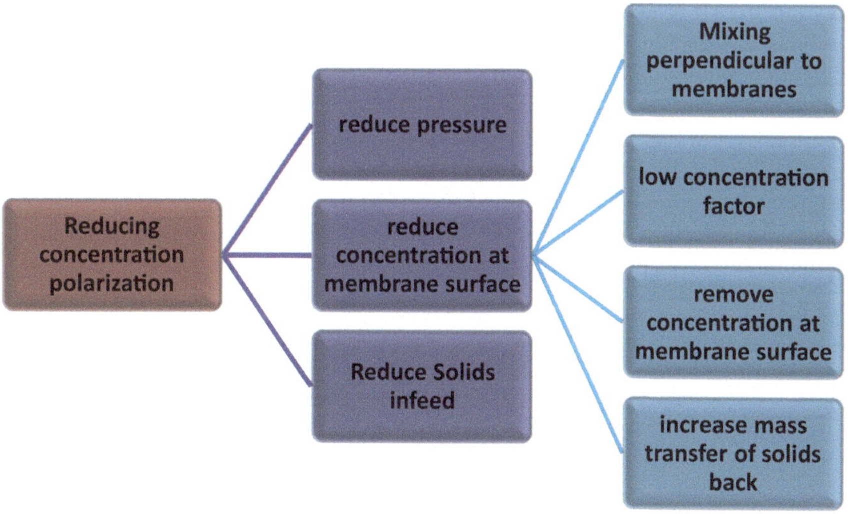

Fig. 8 Reducing concentration polarization effects

5.3 *Categories of Cleaning Methods*

The membranes have been reported to be cleaned using a number of techniques. According to Fig. 9, there are five broad groups into which these methodologies might be divided. Once the type of fouling has been identified, cleaning treatments can be employed to remove it from the membrane and restore its integrity.

The cleaning solutions to be used with the membranes are often provided by the membrane manufacturers. The pH restriction and oxidative impacts are two things to be aware of while using cleaning products. Figure 9 shows the five different types of regularly employed cleaning techniques for membranes, including NF and UF membranes in the water business [39].

6 Membrane Materials

Gas and oil separation, water production, wastewater treatment, and catalyst recovery are all accomplished using membranes. The intended application of the membrane material should be taken into account when selecting it. Chemical stability, permeability, fouling resistance, mechanical strength, pore features, casting ability, flux, and other parameters should therefore be taken into account. It's important to take into account additional elements including cost, selectivity, and separation issues [32]. Natural, synthetic, and biological membrane materials are categorized. Different types of synthetic membranes include inorganic, liquid, organic, composite, nanomaterials-polymeric, and hybrid membranes. Organic polymeric membranes are restricted to low temperatures uses and the separation of blends that are chemically inactive. It is advised to employ inorganic membranes with chemically active combinations and in high-temperature applications composed of materials such as ceramic, aluminium, glass, high-grade steel, and fiber-reinforced carbon. Both inorganic and polymeric membranes fall under the composite membrane category, commonly known as a hybrid or mixed matrix membrane. Only certain substances can pass through the membrane, which serves as the real barrier. It is layered on a porous backup. Membranes classified according to several classification whether it is based on materials, or on the structure and morphology, or according to geometry configuration, or separation process and lastly according to applications [32].

Fig. 9 Cleaning membranes main methods for fouling control

During the MBR wastewater treatment process, the particles and liquids are separated using MF or UF membranes. It is common practise to employ a membrane, which is a two-dimensional material, to separate fluid components according to their respective sizes or electrical charges. Permselectivity is the capability to allow only specific molecules to pass. The components separate during this physical phase without affecting their chemical structure. Retentate describes chemicals that are rejected while permeate describes substances that flow through a membrane's pores [4]. For MBR, there are numerous membrane configuration options, including plate-and-frame, pleated filter cartridge, spiral-wound, hollow fibre (HF), and tubular.

7 Emerging Membrane Technology for Sustainable Water Treatment

The benefits of membrane technologies are frequently acknowledged to include ease of use, high stability under operating conditions, low energy consumption, high eco-friendly, simplicity of control, scalability, and excellent adaptability. Many membrane activities can now be combined into a single process to increase efficiency in terms of product quality, plant compactness, energy use, and environmental effect [43]. Membrane-based techniques for creating potable water are mostly based on waste water treatment and also desalination. Several studies have been conducted over the last decade to improve the effectiveness of conventional membrane science and water treatment technologies, such as the reuse of waste sludge or brine, membrane reuse, waste energy harvesting, and the use of membrane antifouling techniques to reduce waste. This chapter discusses several emerging trends in desalination and wastewater treatment, as well as the issues they provide. Figure 10 summarizes the membrane science and technology emerging trends for sustainable water treatment.

7.1 Desalination

Desalination has shown to be a source of freshwater supply in areas with a lack of it, and there are currently a number of desalination facilities around the globe. Thermal and membrane-based technologies are the two main desalination methods [42]. By separating fresh water from salt water, desalination has shown to be the most effective method for addressing the problem of freshwater scarcity [6, 12]. Due to the enormous advantages that membrane-based technologies offer over thermal-based methods for desalination especially low energy consumption, many countries are increasingly switching to membrane-based technologies like RO. Numerous researches are being done to create improved membranes for desalination applications [42]. Due to their small capital and operating costs, low capital requirements, and low energy usage,

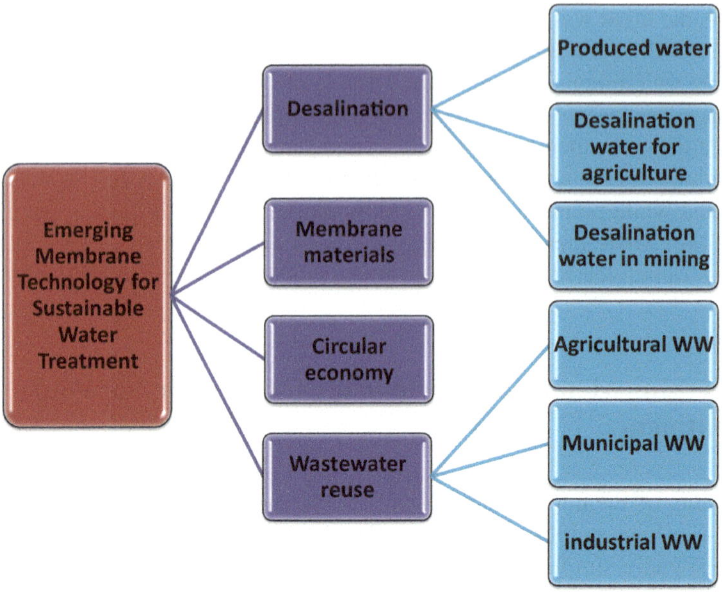

Fig. 10 Membrane science and technology emerging trends for sustainable water treatment

membrane-based techniques like membrane distillation (MD), RO, forward osmosis (FO), and are seen as competitive options.

7.1.1 Produced Water Treatment

A byproduct of the extraction method employed in the oil and gas sector is produced water (PW). In addition to brine, this PW also includes a variety of various organic and inorganic materials. 250 million barrels/day of PW are thought to be produced globally, with a trend towards increased production [42]. PW is challenging to handle because of its complex physicochemical characteristics, which can vary from well to well and over the course of a lifetime. Produced water contains organics and solids that are suspended and dissolved. To acquire fresh water, it is feasible to desalinate the PW by eliminating the pollutants and other dissolved salts that are present in it [5]. Reused water quality and the applicable membrane processes may differ depending on the intended use of the produced water. Other than for reinjection, external reuse applications demand substantially greater water quality. Historically, ultra-low pressure NF and RO were employed to treat generated water to meet potable water standards and for irrigation. However, MF and UF processes, either standalone or in combination, have been effective in separating suspended particles, macromolecules, and oil. Numerous studies have addressed the management of PW membrane-based technologies since PW has been rising each year.

(A) **Reverse Osmosis**

RO is the most widely utilized desalination method worldwide, with the lowest cost of producing fresh water [43]. Over the past 40 years, several technological advancements in both membrane and system design have led to RO's current dominant position [16]. Salt passage has decreased sevenfold during the last 30 years as a result of complex pre-treatment processes, enhanced biological, mechanical, and chemical robustness of RO membranes, and increased membrane permeability. Additionally, these advancements have more than ten times reduced the cost of the membrane per unit volume of produced water [1]. In addition, the combined impact of reduce fouling, decreased energy use, the polarization effect, and improved energy recovery. Recent developments in polymeric membrane materials, which are phased out in three stages, have contributed greatly to the advancement of RO. Early polymer testing by trial and error, using an appropriate polymerization procedure, and studying membrane shape with the use of advanced characterization tools are all examples of early polymer testing. The difficulty of handling this enormous amount of brine that is discharged from RO plants is still being researched. Thermal efficiencies, freshwater costs, generated water quality, configurations, water recovery parameters, and energy consumption all affect the performance and application of RO plants.

(B) **Membrane Distillation:**

An emerging technology for desalinating saltwater is MD. The partial vapour pressure differential caused by the temperature gradient across the membrane drives the MD process. A hybrid process called MD combines thermal separation methods and membrane. In a direct contact membrane distillation (DCMD) setup, freshwater passes through a microporous hydrophobic membrane's permeate/coolant side at a lower temperature while the feed side is heated and contains salt [8]. The membrane's thickness, porosity, liquid entry pressure, pore size, thermal conductivity, roughness, membrane surface area, and, finally, the membrane material itself all have an impact on how well MD performs. Membranes created for the MD process must have the appropriate structural chemistry to ensure optimal performance. Diverse research teams have worked to develop novel membranes using novel fabrication techniques and appropriate nanomaterials in an effort to address these problems. These efforts have given them control over the membrane structural chemistry, which is essential for improving membrane desalination performance [25].

7.1.2 Desalination in Agriculture

Direct freshwater sources used to be affordable, eliminating the need for more expensive desalinated water in agriculture. Desalinated water is becoming more competitive, however, as its cost falls while that of surface water and groundwater rises. Desalinated water is being used in agriculture all over the world more and more.

Since it is now obvious that using groundwater is not sustainable, new laws are anticipated to reduce its use in favor of desalinated water [24].

Desalination removes ions necessary for plant growth as well as undesirable salts from water. When used in agriculture, desalinated water has additional benefits since it provides enough quantities of nutrients including calcium, magnesium, and sulphate. Desalination for agriculture differs in that it requires less manpower, no post-treatment because the water produced will be nutrient-rich, fewer civil engineering projects, and lower energy expenses. The production of water for agriculture is dominated by two desalination technologies: brackish water and seawater RO. However, compared to thermal desalination systems, SWRO has become the most innovative and dominant technique due to superior product water quality, less energy use, and consequently lower water costs [18].

7.1.3 Desalination in Mining

In its early stages of development, desalination was only employed to supply domestic and industrial needs. Once this technology had been enhanced, it was used in various industries, particularly mining. Mining industry consumes large amount of water a wide range of operational tasks, including the transportation of ore and waste in slurries and suspension, the chemical and physical separation of materials, the control of dust, the cleaning of equipment and, the cooling systems used in power generation [27]. Due to the fact that mines are frequently found in isolated, arid areas, there are water difficulties in the mining industry. In addition, as spent mine water is frequently contaminated with explosive residue, salts, and acids, it must be treated before it can be recycled or disposed of. For isolated and dry mines, the main water sources are seawater and groundwater, both of which have a high salinity. However, Desalinated water, which is made from seawater and groundwater, is preferred because it can prevent damage to surrounding aquifers, protect miners' health, and limit corrosion of underground mining equipment [24, 27]. RO, and Electrodialysis are the main desalination technology used to treat mine waters. modified RO processes including seeded RO (SRO) and the slurry precipitation and recycle RO (SPARRO) technology have been proposed [24].

7.2 Wastewater Reuse

Using membrane technology to recover water for reuse from different WW sources has grown in popularity. Waste water (WW) reuse and sustainable water recycling are urgently needed for rising water demand of human activities and sever water scarcity. After proper treatment, WW reuse can successfully contribute to the resolution of this pressing issue.

7.2.1 Agricultural Wastewater Treatment

Agriculture and aquaculture WW have a long history and have been widely used in a variety of geographic, environmental, and geopolitical contexts. The use of treated wastewater in aquaculture and agriculture has proven to be great despite its multiple socioeconomic, environmental, and health effects. It is necessary to develop WW treatment procedures in order to improve the quality of the treated water for reuse [18]. To reduce quantities of solids and different organic and inorganic pollutants, biological, chemical, and physical processes are used in traditional WW treatment facilities. These procedures use a variety of techniques, such as flotation and sedimentation, to efficiently remove both organic and inorganic particles, including oil and grease. The membrane technology appears to be the most effective in WW reuse with better perspective of an appropriate degree of WW treatment efficiency when compared to the standard filtration, coagulation followed by flocculation, and biochemical treatment systems [11]. MBRs, membrane filtration, RO, and electrodialysis and super advanced disinfection processes such as chlorination process all used for WW reuse. A proper characterization of urban WW quality and analysis of the results are necessary for choosing better treatment and reuse methods for agricultural WW. Agriculture irrigation using treated WW must adhere to sanitary, environmental, and technological criteria. The main benefits of using WW for irrigation in agriculture and aquaculture, are food security employment, and increased income [18].

7.2.2 Municipal Wastewater Treatment

Sewage is the primary WW created by human activities in daily life and is referred to as municipal WW [24]. Due to the significant contribution of human faeces, municipal WW comprises a large percentage of unutilized residuals of ingested food and their degradation products. Municipal WW contains compounds containing carbon, nitrogen, phosphorous, potassium, sulphur, and other possible nutritional sources. The same is true for recovering metals and heavy metals from municipal streams. Along with the iron, manganese, nitrate, and sulphate added by geogenic sources, varying salts can also be detected depending on water consumption/utilization and source type [20]. Organic matter, suspended particles, and pathogens comprise the majority of the components in municipal WW. To avoid eutrophication, effluent that contains nitrogen must be thoroughly treated before being released [9]. Due to their benefits over traditional activated sludge systems, which are listed in this chapter's section on the advantages of MBR, membrane bioreactors are frequently used in municipal WW treatment systems for the advanced waste treatment [9]. This is mostly due to growing worries about trends in water contamination and quality in relation to more difficult global issues like increased urbanization and rising domestic water demands [36]. MBR conduct the sludge separation by integrating either a MF or an UF membrane into a standard activated sludge (CAS) reactor, that remove efficiently high amounts of TOC and biochemical oxygen demand (BOD) and completely removes total suspended solids (TSS) with advantage of inexpensive and easy to operate, Also, MF typically is used as a pretreatment stage for other pressure-driven membrane processes like RO and NF to reduce membrane fouling [24].

7.2.3 Industrial Wastewater Reuse

Industrial wastewater arises from a variety of industries, such as the food, mining, steel and, iron pulp, and paper pharmaceutical, textile, and chemical sectors. Most industrial effluent can be handled effectively by membrane technology similar to municipal wastewater treatment [24]. In comparison to conventional treatment techniques, which often combine physical, chemical, and biological processes, membrane technology offers more opportunities for reuse together with improved removal efficiency and selectivity. Additionally, due to the phenomenon of fouling, certain physicochemical factors, such as turbidity, total suspended particles, TOC, and COD, directly affect the application of membranes [15]. The RO has been recognized, among other membrane processes, for its exceptional efficacy in the separation of monovalent ions, such as sodium and chloride ions, with percentages above 99.5%, and microscopic particles, such as microorganisms. Usually, the RO membranes were used for filtration after the NF, MF, and UF membranes were used as a pretreatment. Preventing fouling of the RO membranes and maintaining a steady flow of treated water are the intended outcomes [15].

7.3 Membrane Materials (Green and Sustainable Membranes)

Traditional membranes frequently include toxic materials and become fouled with pollutants, posing safety and environmental problems. Therefore, non-toxic materials with desired efficiency should take the place of petroleum-based membranes [34]. A promising strategy to reduce the use of membranes made of petroleum, which have risky properties and are expensive, is to apply green chemistry to the manufacturing of nano-membranes [35]. Nano-membranes have recently achieved a strong point in this technical arena, despite the fact that membrane-based water purification plants are costly, and environmentally unfavorable. Green chemistry and synthesis should be integrated with nanotechnology to fulfill the aforementioned needs [31]. Such issues in membrane technology have been proved to be amenable to new technologies like nanofabrication and nanoparticles [46]. Furthermore, manufacturers of membranes are inevitably concerned about disposal and recycling of synthetic nano-membranes [34]. Over the past twenty years, substantially more focus and funding have been placed on green membranes and components made from green nanomaterials of various types [47]. As it a promising for removal of contaminants but some challenges to be considered about green technology regarding to,

- Green nano-membranes should be designed with sustainability, toxicity, and biosafety standards in mind in order to achieve manufacturing at industrial and commercial sizes;
- Cost-effectiveness studies are focused on the production of green nanomembranes should be considered and compared with conventional traditional membranes.

In comparison to conventional materials like polymeric membranes, using natural components in membrane technology has various benefits as bioavailability and biocompatibility which make it more environmentally friendly. In order to improve their qualities and functionality, natural materials can also be mixed with other substances like metal nanoparticles, graphene, and carbon nanotubes. With multiple current research projects targeted at enhancing their functionality and extending their applications, it appears that green or natural component-based membranes have a bright future in water treatment and other applications [34].

7.4 Circular Economy and Membrane Technology

Natural resources are restored and materials are never wasted in a circular economy. In this type of economy, processes including maintenance, remodelling, reuse, recycling, remanufacturing, and composting keep resources and goods in use. By separating economic activity from the use of finite resources, the circular economy responds to issues like pollution and climate change [21]. Wastes generated during desalination and WW treatment utilising membrane-based techniques must be disposed of properly. These wastes include waste heat, waste sludge, spent cooling water, brine, and process materials that have reached the end of their useful life. The most convenient methods of disposing of membranes are burning and landfilling, nevertheless, these methods are considered to be unsustainable practises. Waste minimization must be ensured in the circular economy's future directions for membrane-based desalination systems. Although membrane reuse, recycle has been recognised as one of the practical options, it also has certain drawbacks, most notably bio-fouling. Membrane Biofilm Reactors (MBfRs) have discovered uses for fouled RO membranes that can't be recycled as NF or UF membrane [50]. Because recovered membranes typically still go through the production step, where additional energy and resources are used, recycling is a less preferable circular economy idea than reuse. Recycling materials upholds the ideals of the circular economy better than incineration or the least preferred alternative, disposal, even if it is not as effective as reuse. In order to increase membrane longevity and lower the rate of membrane replacement, the antifouling capabilities of the membrane have also been improved [50].

Waste from one membrane process, such as reject brine, cooling water, and waste heat, can be reused in another process to reduce overall energy consumption. Problematic and stubborn types of various chemicals in the solution could lead to the handling of the brine streams. Traditional procedures are purportedly regarded as appropriate processes after the removal of organic matter because they effectively remove and detoxify organic matter. The oxidation method's high energy consumption and the coagulation method's sludge generation are both major obstacles to their use in these applications. An additional concern related to oxidative reactions is the production of by-product chemicals. However, the traditional methods are utterly ineffective at removing additional compounds from the concentrate solution. Membrane technology offers an effective alternative for handling concentrates that may avoid the risk of often using removal and elimination procedures to manage concentrates. When the environment, geographic conditions, and climatic conditions are taken into account, traditional treatment methods are only effective in eliminating the organic matter, this is more advantage membrane technology over the traditional treatment techniques. The employment of membrane-based techniques for concentrate disposal, such as MD, FO, and Electrodialysis, is fascinating, trustworthy, and ecologically advantageous because they provide high-quality effluent while recovering resources. Compared to other treatment procedures, the FO method requires low energy consumption. Existence of a waste heat source to warm the feed water to the MD process or warm the draw solution in the FO process is a significant advantage of both processes [3]. Recently, it have seen a significant increase in the usage of brine reuse in zero-liquid discharge (ZLD) technologies. ZLD is a strategic WW management system that makes sure there won't be any environmental discharge of industrial WW. It is accomplished by first recycling WW, followed by recovery and reuse for industrial purposes. ZLD attempts to recover 100% treated water by avoiding liquid waste from being discharged. ZLD may improve the water economics of treatment by increasing treated water output while decreasing concentrated discharge production from water producing plants. ZLD techniques are relatively expensive, and some of them, such evaporation for brine minimization, still need a lot of solar energy to operate. ZLD techniques are therefore rarely applied in the desalination industry [50].

8 Hybrid Membrane Processes (HMP)

A membrane filtration unit (MF/UF/NF) and additional processes, such as adsorption, coagulation, and ion exchange, are combined in an integrated or hybrid membrane system. It is believed that the combination of treatments utilized in hybrid membrane processes will allow for the possibility of going beyond the limitations of conventional techniques. The following is a list of the advantages: HMP can be utilized to boost permeate flux and reduce membrane fouling, which improves removal efficiency and results in cleaner products [37]. The industry has extensively used membrane technologies, such as MD, RO, FO, osmotically enhanced dewatering-reverse osmosis (OED), and crystallisation systems, to create high-quality water from the feed water extraction process. The primary purpose of hybridization is to eliminate pollutants, which involves a pretreatment stage to remove big oil droplets, particles, and gas bubbles. This is done to guarantee the membrane's effectiveness and long-term use in the treatment of WW. The industry has made extensive use of membrane technologies to produce high-quality water from the process of extracting feed water [3]. The economic viability of hybrid membrane systems in water treatment and desalination processes is contingent upon several important and crucial criteria, one of which is cost.

9 Future Scope and Conclusions

The emerging trends in membrane technology and its application have been discussed in this chapter with its challenges, and its reported achievements and innovative approaches in the last decade. These are reported as the study's main conclusions.

1. Membrane separation is an energy-efficient, clean technology that can substitute conventional processes like ion exchange, distillation, filtration, and chemical treatment systems. Other advantage comprises its capacity for high-quality product production and adaptability in system design.
2. Membrane technology has a broader range of industrial and environmental applications because of its multidisciplinary nature, including water treatment for domestic and industrial wastewater supply, desalination, biotechnology industry, sludge treatment, aquaculture, water reuse and beverage, food, pharmaceutical industry and metallurgical separation processes.
3. The membrane bioreactors produce high quality effluent with less sludge generation, which may be reused or recycled right away.
4. Membrane bioreactors can also be used to develop hybrid processes by combining it with various separation techniques. Membrane fouling, concentration polarisation, short membrane lifetime, poor selectivity, poor flow, and high cost are the main methods drawbacks.
5. Membrane fouling and concentration polarization are the main challenges facing membrane industry. Numerous membrane types are employed in the production of drinking water, desalination, wastewater reclamation, and reuse. New membrane development and inventive operational methods gave rise to a promising sector. The primary obstacles to the development of membrane technology typically have to do with economic concerns: membranes are costly, and membrane systems are still energy-intensive, both of which have a negative influence on operating costs.
6. Desalination and wastewater reuse are the mainstays of emerging membrane-based methods.

7. RO, MD and FO are the main approaches being done to create improved membranes for desalination applications, especially in the management of PW membrane-based technologies since PW has been rising each year.
8. Waste water (WW) reuse and sustainable water recycling are urgently needed for rising water demand of human activities and sever water scarcity. Agricultural, aquaculture, municipal and industrial wastewater reuse are main applications.
9. Using green chemistry to make nano-membranes is a promising solution to reduce the usage of petroleum-based membranes, which are dangerous and expensive. To achieve the aforementioned requirements, green chemistry and synthesis must be integrated with nanotechnology.
10. In order to improve the properties and functionality of natural materials, they can also be coupled with additional substances like graphene, carbon nanotubes, and metal nanoparticles. With multiple ongoing research projects targeted at enhancing their performance and broadening their applications, the future of green or natural component-based membranes in water treatment and other applications seems promising.
11. The discharge of wastes leaving the activities is required for desalination and WW treatment using membrane-based techniques. These wastes include process materials that have reached the end of their useful lives as well as waste sludge, heat, brine, and spent cooling water, among others.
12. Future directions for membrane-based desalination systems in the circular economy must assure waste minimization, although membrane recycling and reuse have been acknowledged as possible options.
13. It is anticipated that the hybrid membrane processes' combination of treatments will make it possible to overcome the limits of traditional methods. Cost is one of the most critical factors in determining whether hybrid membrane systems are economically viable for use in desalination and water treatment processes.

Given the promising future of membrane technology, the development of more sophisticated and target-specific membranes will necessitate open-minded methods.

References

1. Abdelrasoul A, Doan H, Lohi A, Sustainable Water Technology and Water-Energy Nexus (2017) Biomimetic and bioinspired membranes for new frontiers in sustainable water treatment technology. InTech, London, UK
2. Abdullah N, Rahman MA, Dzarfan Othman MH, Jaafar J, Ismail AF (2018) Membranes and membrane processes. In: Current trends and future developments on (bio-) membranes. Elsevier, Amsterdam, pp. 45–70. https://doi.org/10.1016/B978-0-12-813549-5.00002-5
3. Ahmad NA, Goh PS, Yogarathinam LT, Zulhairun AK, Ismail AF (2020) Current advances in membrane technologies for produced water desalination. Desalination 493:114643. https://doi.org/10.1016/j.desal.2020.114643
4. Al-Asheh S, Bagheri M, Aidan A (2021) Membrane bioreactor for wastewater treatment: a review. Case Stud Chem Environ Eng 4:100109. https://doi.org/10.1016/j.cscee.2021.100109

5. Al-Ghouti MA, Al-Kaabi MA, Ashfaq MY, Da'na DA (2019) Produced water characteristics, treatment and reuse: a review. J Water Process Eng 28:222–239. https://doi.org/10.1016/j.jwpe.2019.02.001
6. AlZainati N, Saleem H, Altaee A, Zaidi SJ, Mohsen M, Hawari A, Millar GJ (2021) Pressure retarded osmosis: advancement, challenges and potential. J Water Process Eng 40:101950. https://doi.org/10.1016/j.jwpe.2021.101950
7. Baker RW (2012) Membrane technology and applications, 3rd edn. Wiley, Hoboken
8. Boukhriss M, Timoumi M, Bacha HB (2023) Experimental of membrane distillation unit coupled with a DCMD using solar energy. Solar Compass 7:100055. https://doi.org/10.1016/j.solcom.2023.100055
9. Deng J, Xiao X, Li Y-Y, Liu J (2023) Low-carbon nitrogen removal from power plants circulating cooling water and municipal wastewater by partial denitrification-anammox. Biores Technol 380:129071. https://doi.org/10.1016/j.biortech.2023.129071
10. Galinha CF, Sanches S, Crespo JG (2018) Membrane bioreactors. In: Fundamental modelling of membrane systems. Elsevier, Amsterdam, pp 209–249. https://doi.org/10.1016/B978-0-12-813483-2.00006-X
11. Gao C, Wang Z, Ji X, Wang W, Wang Q, Qing D (2023) Coupled improvements on hydrodynamics and water quality by flowing water in towns with lakes. Environ Sci Pollut Res 30(16):46813–46825. https://doi.org/10.1007/s11356-023-25348-3
12. Ghazi ZM, Rizvi SWF, Shahid WM, Abdulhameed AM, Saleem H, Zaidi SJ (2022) An overview of water desalination systems integrated with renewable energy sources. Desalination 542:116063. https://doi.org/10.1016/j.desal.2022.116063
13. Guillen-Burrieza E, Ruiz-Aguirre A, Zaragoza G, Arafat HA (2014) Membrane fouling and cleaning in long term plant-scale membrane distillation operations. J Membr Sci 468:360–372. https://doi.org/10.1016/j.memsci.2014.05.064
14. Hilal N, Khayet M, Wright CJ (2012) Membrane modification: technology and applications. CRC Press, Boca Raton
15. Issaoui M, Jellali S, Zorpas AA, Dutournie P (2022) Membrane technology for sustainable water resources management: challenges and future projections. Sustain Chem Pharmacy 25:100590. https://doi.org/10.1016/j.scp.2021.100590
16. Joshi M, Swaminathan J (2023) Reverse osmosis membrane and module improvement roadmap for maximum impact. Desalination 554:116511. https://doi.org/10.1016/j.desal.2023.116511
17. Kabay N, Shirazi MMA, Güler E, Bryjak M (2022) Grand challenges in membrane modules and processes. Front Membr Sci Technol 1. https://doi.org/10.3389/frmst.2022.913597
18. Karimi-Maleh H, Darabi R, Karimi F, Karaman C, Shahidi SA, Zare N, Baghayeri M, Fu L, Rostamnia S, Rouhi J, Rajendran S (2023) State-of-art advances on removal, degradation and electrochemical monitoring of 4-aminophenol pollutants in real samples: a review. Environ Res 222:115338. https://doi.org/10.1016/j.envres.2023.115338
19. Kelly ST, Zydney AL (1997) Protein fouling during microfiltration: comparative behavior of different model proteins. Biotechnol Bioeng 55(1):91–100. https://doi.org/10.1002/(SICI)1097-0290(19970705)55:1%3c91::AID-BIT11%3e3.0.CO;2-6
20. Kharraz JA, Khanzada NK, Farid MU, Kim J, Jeong S, An AK (2022) Membrane distillation bioreactor (MDBR) for wastewater treatment, water reuse, and resource recovery: a review. J Water Process Eng 47:102687. https://doi.org/10.1016/j.jwpe.2022.102687
21. Kowalik-Klimczak A, Gajewska-Midziałek A, Buczko Z, Łożyńska M, Życki M, Barszcz W, Ciciszwili T, Dąbrowski A, Kasierot S, Charasińska J, Gorewoda T (2023) Circular economy approach in treatment of galvanic wastewater employing membrane processes. Membranes 13(3):325. https://doi.org/10.3390/membranes13030325
22. Krzeminski P, van der Graaf JHJM, van Lier JB (2012) Specific energy consumption of membrane bioreactor (MBR) for sewage treatment. Water Sci Technol 65(2):380–392. https://doi.org/10.2166/wst.2012.861
23. Ladewig B, Al-Shaeli MNZ (2017) Fundamentals of membrane processes, pp 13–37. https://doi.org/10.1007/978-981-10-2014-8_2

24. Le NL, Nunes SP (2016) Materials and membrane technologies for water and energy sustainability. Sustain Mater Technol 7:1–28. https://doi.org/10.1016/j.susmat.2016.02.001
25. Liao Y, Wang R, Fane AG (2013) Engineering superhydrophobic surface on poly(vinylidene fluoride) nanofiber membranes for direct contact membrane distillation. J Membr Sci 440:77–87. https://doi.org/10.1016/j.memsci.2013.04.006
26. Liu W, Song X, Huda N, Xie M, Li G, Luo W (2020) Comparison between aerobic and anaerobic membrane bioreactors for trace organic contaminant removal in wastewater treatment. Environ Technol Innov 17:100564. https://doi.org/10.1016/j.eti.2019.100564
27. Macedonio F, Drioli E (2018) Special issue of desalination journal on "membrane engineering for desalination in the mining and extraction industry." Desalination 440:1. https://doi.org/10.1016/j.desal.2018.04.025
28. Mannina G, Alliet M, Brepols C, Comas J, Harmand J, Heran M, Kalboussi N, Makinia J, Robles Á, Rebouças TF, Ni B-J, Rodriguez-Roda I, Victoria Ruano M, Bertanza G, Smets I (2021) Integrated membrane bioreactors modelling: a review on new comprehensive modelling framework. Biores Technol 329:124828. https://doi.org/10.1016/j.biortech.2021.124828
29. Moneer AA (2023) The potential of hybrid electrocoagulation-membrane separation processes for performance enhancement and membrane fouling mitigation: a review. Egypt J Aquat Res. https://doi.org/10.1016/j.ejar.2023.08.007
30. Mulder M (1996) Basic principles of membrane technology. Kluwer Academic Publishers, Boston
31. Nasrollahzadeh M, Sajjadi M, Iravani S, Varma RS (2021) Green-synthesized nanocatalysts and nanomaterials for water treatment: current challenges and future perspectives. J Hazard Mater 401:123401. https://doi.org/10.1016/j.jhazmat.2020.123401
32. Nath K (2011) Membrane separation processes. PHI Learning Private Limited, New Delhi
33. Pandey A, Singh R (2014) Industrial waste water treatment by membrane bioreactor system. Elixir Chem Eng 70:23772–23777
34. Rabiee N, Sharma R, Foorginezhad S, Jouyandeh M, Asadnia M, Rabiee M, Akhavan O, Lima EC, Formela K, Ashrafizadeh M, Fallah Z, Hassanpour M, Mohammadi A, Saeb MR (2023) Green and sustainable membranes: a review. Environ Res 231:116133. https://doi.org/10.1016/j.envres.2023.116133
35. Rethinam S, Basaran B, Vijayan S, Mert A, Bayraktar O, Aruni AW (2020) Electrospun nano-bio membrane for bone tissue engineering application—a new approach. Mater Chem Phys 249:123010. https://doi.org/10.1016/j.matchemphys.2020.123010
36. Sadr SMK, Saroj DP (2015) Membrane technologies for municipal wastewater treatment. In: Advances in membrane technologies for water treatment. Elsevier, Amsterdam, pp 443–463. https://doi.org/10.1016/B978-1-78242-121-4.00014-9
37. Saha P, Jampa SS, Sinha MK, Khuntia S (2023) Hybrid membrane process for water treatment: a short review. AQUA Water Infrastruct Ecosyst Soc 72(5):608–622. https://doi.org/10.2166/aqua.2023.097
38. Saleh TA, Gupta VK (2016a) Membrane classification and membrane operations. In: Nanomaterial and polymer membranes. Elsevier, Amsterdam, pp 55–82. https://doi.org/10.1016/B978-0-12-804703-3.00003-6
39. Saleh TA, Gupta VK (2016b) Membrane fouling and strategies for cleaning and fouling control. In: Nanomaterial and polymer membranes. Elsevier, Amsterdam, pp 25–53. https://doi.org/10.1016/B978-0-12-804703-3.00002-4
40. Santos AD, Martins RC, Quinta-Ferreira RM, Castro LM (2020) Moving bed biofilm reactor (MBBR) for dairy wastewater treatment. Energy Rep 6:340–344. https://doi.org/10.1016/j.egyr.2020.11.158
41. Scholes E, Verheyen V, Brook-Carter P (2016) A review of practical tools for rapid monitoring of membrane bioreactors. Water Res 102:252–262. https://doi.org/10.1016/j.watres.2016.06.031
42. Shahrim NA, Abounahia NM, El-Sayed AMA, Saleem H, Zaidi SJ (2023) An overview on the progress in produced water desalination by membrane-based technology. J Water Process Eng 51:103479. https://doi.org/10.1016/j.jwpe.2022.103479

43. Shehata N, Egirani D, Olabi AG, Inayat A, Abdelkareem MA, Chae K-J, Sayed ET (2023) Membrane-based water and wastewater treatment technologies: Issues, current trends, challenges, and role in achieving sustainable development goals, and circular economy. Chemosphere 320:137993. https://doi.org/10.1016/j.chemosphere.2023.137993
44. Shirazi S, Lin C-J, Chen D (2010) Inorganic fouling of pressure-driven membrane processes—a critical review. Desalination 250(1):236–248. https://doi.org/10.1016/j.desal.2009.02.056
45. Song X, Luo W, Hai FI, Price WE, Guo W, Ngo HH, Nghiem LD (2018) Resource recovery from wastewater by anaerobic membrane bioreactors: opportunities and challenges. Biores Technol 270:669–677. https://doi.org/10.1016/j.biortech.2018.09.001
46. Vatanpour V, Mansourpanah Y, Mousavi Khadem SS, Zinadini S, Dizge N, Ganjali MR, Mirsadeghi S, Rezapour M, Saeb MR, Karimi-Male H (2021) Nanostructured polyethersulfone nanocomposite membranes for dual protein and dye separation: lower antifouling with lanthanum (III) vanadate nanosheets as a novel nanofiller. Polym Testing 94:107040. https://doi.org/10.1016/j.polymertesting.2020.107040
47. Wang Y, Wang S, Wang T, Song T, Wu X, Guo L, Xie W, Qiu P, Dong Q, Li Q (2023) A green nanocomposite membrane for concrete moisturizing, with excellent barrier properties and aging resistance. Mater Today Commun 35:105553. https://doi.org/10.1016/j.mtcomm.2023.105553
48. Werkneh AA (2022) Application of membrane-aerated biofilm reactor in removing water and wastewater pollutants: current advances, knowledge gaps and research needs—a review. Environ Challenges 8:100529. https://doi.org/10.1016/j.envc.2022.100529
49. Yadav G, Mishra A, Ghosh P, Sindhu R, Vinayak V, Pugazhendhi A (2021) Technical, economic and environmental feasibility of resource recovery technologies from wastewater. Sci Total Environ 796:149022. https://doi.org/10.1016/j.scitotenv.2021.149022
50. Yusuf A, Sodiq A, Giwa A, Eke J, Pikuda O, De Luca G, Di Salvo JL, Chakraborty S (2020) A review of emerging trends in membrane science and technology for sustainable water treatment. J Clean Prod 266:121867. https://doi.org/10.1016/j.jclepro.2020.121867
51. Zhang TC, Surampalli RY, Vigneswaran S, Tyagi RD, Ong SL, Kao CM (2012) Membrane technology and environmental applications. American Society of Civil Engineers, Reston

Sequential Batch Reactor Processes and Membrane Technologies

Sequential Batch Reactor to Purify Wastewater from Olive Oil Mills, Mixed with Municipal Wastewater

Khadija Zahidi, Latifa Mouhir, Chaymae Haddaji, Zineb Ettaloui, Oussama Hartal, Amina Khalidi Idrissi, Abdelaziz Madinzi, Khalid Digua, Roukaya Bouyakhsass, Aysegul Pala, Tonni Agustiono Kurniawan, Abdelkader Anouzla, Salah Souabi, and Safaa Khattabi Rifi

Abstract Creating wastewater from olive oil mills (OMW) represents a real environmental challenge in food manufacturing. These wastewater effluents contain several complex components. This study investigated the capacity of a sequential batch reactor (SBR) to treat diluted olive oil mill wastewater in conjunction with domestic sewage. The study examined two OMW dilutions. The D1 dilution increased OMW biodegradability and reduced pollution. The efficacy of each reactor in reducing pollution from both dilutions was monitored throughout the experiment. Both systems operated on a continuous 24-h repeating procedure. The SBR D1 showed considerable reductions in COD, polyphenols, nitrates, ammonium, and phosphorus, reaching 68.9%, 67.1%, 75.3%, and 95.3%, respectively. In contrast, 40.2%, 47%, 61.2%, 61.1%, and 94.1% of COD, polyphenols, nitrates, ammonium, and phosphorus, respectively, were removed by SBR D2. The proposed treatment method represents a simple, cost-effective process that could be used in the olive oil sector.

K. Zahidi · L. Mouhir · C. Haddaji · Z. Ettaloui · O. Hartal · A. K. Idrissi · A. Madinzi · K. Digua · R. Bouyakhsass · A. Anouzla · S. Souabi (✉) · S. K. Rifi
Laboratory of Process Engineering and Environment, Faculty of Science and Technology, University Hassan II, Mohammedia, Morocco
e-mail: salah.souabi@gmail.com

C. Haddaji
e-mail: chaymae.haddaji@etu.fstm.ac.ma

K. Digua
e-mail: k.digua@univh2m.ac.ma

A. Pala
Environmental Research and Development Center (CEVMER), Dokuz Eylul University, Izmir, Turkey
e-mail: aysegul.pala@deu.edu.tr

T. A. Kurniawan
College of Ecology and the Environment, Xiamen University, Xiamen 361102, Fujian, China

S. Yadav et al. (eds.), *Emerging Technologies in Biological and Hybrid Wastewater Treatment: Lessons from Developed to Enhancing Practices in Developing Countries*, Earth and Environmental Sciences Library,
https://doi.org/10.1007/978-3-031-74515-7_4

Keywords Olive oil mill wastewater · Municipal wastewater · SBR · Polyphenols · Nitrification · Denitrification

1 Introduction

Olive cultivation is one of humanity's earliest agricultural practices. Olive oil is crucial in Mediterranean diets due to its high nutritional value and multiple health advantages. However, they are also grown in Europe, the Middle East, the United States, Argentina, and Australia. Olive oil production is expanding worldwide, owing to its numerous virtues and usefulness as a health-promoting diet [1]. Olive oil effluent, often called OMW (Olive Mill Wastewater), combines solids and liquids and is generally slightly acidic in pH. These effluents have distinctive characteristics such as a pronounced odor, a dark brown color, and suspended solids. They consist mainly of water, representing between 80 and 95% of their total weight, and organic elements such as lipids, polyphenols, carbohydrates, long-chain fatty acids, and inorganic components. Because of its high salinity and high concentration of solid and organic contaminants, such as phenolic compounds, direct discharge of OMW without previous treatment can result in soil and water pollution as well as deterioration of its physical, chemical, and biological properties [2].

Spreading wastewater in evaporation ponds is an economical solution for minimizing environmental pollution, replacing direct discharge into rivers, due to the natural partial decomposition of biodegradable matter. However, the use of this method can give rise to certain environmental problems associated with clogging, flooding, filtration and evaporation of heavily polluted wastewater. The SBR (Sequencing Batch Reactor) is an integrated wastewater treatment system that uses an anaerobic-aerobic mode. In these anaerobic-aerobic bioreactors, wastewater is managed by a fill-and-pull method. The typical SBR process consists mainly of five phases, carried out in the following order: filling, reaction, settling, reaction, and emptying. When compared to traditional wastewater treatment methods, the sequential batch reactor (SBR) stands out for its ease of use, speed, high level of functionality, and capacity to respond efficiently to varied pollutants. SBR systems have a variety of advantages, including reduced operating costs, and the ability to merge nitrification and denitrification processes in a single reactor. Furthermore, they have proven good performance in eliminating nitrogen, phosphorus, and chemical oxygen requirements.

The main objective of this study is to measure the efficiency of this treatment process. The extent to which the pollutants contained in the OMW are eliminated or reduced is assessed. The ultimate goal is to lower the OMW's pollutant load to an environmentally acceptable level, hence minimizing the negative effects on aquatic ecosystems and water resources. This pollutant reduction evaluation will help determine the efficiency of the SBR reactor in the context of diluted OMW treatment.

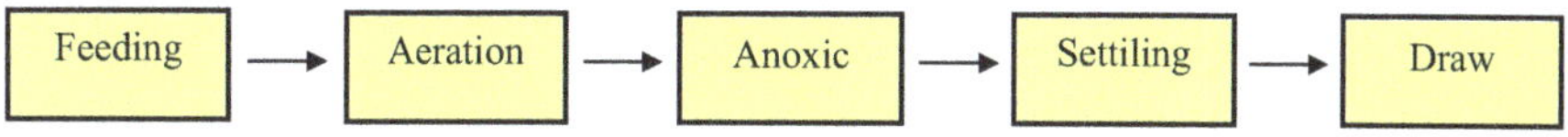

Fig. 1 Cycles used for the two SBRs during the experimentation period

2 Materiels et Methodes

2.1 Wastewater and Seeding

A wastewater sample from a municipal wastewater treatment facility in Benslimane, Morocco, was used as the inoculant to initiate the sequencing batch reactor (SBR). In December 2020, a 30-L sample of freshly obtained, untreated OMW was collected using an olive oil mill fitted with batch-pressure extraction equipment.

2.2 Sequencing Batch Reactor (SBR)

First, two SBR reactors (D1 and D2) received a 2-L mixture of diluted OMW. Dilution was accomplished by mixing 50% OMW with 50% wastewater in one reactor and 75% OMW with 25% wastewater in the other. The 50 and 25% dilution percentages were chosen to improve OMW biodegradability and facilitate the treatment of a large quantity of OMW during the treatment period.

Before initiating the SBR process, a mixture of wastewater and OMW, comprising 1 L and 1 L of OMW, underwent continuous aeration with agitation for 72 h. This step intended to increase bacterial biomass. After allowing the organisms to settle, the clear supernatants were carefully removed, and the reactors were then filled with diluted OMW to a total capacity of 2 L.

These stages included 30 min of effluent addition, followed by 17 h of aeration, 4 h of anoxia, 2 h of anaerobic conditions, and finally a 60-min decantation phase. Figure 1 illustrates the recommended sequence of steps.

The oxygen concentration was carefully regulated to remain around 2 mgO_2/L throughout the experiment. No additional carbon source was introduced during the study. Mechanical agitation was employed during all phases except for the decantation and emptying stages. During the decantation and emptying phases, a peristaltic pump was utilized to extract the effluent (Fig. 2).

2.3 Pysicochemical Analyses

Using an "Accumet basic ab15 pH meter" by standard NF T 90-008 February 2001 (T 90-008), the pH of the various samples was measured.

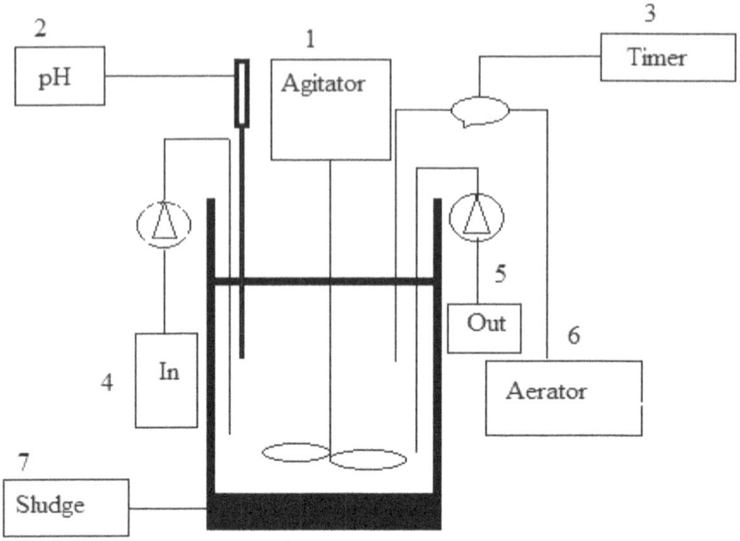

Fig. 2 Schematic SBR reactor. 1. Mixer motor. 2. pH probe. 3. Timer. 4. Inlet feed tank. 5. Outlet discharge tank. 6. Aerator. 7. Sludge

A Model 2100N Turbidimeter-HACH turbidimeter was used to measure turbidity by NF EN ISO 7027 March 2000 (T 90-033). The turbidimeter is not greater than 4000 NTU. NF T 90-105, published in January 1997, establishes the suspended matter (TSS) (T 90-105). Chemical oxygen demand was analyzed in accordance with standard NF T90-101 (T90-101). The biological oxygen demand (BOD_5) was calculated using the manometric method NF EN 1899 May 1998 (T90-103).

The Folin-Ciocalteu reagent was used in the colorimetric method to evaluate the amount of phenolic compounds.

Since January 1997 (T 90-045), the NF EN ISO 78-90 standard has assessed nitrates.

NF T 90-015 evaluated the ammoniacal nitrogen concentration in January 1997.

Total Phosphorus was considered using the NF T 90-023 method from January 1997.

3 Results and Discussion

3.1 Physical–Chemical Analysis

Table 1 summarizes the physicochemical composition of wastewater, raw OMW, and diluted OMW at concentrations of 25 and 50%. Raw and diluted OMW effluents have a dark brown color and an acid pH. Acidity can be attributed to the organic acids present in the effluent. Moreover, they are characterized by a high concentration of toxic substances, mainly phenolic compounds. Specifically, the raw OMW contains approximately 4430 mg/L of phenolic compounds.

The diluted OMW at 25% and 50% concentrations contain approximately 3591 mg/L and 3117 mg/L of phenolic chemicals, respectively. Krastanov et al. found lower quantities of polyphenols in diluted OMW samples, indicating that dilution reduced the polyphenol load. Polyphenols are poisonous to a wide range of bacteria at high doses. OMW effluents contain a wide range of phenolic chemicals, including cinnamic acid derivatives, tyrosol, and benzoic acid. Babić et al. found these chemicals as the main phenolic components in OMW. In terms of suspended solids, the raw OMW content is around 18.3 g/L, mainly in the insoluble fractions of the olive pulp. The average OMW load is around 18.3 g/L, mainly in the insoluble fractions of olive pulp. OMW is also characterized by a high level of organic matter quantified in terms of COD, which amounts to around 115,200 mg O_2/L.

Nevertheless, the effluents of OMW hold the potential for agricultural valorization through direct application on the soil or after storage due to their high nitrogen content. OMW's nutritional profile varies from one olive growing season to the next, depending on factors such as fertilizer treatment. When diluted to 50%, OMW decreased nitrates by 53.5%, phosphorus by 60.2%, and ammonium by 65.7%. When the BOD5/COD ratio decreases between 0.4 and 0.8, the wastewater is considered biodegradable. The biodegradability index of raw OMW, as shown in Table 1, is approximately 0.33. This figure shows that raw OMW is poorly biodegradable. However, diluting OMW with urban wastewater results in a considerable increase in biomass, which improves biodegradability and reduces toxicity.

Table 1 Physicochemical analysis of untreated and diluted OMW

	Crude OMW	OMW (25%)	OMW (50%)
pH	4.43	4.36	4.47
Turbidity (NTU)	> 4000	> 4000	> 4000
Polyphenols (mg/L)	4430 ± 0.07	3591 ± 0.04	3117 ± 0.03
Total suspended solids (TSS) (g/L)	18.3 ± 1.05	15.2 ± 1.02	10.9 ± 0.9
BOD_5 (mgO_2/L)	38,200 ± 3	40,500 ± 1.09	34,400 ± 2.04
COD (mgO_2/L)	115,200 ± 1.09	96,000 ± 1.02	62,400 ± 1.07
BOD_5/COD	0.33	0.42	0.55
Nitrate (mg/L)	1983 ± 0.06	1321 ± 0.09	921 ± 0.02
NH_4^+ (mg/L)	20.8 ± 1.07	16 ± 1.16	7.1 ± 0.06
Total phosphorus (mg/L)	980 ± 1.05	520 ± 2.01	390 ± 2.00

3.2 SBR Effect on OMW

3.2.1 pH and COD Variation

Figure 3 illustrates the pH variation during SBR treatment.

The pH level is a critical factor in determining the efficiency of the biological purification operation, particularly in the case of SBR systems. It has a substantial effect on the metabolic activity of microbial substrates, influencing different species of bacteria to varying degrees. Autotrophic bacteria are sensitive to pH variations [4].

Gram-positive bacteria are more tolerant to acidic conditions than Gram-negative bacteria. Gram-positive bacteria can survive higher acidity levels than Gram-negative bacteria. Gram-positive bacteria can manage with acidic pH using a number of adaptive strategies. These include modifying the composition of cell membranes, creating alkaline substances, altering metabolic pathways, and protecting macromolecules from acid damage. Firmicutes bacteria were found in the microbial community of OMWs. The experimental findings revealed a gradual decrease in pH levels over time, with the pH decreasing from 4.4 to 3.7 for the 25% sample and from 4.5 to 3.9 for the 50% sample (Fig. 3). In contrast, pH levels tend to rise during denitrification during the cycle's anoxia phase. Nitrification reactions create H^+ ions, leading to lower pH levels.

The SBR system represents a more pragmatic option for wastewater treatment, particularly when the organic load is significant [3]. Figure 4 demonstrates the variation in COD levels. The findings indicate significant fluctuations in COD concentrations in the initial days of treatment. These variations are indicative of the adaptation period. Microorganisms can increase their resistance to various harmful substances during this acclimatization phase. This procedure enhances their ability to properly handle and destroy these chemicals. The reactor's adaptation phase was relatively brief, due primarily to dilution and the pre-feeding adaption period. By the ninth day, the inoculum had successfully evolved, resulting in an increase in elimination rates. At the end of the treatment, the COD removal efficiency was 40.2% for

Fig. 3 pH variation during SBR treatment

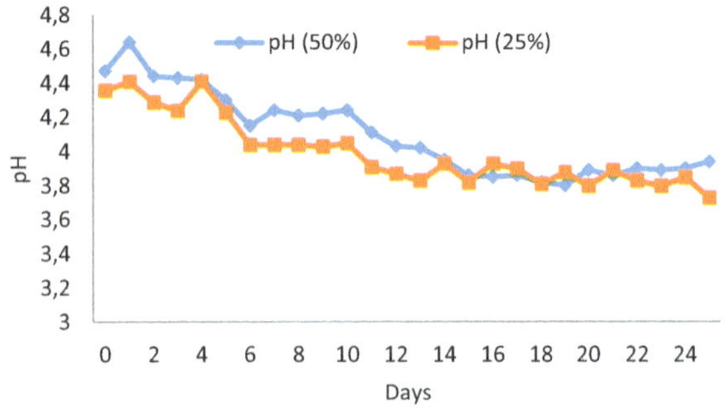

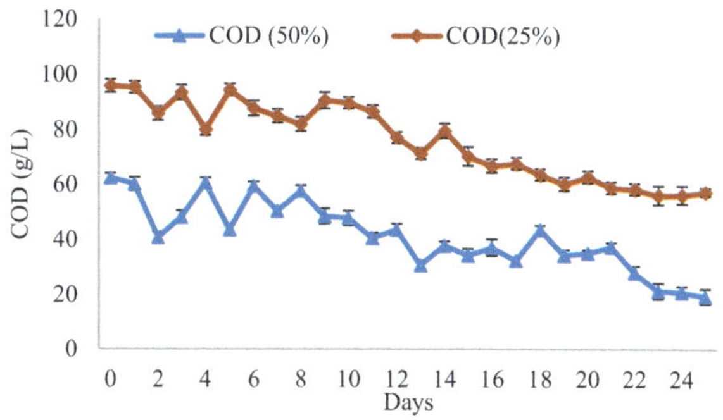

Fig. 4 Variation in COD during treatment

the 25% mixture and 68.9% for the 50% mixture (Fig. 4). These findings emphasize the considerable reduction of organic matter, as indicated by COD, resulting from heterotrophic bacteria growth and metabolism during the aeration phase. The decrease in COD can be ascribed to nitrifying bacteria consuming soluble organic materials, which promotes the increase of bacterial biomass.

During the treatment process in the SBR, the remaining fraction of COD is believed to be associated with non-biodegradable substances. The elimination of COD happens predominantly during the SBR's aeration phase. This phase is particularly successful at reducing the proportion of biodegradable organic components in wastewater. However, certain non-biodegradable compounds may survive and contribute to the remaining COD percentage that is not removed.

3.2.2 Polyphenols Variation

Figure 5 depicts the temporal changes in polyphenol concentration for the two samples.

Initial polyphenol concentrations were 3591 and 3117 mg/L for the 25% and 50% dilutions respectively. Elimination percentages ranged from 6.7% to 67.8%

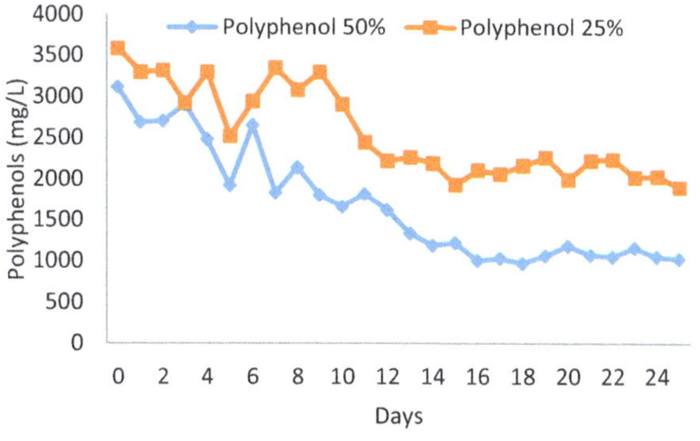

Fig. 5 Polyphenol degradation during SBR treatment

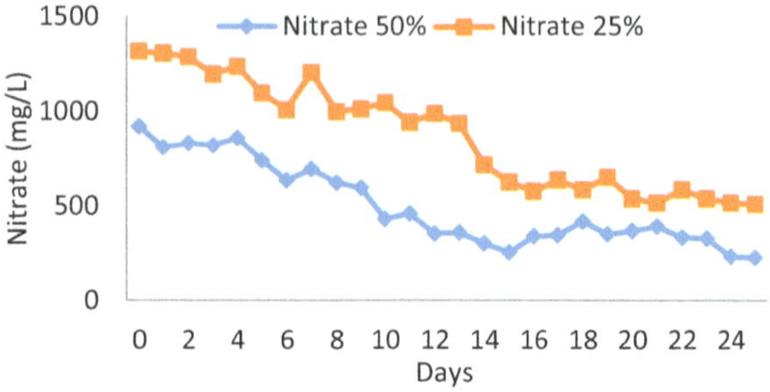

Fig. 6 Nitrate change during SBR process

for OMW diluted to 50%, and from 6.6% to 47% for samples diluted to 25%. Phenolic compound types, chemical structures and concentrations affect their biological degradation. OMW contains a variety of microorganisms, including fungus, beta-proteobacteria. Farabegoli et al. reported that the SBR worked well in polyphenol degradation, with an average output of approximately 76% for diluted OMW (1:25).

3.2.3 Nutrient Variation

The results demonstrate that after 10 days of operation, nitrate removal efficiency increased to 75.3% for 50% diluted OMW and 61.2% for 25% diluted OMW (Fig. 6). The procedure resulted in the nearly full oxidation of ammonium. Denitrification occurs directly after ammonium oxidation to nitrite, as Lemaire et al. stated. However, other bacteria, including *Rhodococcus* sp., can nitrify and denitrify simultaneously. When grown in aerobic environments, these bacteria transform Ammonium (NH_4^+) into nitrogen gas aerobically by using organic substrates as carbon and energy sources. *Rhodococcus* sp. was among the microorganisms El-Khateeb et al. showed to be present in OMW.

Ammonium concentrations were significantly lowered by the SBR reactor, reaching 7.1 mg/L and 16 mg/L for the 50% and 25% mixes, respectively (Fig. 7). During the first ten days of reactor operation, the effluent concentrations varied between 5.71 mg/L and 4 mg/L for the 50% OMW and between 5.9 mg/L and 41.7 mg/L for the 25% OMW. This period was considered as an adaptation phase for the microbial community. The ammonium removal efficiencies increased from 53 to 86% for the 50% mixture and from 38 to 61% for the 25% OMW by day 25.

3.2.4 Phosphorus Variation

The change in phosphorus concentration over the treatment period is depicted in Fig. 8.

Phosphorus removal is a crucial aspect as it contributes to the eutrophication of river waters. The Sequential Batch Reactor (SBR) is widely used to remove nutrients from wastewater, including nitrogen and Phosphorus. Figure 8 demonstrates that both

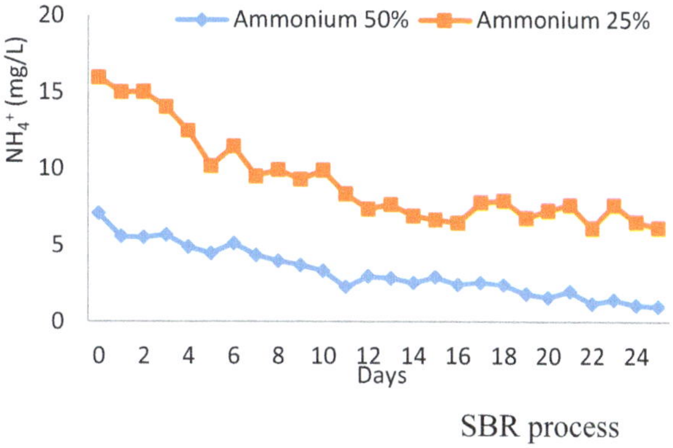

Fig. 7 Ammonium change during SBR process

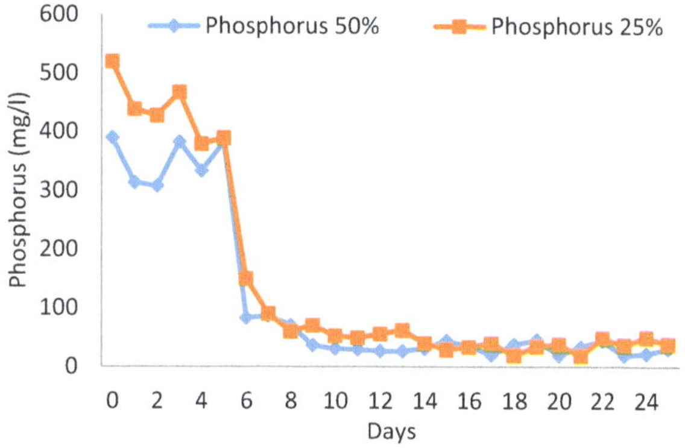

Fig. 8 Phosphorus variation during SBR process

reactors had phosphorus removal rates greater than 80%, with values for the 50% mixture reaching 95.3% and the 25% mixture reaching 94.1%. The reason behind these elevated clearance percentages is the presence of phosphorus-accumulating organisms (PAOs). Phosphorus concentrations in both reactors dramatically dropped on the sixth day of operation, demonstrating the effectiveness of PAOs in absorbing Phosphorus. The significant uptake of Phosphorus by PAOs is facilitated by the alternating anaerobic, aerobic, and occasionally anoxic phases in the SBR process, according to Fernandes et al. In anaerobic environments, bacteria produce poly-β-alkanoates (PHA) through the internal hydrolysis of polyphosphates, releasing Phosphorus. Previous studies have demonstrated that SBR treatment with a forced anoxic phase achieved phosphorus removal exceeding 80% in municipal wastewater.

4 Conclusion

The current study explored the efficacy of SBR in the treatment of OMW. The results revealed that the OMWs had high levels of polyphenols and COD. As a consequence, BOD5/COD ratios of 0.42 for D2 and 0.55 for D1 demonstrate that diluting OMW with 25% and 50% municipal wastewater lowers pollution while enhancing effluent biodegradability. The D1-fed reactor produced the most promising results, removing COD, polyphenols, ammonium, nitrate, and phosphorus at rates of 68.9%, 67.1%, 86%, 75.3%, and 95.3% respectively. Olive oil mill effluent can be successfully cleaned by biological treatment of diluted OMW with SBR, allowing for enormous flexibility in pollutant loading. This type of care is a simple and effective approach of resolution.

5 Recommendations

Several industries use polyphenols. The food industry uses polyphenols as natural food additives for their antioxidant properties. The valorization of the polyphenols present in OMW can therefore offer economic and environmental benefits.

References

1. Fakhfakh F, Raissi S, Kriaa K et al (2024) Modeling and optimization of a green process for olive mill wastewater treatment. Water 16:327
2. Khalil J, Habib H, Bouguerra S et al (2024) The impact of olive mill wastewater on soil properties, nutrient and heavy metal availability—a study case from Syrian vertisols. J Environ Manage 351:119861
3. Ramírez XMV, Mejía GMH, López KVP et al (2012) Wastewater treatment from biodiesel production via a coupled photo-Fenton–aerobic sequential batch reactor (SBR) system. Water Sci Technol 66:824–830
4. Yan L, Liu Y, Ren Y et al (2013) The effect of pH on the efficiency of an SBR processing piggery wastewater. Biotechnol bioprocess Eng 18:1230–1237

Study on the Effectiveness of the Sequential Batch Reactor on the Reduction of Wastewater Pollution by Vegetable Oils from Oil Refining

Khadija Zahidi, Latifa Mouhir, Safaa Khattabi Rifi, Chaymae Haddaji, Zineb Ettaloui, Roukaya Bouyakhsass, Abdelaziz Madinzi, Khalid Digua, Mohammed Chatoui, Tonni Agustiono Kurniawan, Abdelkader Anouzla, Aysegul Pala, and Salah Souabi

Abstract Industrial wastewater commonly contains a lot of pollutants, which contributes to eutrophication in aquatic systems. To improve the treatment of vegetable oil refining wastewater, we investigated the potential of a biological treatment, precisely an aerobic-anoxic sequencing batch reactor system. The reactors were operated on a 24-h cycle, consisting of 20 min for filling, 17 h for aeration, 4 h for anoxic conditions, 2 h for settling, 20 min for draw, and 20 min of idle time. Results showed removal efficiencies of 83%, 94%, 80%, 93%, 96%, and 91% for COD, BOD_5, turbidity, nitrate, ammonium, and phosphate, respectively. Despite its

K. Zahidi · L. Mouhir · S. K. Rifi · C. Haddaji · Z. Ettaloui · R. Bouyakhsass · A. Madinzi · K. Digua · M. Chatoui · A. Anouzla · S. Souabi (✉)
Laboratory of Process Engineering and Environment, Faculty of Science and Technology, University Hassan II, Mohammedia, Morocco
e-mail: salah.souabi@gmail.com

C. Haddaji
e-mail: chaymae.haddaji@etu.fstm.ac.ma

A. Madinzi
e-mail: abdelaziz.madinzi@fstm.ac.ma

K. Digua
e-mail: k.digua@univh2m.ac.ma

A. Anouzla
e-mail: aanouzla@gmail.com

T. A. Kurniawan
College of Ecology and the Environment, Xiamen University, Xiamen 361102, Fujian, China

A. Pala
Environmental Research and Development Center (CEVMER), Dokuz Eylul University, Izmir, Turkey
e-mail: aysegul.pala@deu.edu.tr

S. Yadav et al. (eds.), *Emerging Technologies in Biological and Hybrid Wastewater Treatment: Lessons from Developed to Enhancing Practices in Developing Countries*, Earth and Environmental Sciences Library,
https://doi.org/10.1007/978-3-031-74515-7_5

complexity, the SBR biological reactor is considered a promising solution for industrial wastewater treatment. The O/A-SBR system is an effective treatment process for the vegetable oil refining industry.

Keywords Aerobic-anoxic · Industrial wastewater · Cost-effectiveness · Pretreatment · Sequencing batch reactor · Vegetable oil refining wastewater

1 Introduction

The vegetable oil industry faces environmental problems related to refining [5]. Vegetable oil refinery is a big consumer of water and, hence, a significant wastewater source. The manufacture of vegetable oils begins with crude oil extraction from oil seeds, followed by refining. This refining process typically produces substantial amounts of wastewater at various stages [11, 13]. These wastewaters have high chemical oxygen demand (COD) and contain a lot of sodium salts from free fatty acids, grease, soap stocks, oil, sulfates, and phosphates [18]. The direct discharge of untreated VORW into water bodies decreases oxygen concentrations and hinders sun radiation penetration, resulting in the extinction of flora and wildlife. VORW therefore causes gastrointestinal issues, skin irritation, and respiratory difficulties [17].

In the vegetable oil industry, wastewater treatment is essential to protect the environment and safeguard public health. Generally, innovative technologies are available to minimize the impact of these discharges on the environment [18]. Various methods have been used to treat this type of wastewater. The appropriate treatment method will depend on the specific characteristics of the effluent and technical and financial constraints [35]. Physical procedures for separating solid particles and pollutants include adsorption, membrane filtration, air flotation, and reverse osmosis [10, 19, 32, 36]. Chemical processes including coagulation-flocculation and electrocoagulation need processing reagents [12]. Biological treatments employ microorganisms to degrade organic pollutants in systems such as sequential batch reactors, moving bed biofilm reactors, and up-flow anaerobic sludge beds [9, 21, 27].

Biological treatment is generally adequate for wastewater from food processing industries. These effluents contain typically organic and fatty constituents and have excellent biodegradability. Biological degradation is more effective at breaking down vegetable and animal oils than physicochemical processes [4]. The sequential batch reactor (SBR) is noted for its capacity to treat a wide range of effluent. It is resistant to variations in water quality and provides a high level of treatment [34]. This study assesses the efficacy of biological treatment as an aerobic-anoxic batch reactor with alternate working cycles for carbon, nitrogen, and water removal.

2 Materials and Methods

2.1 Sampling

The work was carried out on crude samples collected from the LESIEUR CRISTAL company, Casablanca-Morocco, which specializes in vegetable oil refining. Samples were stored at the laboratory in 20 L plastic containers at 4 °C [1].

2.2 SBR Setup

Treatment of wastewater by the SBR is performed on a 24-h cycle, with each cycle primarily comprising a 20-min charging phase and a 17-h aeration phase. The reactor mixture is stirred at 100 revolutions per minute (rpm). A 4-h anoxic phase followed this. This phase plays a crucial role in facilitating nitrite and nitrate denitrification, a process carried out under conditions of low aeration (at levels between 0.1 and 0.3 mg ± 0.1 O_2/L) and with a stirring rate of 50 rpm, a 2-h settling phase, a 20-min drawdown phase, and a 20-min inactivity phase. After the completion of the reaction phase, the settling phase begins. Drawdown is the final stage of biological treatment, removing the treated effluent. During the inactivity phase, excess sludge is removed from the bottom of the reactor while preserving the quantity of sludge for the following cycle. Samples were taken at the end of each cycle for analysis. Figure 1 shows the SBR plant used in this work.

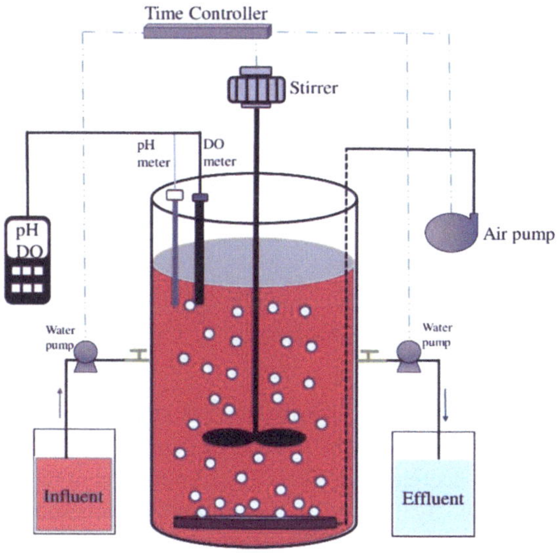

Fig. 1 Experimental setup of the SBR

2.3 Characterization Methods

Turbidity was analyzed using a 'Hach 2100 N' turbidimeter. pH was determined by standard NF T 90-008 (February 2001). Electrical conductivity was measured with a "BASIS Jenco 3177M" conductivity meter by NF EN 27888 (January 1994). AFNOR standard T90-101 (February 2001) was used for chemical oxygen demand (COD) analysis. Biochemical oxygen demand was measured according to APHA [1] in an incubator for five days at 20 °C in conditions of darkness. Phenolic compounds were analyzed spectrometrically in the presence of Folin-Ciocalteu reagent. Nitrate content was assessed spectrometrically by EN ISO 78-90 (January 1997), designated T 90-045. Ammonium concentration was measured using standard NF T 90-015-2. NF T 90-023 (January 1997) determined total phosphorus content spectrometrically. Suspended solids (SS) were quantified by AFNOR standard NF T 90-105-2 centrifugation. Analysis of oil and fat content followed EPA standard method 1664.

3 Result and Discussion

3.1 Samples Characterization

A pH of 6.9 ± 0.01 characterizes the samples. An estimated oil and fat concentration of 0.7 ± 0.1 g/L. The pH is, therefore, an essential factor in maintaining the stability of the oily emulsion. The results show that turbidity is of the order of 3924 ± 6 NTU. Suspended solids (SS) are present at a 6.35 ± 0.2 g/L concentration. VORWs are characterized by a chemical oxygen demand (COD) of 11 ± 0.3 g/L and a biochemical oxygen demand (BOD_5) of 5 ± 0.2 g/L. In this instance, the high BOD_5/COD ratio signifies the excellent biodegradability of the samples. Similarly, it exhibits a notable concentration of chemical contaminants, including ammonium at 42.1 ± 0.4 mg/L, nitrate ions at 40.9 ± 0.3 mg/L, phosphate at 48 ± 0.6 mg/L, and phenol at 30.8 ± 0.2 mg/L. Analyses show that VORW exceeds indirect discharge standards with high levels of organic and chemical pollutants. This means they must be treated before being discharged into the receiving environment [29].

3.2 Turbidity Variation

Significant fluctuations were observed during the first few days of treatment. These variations are due to the microorganisms' adaptation phase (Fig. 2). Then, turbidity gradually decreased to 620 NTU. Elimination of turbidity exceeded 80%. This result is explained by the efficient separation of solids and liquid during bioreactor settling.

Several researches have emphasized an association between turbidity, total suspended solids, and color [25, 26]. The results agree with Zalaghi et al. [39],

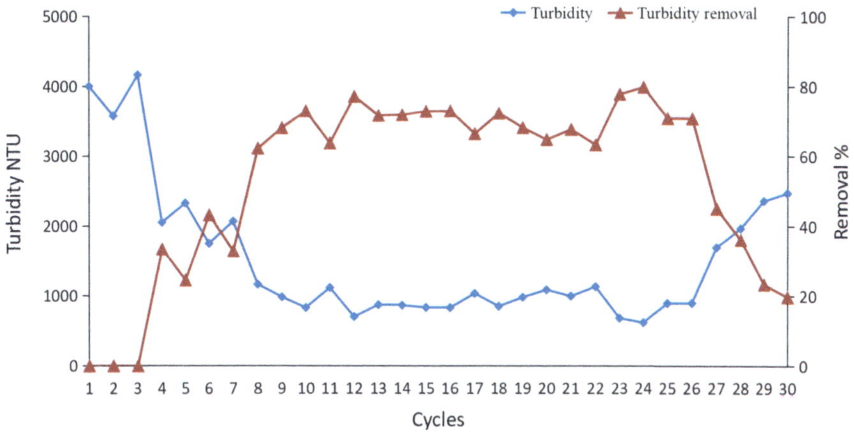

Fig. 2 The efficiency of turbidity removal in the SBR

who found that SBR removed 86% of the TSS from the leachate discharge. Mahvi [23] also reported a reduction in total TSS concentration to 61.25%. Yong et al. [38] also reduced TSS after treatment with SBR, with an efficiency of 55%. With the increase in aeration rate, color removal efficiency also increased to 38%.

Farraji et al. [8] have shown that bentonite improves the efficiency of the SBR process. It achieved remarkable removal rates of 94%, 90%, and 46% for turbidity, total suspended solids, and color, respectively.

The effectiveness of SBR treatment is influenced by several operational factors, such as dissolved oxygen (DO) concentration, temperature, and microbial activity [2]. A high concentration of DO promotes turbidity reduction by allowing better adsorption and aggregation of flocs. Aerobic conditions also favor greater adsorption capacity of colloidal materials. In addition, a high concentration of DO reduces the presence of filamentous bacteria and prevents sludge aggregation. Lower temperatures can lead to reduced microbial activity. Keeping optimal DO levels and appropriate temperature ranges is critical to improving the performance of SBR systems in reducing turbidity [37].

3.3 Reduction of COD and BOD_5 by SBR

Figures 3 and 4 illustrate the reduction in chemical and 5-day biological oxygen demand concentrations over the operating period.

As depicted in Fig. 3, the results indicate that the COD initially went to a higher value during the early days of the treatment. From the start to the 5th day, the therapy displayed instability with fluctuations in COD concentration. This rise in COD may be attributed to the microorganism's mortality due to the acclimatization period, indicating their ongoing adaptation to the new conditions [31].

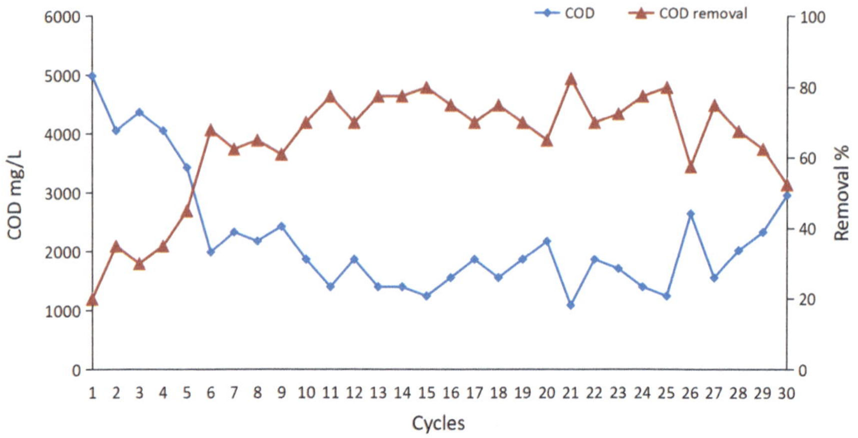

Fig. 3 The efficiency of COD removal in the SBR

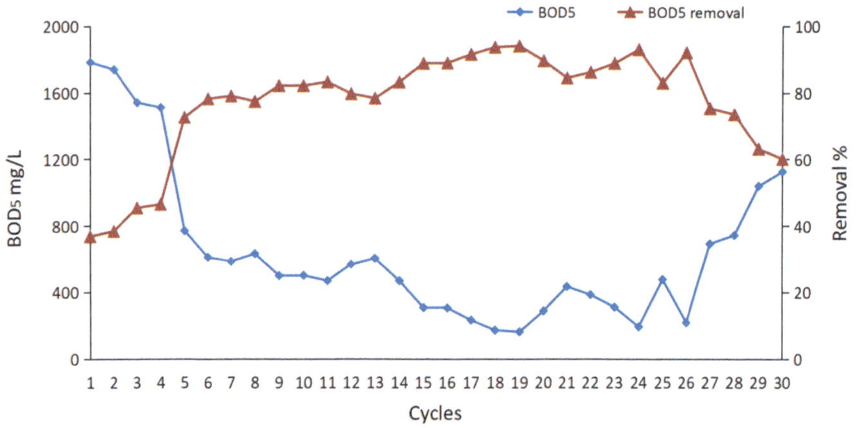

Fig. 4 The efficiency of BOD_5 removal in the SBR

The increase in the COD can likely be attributed to bactericidal minerals and organic micropollutants. Industrial wastewaters often require an acclimatization period for the microbial community to gradually adapt to potentially toxic compounds. Within this timeframe, the emergence of particular enzymatic genes facilitates biodegradation [6]. COD removal efficiency in the reactor started to increase from day five onwards, signaling the gradual adaptation of the microorganisms to the SBR system. This treatment removed 83% of the COD.

This study's observed COD removal efficiency mirrors the findings of Al-Abadi and Al-Sulaiman, who reported an 86% removal rate when utilizing the SBR treatment process. These findings are higher than those published by Rifi et al. [30]. The reactor's continuous operation during a 24-h cycle achieved a 69% COD reduction when treating olive oil mill wastewater mixed with municipal sewage. Jena et al. [14] used an anoxic-aerobic SBR to remove COD from highly concentrated wastewater, with an efficiency of 72%. Alagha et al. [3] found that their results showed that. In addition, lengthening the aeration cycle time improved treatment efficiency by 5%, as Kulikowska et al. [20] noted.

BOD_5, a metric for assessing the effectiveness of biological processes, underwent significant removal efficiencies after the SBR cycles. The reactor eliminated 94% of the BOD_5 (Fig. 4). This reduction is explained by the efficiency of the sludge used, which effectively decomposes biodegradable organic matter in the VORW. Cárdenas et al. [7], who recorded a BOD_5 removal rate of 91%.

Bacteria require energy and carbon for their growth and metabolism. When oxygen is available, organic substances can undergo transformation into various states, assimilated by the biomass through respiratory oxidation or oxidized by the bacteria into CO_2 and nutrients such as nitrogen, phosphorus, and sulfur [40]. This way, the organic carbon used for denitrification is oxidized into carbon dioxide (CO_2) and water, reducing COD [16].

3.4 Nutrient Reduction in SBR

Figures 5 and 6 show that the system is unstable during the first phase. Fluctuations in nitrogen compound concentrations mark this instability. The adaptation of the bacterial community in the reactor can explain these variations. After this phase, there is a noticeable reduction in nitrogen concentrations. Throughout the 30-day operational period, ammonium removal efficiency reached 96% and nitrate removal efficiency 93%. Muhamad et al. [24] documented comparable findings in their investigation of wastewater treatment in recycled paper mills using an attached growth-SBR, where they observed significant removal efficiencies for nitrogen compounds ranging from 70 to 94%. Quan and Gogina [28] delved into managing wastewater from Vietnamese municipal wastewater treatment plants in a separate study. They employed a modified SBR reactor that introduced an anoxic phase into the operational cycle, creating conditions conducive to removing ammonia–nitrogen. Remarkably, within a mere 3 h of the anoxic-denitrification phase, all nitrates were transformed into nitrogen gas (N_2), yielding an ammonium removal efficiency ranging from 75 to 80%.

Figure 7 illustrates the initial variations in phosphate removal at the outset of the treatment, followed by stabilization after cycle number 6. Phosphate removal rates are very high, mainly in the final stages of the process. Removal rates ranged from 72 to 91%. The observed drop in phosphate levels can be attributed to a variety of mechanisms, including precipitation, adsorption and degradation by microorganisms as part of the biological treatment process [15].

In a separate investigation, Liu et al. [22] studied the removal of nutrients from wastewater within an anaerobic–aerobic–anoxic system, achieving remarkable removal rates of 97% for COD, 96% for total nitrogen (TN), and 94% for total phosphorus (TP). Showkat and Najar [33] showed that wastewater treatment with

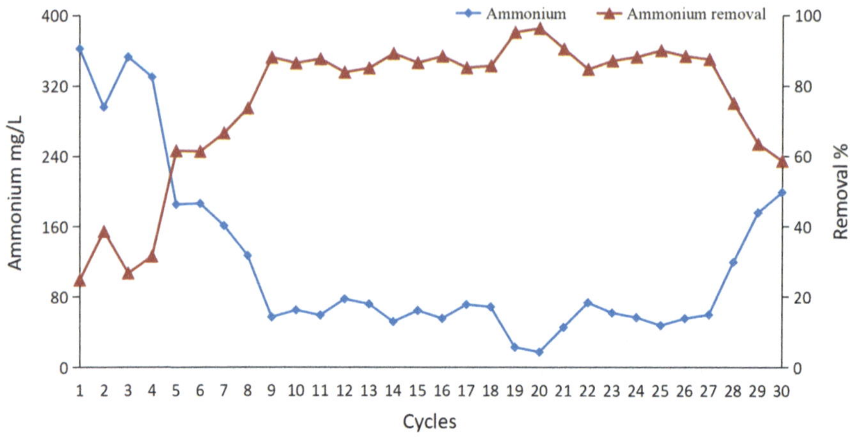

Fig. 5 The efficiency of ammonium removal in the SBR

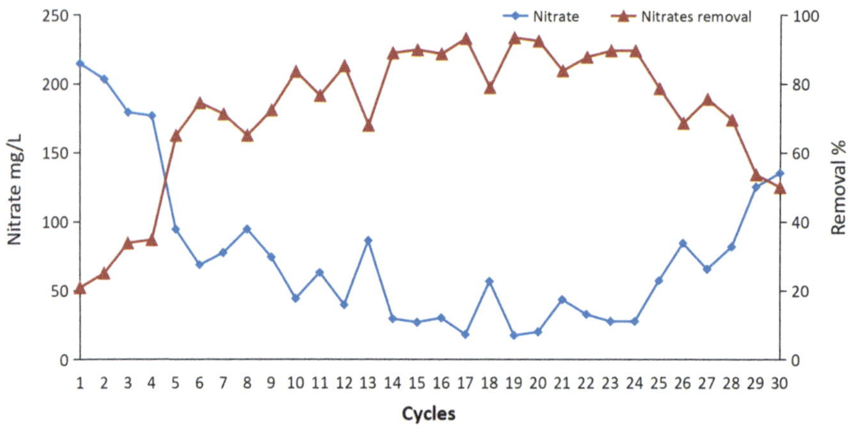

Fig. 6 The efficiency of nitrates removal in the SBR

SBR removed 65% of nitrates and 68% of total nitrogen. SBR treatment of recycled wastewater from paper mills has shown 70–94% for nitrogen removal efficiencies and 42–71% for phosphate [24].

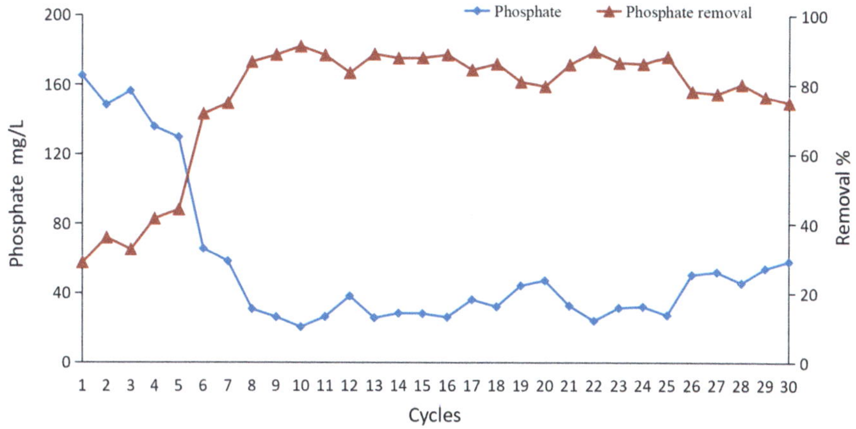

Fig. 7 The efficiency of phosphate removal in the SBR

4 Conclusion

VROW treatment is considered very difficult due to its complex composition of humic substance and organic matter. The evaluation of combined nitrification–denitrification and dephosphatation in an aerobic-anoxic SBR reactor for the treatment of wastewater from vegetable oil refineries was performed. This study identified the optimum aeration time for wastewater treatment from vegetable oil refineries as 17 h, including a 2-h anoxic phase. Under these conditions, the system showed significant removal efficiencies, reaching 83%, 94%, 80%, 93%, 96%, and 91% for COD, BOD_5, turbidity, nitrate, ammonium, and phosphate, respectively. The results show that SBR suits the industrial waste treatment sector well. In addition, the process's ability to adapt to water quantity and quality fluctuations makes it a versatile and advantageous solution for VORW treatment.

5 Recommendations

The perspectives proposed in this study are intended to provide practical guidance for decision-makers, practitioners and researchers to advance the sustainable treatment of industrial wastewater and help protect the environment and contribute to environmental protection:

Ongoing research into new treatment methods: Further research could focus on innovative techniques such as biofiltration, advanced oxidation, or adsorption to further reduce pollutants present in industrial wastewater.

Optimization of treatment processes: It is important to continue efforts to optimize existing treatment processes.

Sustainability and resource reuse: The environmental sustainability of treatment methods should be given particular attention.

References

1. APHA A (1998) Standard methods for the examination of water and wastewater analysis. American Public Heal Association, Washington, DC
2. Al-Attabi AWN (2018) Treatment of petroleum refinery wastewater in an innovative sequencing batch reactor. Liverpool John Moores University (United Kingdom)
3. Alagha O, Allazem A, Bukhari AA, Anil I, Mu'azu ND (2020) Suitability of SBR for wastewater treatment and reuse: pilot-scale reactor operated in different anoxic conditions. Int J Environ Res Public Health 17:1617
4. Ali SI, Moustafa MH, Nwery MS, Farahat NS, Samhan F (2022) Evaluating the performance of sequential batch reactor (SBR & ASBR) wastewater treatment plants, case study. Environ Nanotechnol Monit Manag 18:100745
5. Amir S, Abouelwafa R, Meddich A, Souabi S, Winterton P, Merlina G, Revel J-C, Pinelli E, Hafidi M (2010) PLFAs of the microbial communities in composting mixtures of agro-industry sludge with different proportions of household waste. Int Biodeterior Biodegradation 64:614–621
6. Chan YJ, Chong MF, Law CL (2010) Biological treatment of anaerobically digested palm oil mill effluent (POME) using a lab-scale sequencing batch reactor (SBR). J Environ Manage 91:1738–1746
7. Cárdenas C, Yabroudi SC, Benítez A, Páez K, Perruolo T, Angulo N, Araujo I, Herrera L (2012) Desempeño de un reactor biológico secuencial (RBS) en el tratamiento de aguas residuales domésticas. Rev Colomb Biotecnol 14:111–120
8. Farraji H, Qamaruz Zaman N, Aziz HA, Ashraf MA, Mojiri A, Mohajeri P (2015) Landfill leachate treatment by bentonite augmented sequencing batch reactor (SBR) system. Appl Mech Mater 466–471
9. Gaur RZ, Khan AA, Lew B, Diamantis V, Kazmi AA (2017) Performance of full-scale UASB reactors treating low or medium strength municipal wastewater. Environ Process 4:137–146
10. Gong C, Ren X, Han J, Wu Y, Gou Y, Zhang Z, He P (2022) Toxicity reduction of reverse osmosis concentrates from petrochemical wastewater by electrocoagulation and Fered-Fenton treatments. Chemosphere 286:131582
11. Haddaji C, Ennaciri K, Driouich A, Digua K, Souabi S (2022) Optimization of the coagulation-flocculation process for vegetable oil refinery wastewater using a full factorial design. Process Saf Environ Prot 160:803–816
12. Haddaji C, Khattabi Rifi S, Digua K, Abdelaziz Madinzi A, Chatoui M, Driouich A, Ettaloui Z, Kurniawan TA, Anouzla A, Souabi S (2024) Optimization of coagulation-flocculation process for wastewater treatment from vegetable oil refineries using chitosan as a natural flocculant. Environ Nanotechnol Monit Manag. https://doi.org/10.1016/j.enmm.2024.100973
13. Hartal O, Khattabi Rifi S, Chatoui M, Haddaji C, Madinzi A, Pala A, Souabi S (2024) Combined natural flotation and chemical precipitation for the treatment of vegetable oil refinery wastewater. Int J Environ Sci Technol 1–12
14. Jena J, Kumar R, Saifuddin M, Dixit A, Das T (2016) Anoxic–aerobic SBR system for nitrate, phosphate and COD removal from high-strength wastewater and diversity study of microbial communities. Biochem Eng J 105:80–89
15. Kallali H (2011) Physicochemical and microbial Caracteristics Performency in wastewater treated under aerobic reactor. Am J Environ Sci 7:254–262
16. Kamińska G, Marszałek A (2020) Advanced treatment of real grey water by SBR followed by ultrafiltration—performance and fouling behavior. Water 12:154

17. Khalidi-Idrissi A, Hartal O, Madinzi A, El-Abbadi K, Souabi S (2024) Natural flotation and coagulation–flocculation: a dual approach to refinery wastewater treatment. Euro-Mediterranean J Environ Integr 1–19
18. Khouni I, Louhichi G, Ghrabi A (2020) Assessing the performances of an aerobic membrane bioreactor for textile wastewater treatment: influence of dye mass loading rate and biomass concentration. Process Saf Environ Prot 135:364–382
19. Khouni I, Louhichi G, Ghrabi A, Moulin P (2020) Efficiency of a coagulation/flocculation–membrane filtration hybrid process for the treatment of vegetable oil refinery wastewater for safe reuse and recovery. Process Saf Environ Prot 135:323–341
20. Kulikowska D, Klimiuk E, Drzewicki A (2007) BOD5 and COD removal and sludge production in SBR working with or without anoxic phase. Bioresour Technol 98:1426–1432
21. Li C, Liu S, Ma T, Zheng M, Ni J (2019) Simultaneous nitrification, denitrification and phosphorus removal in a sequencing batch reactor (SBR) under low temperature. Chemosphere 229:132–141
22. Liu S, Daigger GT, Liu B, Zhao W, Liu J (2020) Enhanced performance of simultaneous carbon, nitrogen and phosphorus removal from municipal wastewater in an anaerobic-aerobic-anoxic sequencing batch reactor (AOA-SBR) system by alternating the cycle times. Bioresour Technol 301:122750
23. Mahvi AH (2008) Sequencing batch reactor: a promising technology in wastewater treatment
24. Muhamad MH, Abdullah SRS, Hasan HA (2015) Efficiency of attached-growth sequencing batch reactor in the treatment of recycled paper mill wastewater. J Teknol 74
25. Nikoonahad A, Ebrahimi AA, Nikoonahad E, Ghelmani SV, Mohammadi A (2016) Evaluation the correlation between turbidity and total suspended solids with other chemical parameters in Yazd wastewater treatment effluent plant. J Environ Heal Sustain Dev 1:66–74
26. Ogleni N, Damar Arifoglu Y, Ileri R (2012) Microbiological and performance evaluation of sequencing batch reactor for textile wastewater treatment. Water Environ Res 84:346–353
27. Pal P, Kumar R (2020) Recent advances in biological treatment processes for wastewater and water treatment. Curr Trends Future Dev Membr 41–66
28. Quan TH, Gogina E (2019) Application of anoxic phase in SBR reactor to increase the efficiency of ammonia removal in Vietnamese municipal WWTPs. In: E3S web of conferences. EDP Sciences, p 1017
29. Ramsay IR, Pullammanappallil PC (2001) Protein degradation during anaerobic wastewater treatment: derivation of stoichiometry. Biodegradation 12:247–256
30. Rifi SK, Fels LE, Driouich A, Hafidi M, Ettaloui Z, Souabi S (2022) Sequencing batch reactor efficiency to reduce pollutant in olive oil mill wastewater mixed with urban wastewater. Int J Environ Sci Technol 1–14
31. Rodríguez DC, Pino N, Peñuela G (2011) Monitoring the removal of nitrogen by applying a nitrification–denitrification process in a sequencing batch reactor (SBR). Bioresour Technol 102:2316–2321
32. Saththasivam J, Ogunbiyi O, Lawler J, Al-Rewaily R, Liu Z (2022) Evaluating dissolved air flotation for oil/water separation using a hybridized coagulant of ferric chloride and chitosan. J Water Process Eng 47:102836
33. Showkat U, Najar IA (2019) Study on the efficiency of sequential batch reactor (SBR)-based sewage treatment plant. Appl Water Sci 9:1–10
34. Singh A, Srivastava A, Saidulu D, Gupta AK (2022) Advancements of sequencing batch reactor for industrial wastewater treatment: major focus on modifications, critical operational parameters, and future perspectives. J Environ Manage 317:115305
35. Solayman HM, Hossen MA, Abd Aziz A, Yahya NY, Hon LK, Ching SL, Monir MU, Zoh K-D (2023) Performance evaluation of dye wastewater treatment technologies: a review. J Environ Chem Eng 109610
36. Yaacob N, Sean GP, Nazri NAM, Ismail AF, Abidin MNZ, Subramaniam MN (2021) Simultaneous oily wastewater adsorption and photodegradation by ZrO_2–TiO_2 heterojunction photocatalysts. J Water Process Eng 39:101644

37. Yan L, Liu S, Liu Q, Zhang M, Liu Y, Wen Y, Chen Z, Zhang Y, Yang Q (2019) Improved performance of simultaneous nitrification and denitrification via nitrite in an oxygen-limited SBR by alternating the DO. Bioresour Technol 275:153–162
38. Yong ZJ, Bashir MJK, Ng CA, Sethupathi S, Lim J-W (2018) A sequential treatment of intermediate tropical landfill leachate using a sequencing batch reactor (SBR) and coagulation. J Environ Manage 205:244–252
39. Zalaghi A, Lamchouri F, Merzouki M, Toufik H (2018) Traitement par le procédé SBR (sequencing batch reactor) des lixiviats de la décharge publique non contrôlée de la ville de Taza (Maroc). Int J Innov Appl Stud 23:299–309
40. Zhao Y, Huang J, Zhao H, Yang H (2013) Microbial community and N removal of aerobic granular sludge at high COD and N loading rates. Bioresour Technol 143:439–446

Study on the Effectiveness of the Sequential Batch Reactor on the Reduction of Wastewater Pollution by Wastewater from Fuel Oil Washing

Khadija Zahidi, Latifa Mouhir, Chaymae Haddaji, Abdelaziz Madinzi, Safaa Khattabi Rifi, Roukaya Bouyakhsass, Khalid Digua, Mohammed Chatoui, Tonni Agustiono Kurniawan, Abdelkader Anouzla, Aysegul Pala, Salah Souabi, and Zineb Ettaloui

Abstract The ecosystem is severely harmed by industrial pollution discharges from wastewater from cleaning fuel oils, especially the maritime environment that receives these discharges. The effectiveness of biological treatment of fuel oil washing wastewater using a sequential batch reactor (SBR) is assessed in this work on a laboratory scale. The sequential batch reactor experiment for fuel-washing wastewater lasted 25 days, operating continuously in a 24-h cycle (with 30 min of filling, 17 h of aeration, 4 h of anoxia, 2 h of settling, and a final 30 min of emptying). To evaluate how effectively the fuel oil washing wastewater treatment process worked, key pollution indicators including turbidity, total suspended particles, chemical oxygen demand (COD), biological oxygen demand (BOD5), ammonium, and nitrate were

K. Zahidi · L. Mouhir · C. Haddaji · A. Madinzi · S. Khattabi Rifi · R. Bouyakhsass · K. Digua · M. Chatoui · A. Anouzla · S. Souabi (✉) · Z. Ettaloui
Laboratory of Process Engineering and Environment, Faculty of Science and Technology, University Hassan II, Mohammedia, Morocco
e-mail: salah.souabi@gmail.com

C. Haddaji
e-mail: chaymae.haddaji@etu.fstm.ac.ma

A. Madinzi
e-mail: abdelaziz.madinzi@fstm.ac.ma

K. Digua
e-mail: k.digua@univh2m.ac.ma

T. A. Kurniawan
College of Ecology and the Environment, Xiamen University, Xiamen 361102, Fujian, China

A. Pala
Environmental Research and Development Center (CEVMER), Dokuz Eylul University, Izmir, Turkey
e-mail: aysegul.pala@deu.edu.tr

S. Yadav et al. (eds.), *Emerging Technologies in Biological and Hybrid Wastewater Treatment: Lessons from Developed to Enhancing Practices in Developing Countries*, Earth and Environmental Sciences Library,
https://doi.org/10.1007/978-3-031-74515-7_6

measured. The removal efficiency of COD is 81%, BOD5 91%, TSS 72%, ammonium 100%, and nitrate 52%. The treatment of industrial wastewater from fuel oil washing wastewater of SBR-type discharges could be a potential approach to limit pollution because of its simplicity, low environmental impact, low energy consumption, and lack of requirement for highly qualified workers.

Keywords Environmental pollutants · Pretreatment · Sequencing batch reactor (SBR) · Vegetable oil refineries · Biodegradability · Oil washing wastewater

1 Introduction

Significant volumes of organic waste, including fats, free fatty acids, hydrocarbons, polyphenols, heavy metals, phospholipids, lipoproteins, and phosphates, are typically produced by fuel oil washing. These discharges contain a variety of contaminants that vary over time [23, 29].

There are numerous methods for treating industrial effluents in this setting. These industrial liquid discharge treatment techniques can be categorized into two general categories: physicochemical and biological [16]. Physicochemical methods of industrial wastewater treatment enable removing suspended and dissolved materials and the lowering organic pollutants [25]. These include membrane procedures, radiation, adsorption, electrocoagulation, air flotation, advanced oxidation procedures, coagulation-flocculation, and adsorption [9, 11]. Due to the high costs of physicochemical treatments and the creation of chemical sludge that is more difficult to handle, biological procedures are the most appropriate alternative methods in industrial discharge treatment [17, 28].

SBR can maintain the aerobic and anoxic phases sequentially while biologically removing nitrogen in a single reactor [10]. The SBR methodology is advantageous compared to other conventional methods since it operates in batch mode, one of its main features. The various parts of the batch go one after the other, with one or more tanks being utilized concurrently for the wastewater, this method takes up less space and can be adjusted to handle nearly any pollutant load. Different biological treatment techniques approved to date for oily wastewater include rotating biological contactor (RBC), activated sludge process (ASP), sequencing batch reactor (SBR), anaerobic baffled reactor (ABR), up-flow anaerobic sludge blanket (UASB), anaerobic filter (AF), moving bed bioreactor (MBBR), anaerobic filter membrane bioreactor (AnFMBR), hybrid reactor, etc. [2]. The effectiveness of the treatment process using traditional biological approaches is limited by the variability of some types of wastewater in terms of amount, content, and complexity (changes in manufacturing product, temporary plant operation, washing, etc.) [11, 18].

Alternative approaches, such as the SBR process, appear promising to enable the mineralization of industrial wastewater loaded with toxic compounds. Filling, reaction, settling, and drawing are the steps that usually comprise the process's discontinuous mode of operation [1, 13, 24].

Flexible operating conditions, high processing efficiency achievable for BOD5, COD, TSS, N, and P, and easy reaction to ongoing variations in pollutant concentrations are among the many advantages associated with SBR [5, 12]. It may also be adjusted to any pollution load and only needs a smaller surface.

2 Materials and Methods

2.1 Procedure for Sampling Oil Washing Water

According to the US Environmental Protection Agency, in January 2021, a total of 50 L of wastewater was gathered from an industrial facility situated in the Casablanca region. The collection process involved obtaining samples every hour from 10:00 to 14:00 over a period of five days. To create a composite sample, these individual samples were combined by adding their volumes together. The composite sample, totaling 10 L, was then stored in 20-L plastic containers at a temperature of 4 °C in the laboratory and characterized afterwards (Table 1).

2.2 Biomass Reactor

In the initial stage, it is essential for microorganisms to be present and consistent with the wastewater being treated in the Sequential Batch Reactor (SBR). To fulfill this requirement, a sample of urban wastewater abundant in microorganisms was collected from a domestic wastewater stabilization basin. The biomass necessary for the Sequential Batch Reactor was cultivated through a batch growth process. Consequently, an initial period was required to establish optimal biomass concentration in the SBR. To achieve this, a mixture of 1 L of fuel wash water and 2 L of aerated municipal wastewater was prepared. For 72 h, aeration was used to help the bacteria in the effluent that needed to be treated acclimatize. Using a pump, an aeration rate of 0.5 mg/min was sustained. Dissolved oxygen levels were continuously adequate and unaffected by flow restrictions during the procedure.

2.3 Setup and Operation Conditions

The filling step, which starts the SBR process, involves filling the reactor and agitating and oxygenating it at the same time to start the reaction phase. The aerobic phase and the anoxic phase are the two other divisions of this phase. During settling, no agitation is employed to allow the sludge to settle down. The subsequent step is the

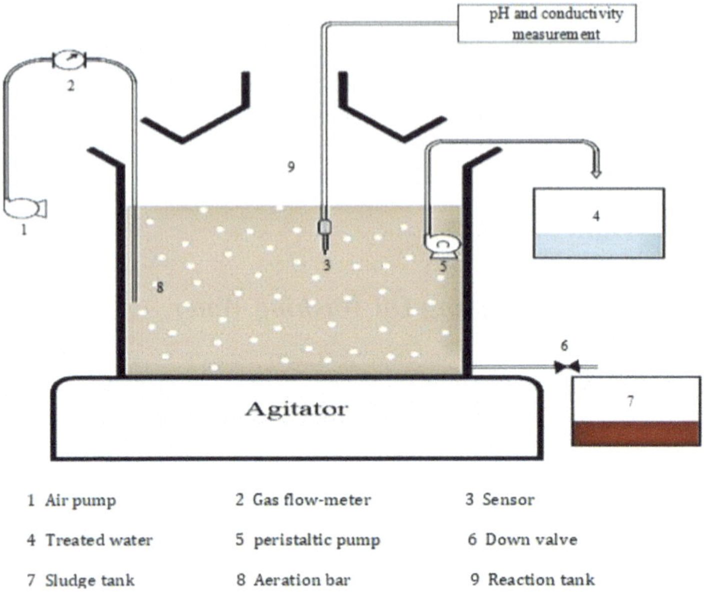

Fig. 1 Experimental equipment of the aerobic-anoxic SBR

draw phase, where the treated effluent is extracted. Following the discharge, a new cycle immediately begins by refilling the reactor.

Figure 1 presents the biological reactor used in this study.

2.4 Analysis Methods

The pH measurements were conducted employing the OHAUS starter 3100 instrument. Turbidity was assessed following NF EN ISO 7027, using the HACH 2100N turbidimeter. According to the NF T90-101 standard, the chemical oxygen demand (COD) was determined. Using the Velp Scientifica sensor system, the Biochemical Oxygen Demand (BOD5) was determined in accordance with the APHA standard. The measurement of nitrate was accomplished through a spectrophotometric method based on sodium salicylate, as specified by STN ISO 7890-3 (75 7455). The ammonium concentration was according to the AFNOR NF T90-015 standards. Based on the methodology outlined by Singleton and Rossi [27], the colorimetric Folin-Ciocalteu method was utilized to quantify phenolic chemicals.

Table 1 Characteristics of oil-washing wastewater

Parameter	Wastewater before mixing	Wastewater after mixing
	Value	
pH	8.54	6.52
Turbidity (NTU)	> 4000	> 4000
TSS (mg/L)	920	1093
NH_4^+ (mg/L)	13.73	0.565
NO_3^- (mg/L)	9.25	11.19
Phenol (mg/L)	21	1252
COD (mg/L)	1527	1828
BOD_5 (mg/L)	590	670
COD/BOD	2.59	2.72
NH_4^+/NO_3^-	1.49	0.05
COD/TSS	1.66	1.76

3 Result and Discussion

3.1 Characterization of Oil-Washing Wastewater

The physicochemical characteristics of the combined urban and industrial wastewater as well as the untreated effluent are displayed in Table 1.

The wastewater mixture has a significant concentration of total suspended solids (TSS) at 1093 mg/L and high turbidity levels (> 4000 NTU). Furthermore, the mixture demonstrates a reduction in phenol concentration of 40.5%, which has the potential to facilitate biomass growth and enhance the biodegradability of the mix. Chemical Oxygen Demand concentration ranges from 1527 to 1828 mg/L, while Biochemical Oxygen Demand (BOD5) varies from 590 to 670 mg/L. An imminent environmental danger is therefore possible considering that these pollutant concentrations go beyond the permitted discharge limits in the receiving environment. Such sewage can contaminate groundwater and surface water, endangering these essential resources when it is released into the environment. Moreover, the marine environment may also potentially be harmed by this wastewater contamination [14].

3.2 Removal of TSS, COD, and BOD5

The assessment of the reactor's performance spanned a duration of 28 days. Figure 2 presents the outcomes related to the elimination of TSS, COD, and BOD5.

The removal efficiency reached 72% for TSS, 80% for BOD5 and 90% for COD. The concentration of TSS exhibited fluctuations until day 15, followed by a decrease

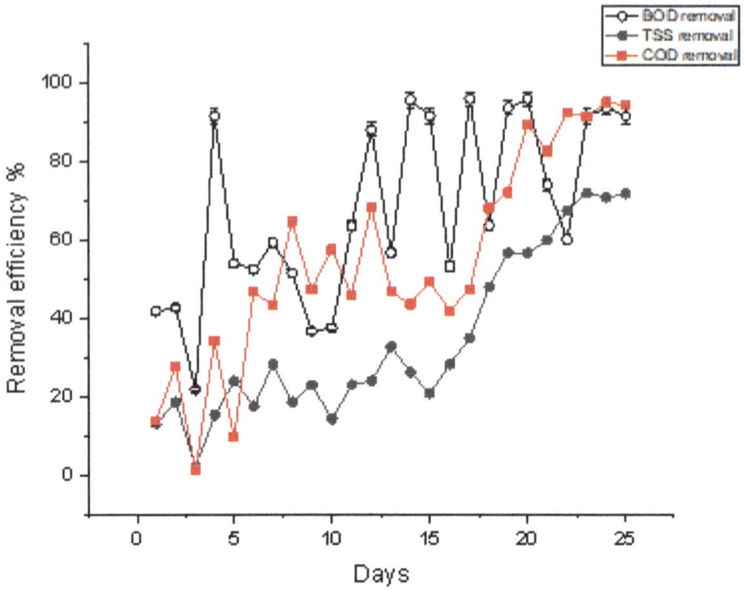

Fig. 2 TSS, COD, and BOD5 removal efficiency

to a stabilized value on the 23th day. These results demonstrate the effectiveness of SBR technology in treating wastewater with high hydrocarbon content, promoting biomass development and efficient degradation of organic matter [3].

The effective elimination of TSS from wastewater depends significantly on the settling of suspended solids, which promotes the proliferation of bacteria that play a vital role in enhancing the biodegradation of organic contaminants, as reported by Metcalf [21]. Sekarani and Hendrasarie [26] additionally verified that the decrease of TSS facilitates the adaptation of biomass for the biodegradation of organic pollutants. The effluent showed COD fluctuation throughout the first treatment period, but after day 15, a progressive drop was noted, leading to a removal effectiveness of about 81%. At day 22, the COD removal content stabilized, suggesting that organic matter was effectively decomposing and that biomass was gradually adapting over time. According to Neisi et al. [22], a batch reactor of the SBR type was used to biodegrade more than 90% of the COD by the mixed microbial mass. On the other hand, Buayoungyuen et al. [6] conducted a study that showcased a remarkable removal efficiency of 90% for COD while maintaining the stability of organic matter and nutrient removal in a reactor that operated on a large scale.

According to the work of Johal et al. [15], aeration and digesting processes are responsible for the observed drop in COD. These researchers found that aerobic digestion of an organic-rich effluent resulted to the removal of 98% of COD. These results highlight how important aeration and digestive processes are to efficiently treating wastewater.

The efficiency of an advanced sequencing batch reactor (SBR) in processing organic materials in an industrial wastewater treatment plant was also been investigated by Diya'uddeen et al. [7]. The findings showed that over a 10-h running cycle, the SBR achieved a high removal efficiency of 93% for BOD (Biochemical Oxygen Demand), with the efficiency rising to 95.2% for BOD removal. Moreover, Talalaj showed that SBR process used as a pretreatment resulted in a BOD5 removal efficiency up to 97.3%.

3.3 *Ammonium and Nitrates Variation*

The variations in ammonium and nitrate ion levels over the duration of the 25-day treatment period are shown in Fig. 3.

The simultaneous nitrification-denitrification process removed NH_4^+-N with an important 98% efficiency. Nearly all ammonia was efficiently removed during the investigation by aeration for 17 h, with oxygen concentrations ranging from 2 to 5 mg/L. Similar findings were documented by Freire et al. [8] for the 30-day treatment of oil wastewater with a (SBR), where SBR removed 95% of the ammonium ions. Furthermore, in the treatment of hydrocarbon-rich wastewater, SBR demonstrated a remarkable reduction in COD of 97%, according to Wei et al. [30]. These researchers also affirmed the increasing adoption of SBR technology as a substitute

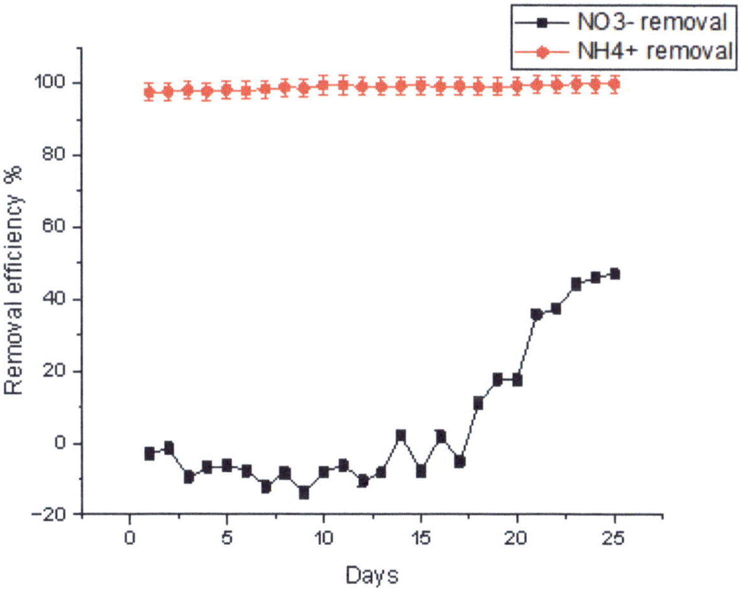

Fig. 3 NO_3^- and NH_4^+ elimination efficiency

for wastewater treatment because of its single-tank configuration and straightforward automation.

Yan et al. [31] conducted a study utilizing SBR for wastewater treatment and achieved removal efficiencies of 97.91 ± 2.04% for NH_4^+-N and 72.28 ± 2.23% for total nitrogen through the nitrite pathway. Importantly, they maintained low dissolved oxygen (DO) levels of 0.7 mg/L in the DO control zone.

Sequential batch reactors (SBRs) have gained significant attention in wastewater treatment because they can perform reactions that would typically require separate tanks in continuous flow systems, such as nitrification and denitrification, all within a single tank using a single sludge biomass. This effectively addresses the issue of NO_x accumulation [32]. Liu et al. [20] demonstrated that an intermittent aeration system is an efficient way to control NO_2 emissions in wastewater treatment plants in comparison with continuous aeration. Nitrification is an aerobic process carried out by litho-autotrophic bacteria, which multiply by obtaining the necessary energy. These nitrifying bacteria use ammonium and nitrite as energy sources and inorganic carbon as their carbon source to sustain cell growth. Ammonia-oxidizing bacteria use two enzymes, ammonium monooxygenase (AMO) and hydroxylamine oxidoreductase, to help oxidize ammonium to nitrite in the first phase [4]. Denitrification is the next phase, and several types of denitrifying bacteria may be engaged in this process.

In their study, Li et al. [19] discovered that operating the SBR reactor for 6 h, both under anaerobic and aerobic conditions, resulted in an impressive 89.6% removal of NT (a specific substance or pollutant). Within the same reactor, they also observed a substantial SND (Simultaneous Nitrification and Denitrification) process with a removal rate of 76.6%. Moreover, they observed a significant decrease in ammonium levels. Through aeration for a duration of 17 h, nearly complete removal of ammonium was achieved, with oxygen concentrations ranging between 2 and 5 mg/L.

3.4 Evolution of the Ratios COD/NH_4^+ et NH_4^+/NO_3^-

Figures 4 and 5 shows the evolution of the DCO/NO_3^- and DCO/NH_4^+, respectively.

The comparative analysis of various ratios provides insights into the changes in parameters such as COD, NO_3^-, and NH_4^+ and their elimination during the Sequential Batch Reactor (SBR) treatment process.

The COD/NH_4^+ ratio's evolution showed that it didn't change throughout the first six days of operation. But following this time, the ratio began to rise. This implies that during the aeration and anoxic stages, nitrification took place, resulting in a decrease in ammonium ion concentration and a rise in nitrate concentration. Reduced organic matter content was the outcome of heterotrophic bacteria's simultaneous nitrification and organic matter breakdown.

The results obtained for the NH_4^+/NO_3^- ratio are presented in Fig. 5.

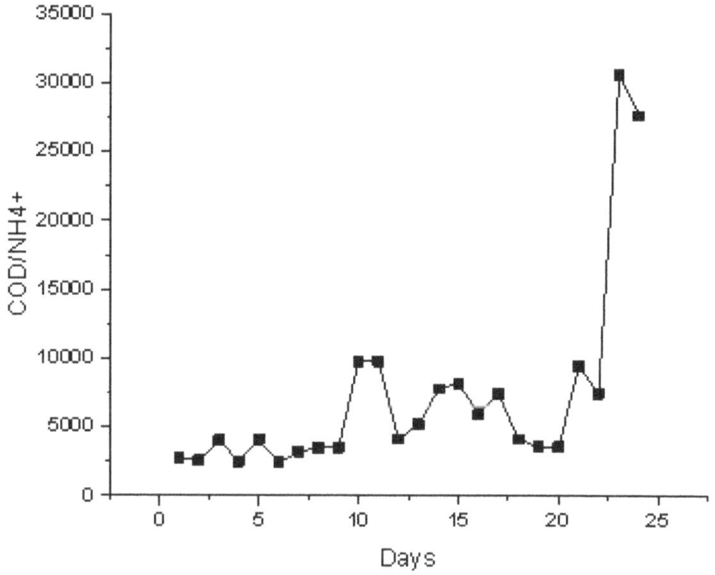

Fig. 4 Evolution of COD/NH_4^+ ratio

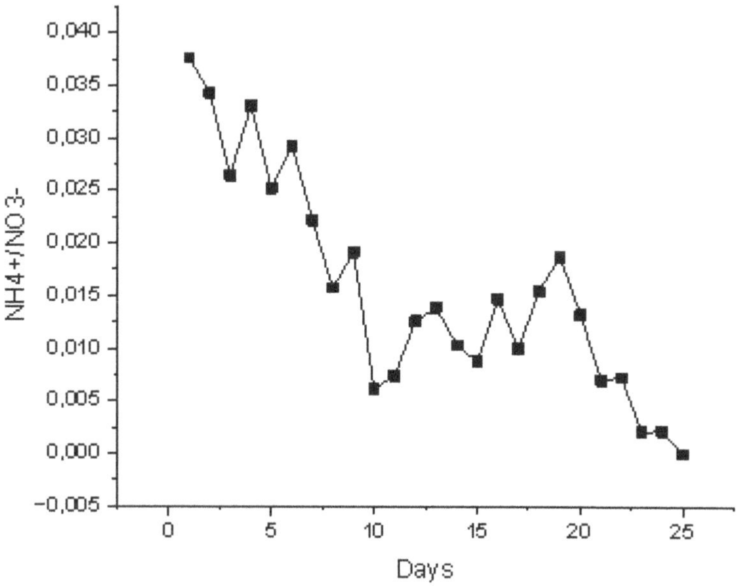

Fig. 5 Ratio evolution NH_4^+/NO_3

The results show that after fluctuating on the first day of reactor operation, the NH_4^+/NO_3^- ratio finally stabilized at 0 over a period of 25 days. This indicates that the ammonia content was successfully removed by nitrification during the aeration period, converting it to NO_3^-. Furthermore, the denitrification mechanism further transformed the NO_3^- into N_2 during the anaerobic phase scheduled inside the 24-h treatment cycle.

4 Conclusion

This study evaluated the Sequential Batch Reactor (SBR) process for treating two types of wastewater generated from fuel oil washing, which exhibited varying pollutant loads over time. The aerobic, anoxic, and sedimentation phases of the treatment were all conducted in turn within the same reactor, with a focus on the long and essential aeration period. The 24-h cycle of the experimental setup showed how well the SBR approach worked to treat effluent from oil washing. Significant removal efficiency were obtained in this study. In fact, a reduction of 81% was observed for COD, 91% for BOD, 72% for TSS, 100% for NH_4^+, and 52% for NO_3^-. These results show that the SBR process is capable of reducing a variety of contaminants found in wastewater. Furthermore, the SBR process offers notable advantages, including flexibility in operating conditions, straightforward automation, low investment costs, and moderate operational expenses. These factors make the SBR process an attractive and promising alternative for the biological treatment of industrial wastewater.

5 Recommendations

Several recommendations are made for future research, policy planners, decision-makers, and stakeholders in the wastewater treatment industry based on the study's findings. Subsequent research efforts need to explore the continuous stability and efficacy of the Sequential Batch Reactor (SBR) procedure across diverse operational scenarios and pollutant concentrations.

Pilot-scale research in several industrial contexts is also advised.

For policy planners and decision-makers, it is crucial to formulate and enforce regulations that encourage the adoption of advanced wastewater treatment technologies, develop incentive programs for industry transition, and establish standards for implementation and monitoring.

The demonstrated cost-effectiveness and operational flexibility of the SBR process should encourage investors and industries to think about funding and using it.

It is also advised to programme technical staff training and community engagement campaigns to increase knowledge of the advantages of this type of treatment process.

These suggestions will help promoting public health and environmental sustainability.

References

1. Akarsu C, Bilici Z, Dizge N (2022) Treatment of vegetable oil wastewater by a conventional activated sludge process coupled with electrocoagulation process. Water Environ Res 94:e10692
2. Al-Dhabi NA, Esmail GA, Alzeer AF, Arasu MV (2021) Removal of nitrogen from wastewater of date processing industries using a Saudi Arabian mesophilic bacterium, *Stenotrophomonas maltophilia* Al-Dhabi-17 in sequencing batch reactor. Chemosphere 268:128636
3. Aslam MT (2013) Settling of solids in raw wastewater-primary settling tanks and storm water tanks. Verlag der Technischen Universität Graz, pp 67–149
4. Bernet V, Spérandio M (2009) Principles of nitrifying process in environmental technologies to treat nitrogen pollution–principles and engineering. Cervantes FJ
5. Bijos JCBF, Pessoa RWS, Queiroz LM, Oliveira-Esquerre KPS (2022) Methane liquid-gas phase distribution during anaerobic sludge digestion: a thermodynamic approach. Chemosphere 134325
6. Buayoungyuen S, Panyapinyopol B, Fongsatitkul P, Patthanaissaranukool W, Warodomrungsimun C (2022) Simultaneous removal of nitrogen, phosphorus, and organic matter from slaughterhouse wastewater using AnA2/O2 SBR and its economic benefits. Environ Asia 15(1)
7. Diya'uddeen BH, Pouran SR, Aziz ARA, Daud WMAW (2015) Fenton oxidative treatment of petroleum refinery wastewater: process optimization and sludge characterization. RSC Adv 5:68159–68168
8. Freire DDC, Cammarota MC, Sant'Anna GL (2001) Biological treatment of oil field wastewater in a sequencing batch reactor. Environ Technol 22(10):1125–1135
9. Ganiyu SO, Martinez-Huitle CA, Oturan MA (2021) Electrochemical advanced oxidation processes for wastewater treatment: advances in formation and detection of reactive species and mechanisms. Curr Opin Electrochem 27:100678
10. Guo J, Yang Q, Peng Y, Yang A, Wang S (2007) Biological nitrogen removal with real-time control using step-feed SBR technology. Enzyme Microb Technol 40:1564–1569. https://doi.org/10.1016/j.enzmictec.2006.11.001
11. Haddaji C, Ennaciri K, Driouich A et al (2022) Optimization of the coagulation-flocculation process for vegetable oil refinery wastewater using a full factorial design. Process Saf Environ Prot 160:803–816
12. He W, Zhang L, Liu H et al (2022) CO_2 sequestration mediated by wollastonite in anaerobic digestion of sewage sludge: from sequence batch to semi-continuous operation. Chemosphere 287:132095
13. Huang T-H, Tung F-T, Chen G-F, Chen W-H (2022) Variations of N concentrations and microbial community in the start-up of anammox using anaerobic heterotrophic sludge: influence of a long reaction-phase time and comparison of the efficiencies of attached-versus suspended-growth cultures. Chemosphere 287:132151
14. Jafarinejad S (2017) Recent developments in the application of sequencing batch reactor (SBR) technology for the petroleum industry wastewater treatment. Chem Int 3:342–350
15. Johal E, Walia BS, Saini MS, Jha MK (2014) Efficiency assessment and mathematical correlation development between BOD and other parameters in Jalandhar Sewage Treatment. Int J Innov Res Sci Eng Technol 3:13088–13096
16. Karimifard S, Moghaddam MRA (2018) Application of response surface methodology in physicochemical removal of dyes from wastewater: a critical review. Sci Total Environ 640:772–797

17. Khan IU, Othman MHD, Iftikhar M, Amin AR, Ahmad R, Muqeet M, Rahman MA (2024) Oily wastewater. In: Advanced ceramics for photocatalytic membranes. Elsevier, Amsterdam, pp 409–431
18. Khouni I, Louhichi G, Ghrabi A (2020) Assessing the performances of an aerobic membrane bioreactor for textile wastewater treatment: influence of dye mass loading rate and biomass concentration. Process Saf Environ Prot 135:364–382
19. Li C, Liu S, Ma T, Zheng M, Ni J (2019) Simultaneous nitrification, denitrification and phosphorus removal in a sequencing batch reactor (SBR) under low temperature. Chemosphere 229:132–141
20. Liu T, Liu S, He S, Tian Z, Zheng M (2021) Minimization of N_2O emission through intermittent aeration in a sequencing batch reactor (SBR): main behavior and mechanism. Water 13(2):210
21. Metcalf E, n.d. AECOM (2014) Wastewater engineering: treatment and resource recovery, 2
22. Neisi A, Afshin S, Rashtbari Y, Babaei AA, Khaniabadi YO, Asadi A, Shirmardi M, Vosoughi M (2018) Efficiency of sequencing batch reactor for removal of organic matter in the effluent of petroleum wastewater. Data Brief 19:2041–2046
23. Preisner M, Neverova-Dziopak E, Kowalewski Z (2020) An analytical review of different approaches to wastewater discharge standards with particular emphasis on nutrients. Environ Manage 66(4):694–708
24. Rajadurai V, Santhi VM, Muthulingam S (2024) Removal of micropollutants from industrial wastewater using sequencing batch biofilm reactor. In: Emerging innovative trends in the application of biological processes for industrial wastewater treatment. Elsevier, Amsterdam, pp 209–235
25. Ray SG, Ghangrekar MM (2019) Comprehensive review on treatment of high-strength distillery wastewater in advanced physico-chemical and biological degradation pathways. Int J Environ Sci Technol 16:527–546
26. Sekarani FA, Hendrasarie N (2020) Reduction of organic parameters in apartment wastewater using sequencing batch reactor by adding activated carbon powder. In: IOP conference series: earth and environmental science, vol 506. IOP Publishing, pp 1–10, 12026.
27. Singleton VL, Rossi JA (1965) Colorimetry of total phenolics with phosphomolybdic-phosphotungstic acid reagents. Am J Enol Vitic 16:144–158
28. Tan X, Acquah I, Liu H et al (2019) A critical review on saline wastewater treatment by membrane bioreactor (MBR) from a microbial perspective. Chemosphere 220:1150–1162
29. Varjani S, Joshi R, Srivastava VK, Ngo HH, Guo W (2020) Treatment of wastewater from petroleum industry: current practices and perspectives. Environ Sci Pollut Res 27:27172–27180
30. Wei Y, Jin Y, Zhang W (2020) Treatment of high-concentration wastewater from an oil and gas field via a paired sequencing batch and ceramic membrane reactor. Int J Environ Res Public Health 17(6):1953
31. Yan L, Liu S, Liu Q, Zhang M, Liu Y, Wen Y, Chen Z, Zhang Y, Yang Q (2019) Improved performance of simultaneous nitrification and denitrification via nitrite in an oxygen-limited SBR by alternating the DO. Bioresour Technol 275:153–162
32. Yilmaz G, Lemaire R, Keller J, Yuan Z (2008) Simultaneous nitrification, denitrification, and phosphorus removal from nutrient-rich industrial wastewater using granular sludge. Biotechnol Bioeng 100:529–541

Nutrient Removal, Constructed Wetlands and Algal-Mediated Treatment Technologies

State of Art Technologies for Nutrient Removal from Wastewater

Poorna Weerarathna Vidanage

Abstract Treating the Wastewater produced in domestic or industrial source has been a practice associated with the human civilization since early stages and continues to be relevant in modern society. The primary focus of wastewater treatment is the removal of nutrients, specifically nitrogen and phosphorus. These nutrients are present in effluents and their presence can lead to eutrophication in water bodies in the natural environment. Nutrient removal techniques have evolved from primitive techniques to the most advanced methodologies historically. The incorporation of scientific knowledge has played a vital role in the modernization of wastewater treatment technologies just like in most modern technologies we enjoy. The objectives of nutrient removal from wastewater have significantly shifted from the sole aim of safeguarding public health and preventing diseases to encompass sustainable initiatives, such as resource recovery from wastewater. This book chapter provides a comprehensive discussion on emerging technologies designed to improve nutrient removal in wastewater treatment systems. Transition from traditional nutrient removal methods to contemporary techniques is also important for any professional who is looking in to learn about the wastewater technologies, which will be thoroughly explored, providing insights into the underlying mechanisms and highlighting the associated advantages and implementation challenges. Additionally, this chapter delves into the investigation of high-rate nutrient removal systems that have proven successful in full-scale wastewater treatment plants. Understanding the notable advancements achieved in conventional wastewater treatment is another important area that will be covered in this chapter. Under the recent advancements, implementation of advanced initiatives, including granular biomass systems, sequencing batch reactors (SBRs), wetland treatment, integrated up-flow anaerobic sludge blanket (UASB) systems, and short aerobic systems will be discussed. Microbial consortia are utilized to remove nutrients in wastewater as an effective strategy. Present trend is to focus on further enhancing the efficiency and economy of biological nutrient removal from

P. W. Vidanage (✉)
Department of Chemical and Process Engineering, University of Moratuwa, Moratuwa 10400, Sri Lanka
e-mail: poornaw@uom.lk

S. Yadav et al. (eds.), *Emerging Technologies in Biological and Hybrid Wastewater Treatment: Lessons from Developed to Enhancing Practices in Developing Countries*, Earth and Environmental Sciences Library,
https://doi.org/10.1007/978-3-031-74515-7_7

wastewater, aligning with sustainable initiatives. Key areas of interest are biological phosphorus removal, biofilm technology for water treatment, and the utilization of microalgae which have been covered in this chapter.

Keywords Wastewater treatment · Nutrient removal · SBR · UASB · Wetland · Microalgae

1 Introduction

Access to clean water is essential to sustain the human civilization. After using water for a variety of domestic or industrial purposes, the disposal of the water becomes mandatory. Traditionally termed as "wastewater," this discharged water is deemed unfit for further utilization due to its inherent impurities and unintended constituents, which often regarded as pollutants, are responsible for distinguishing wastewater from its clean water counterpart.

During the early stages of human civilizations, the management of wastewater was not concerned by society. In the early wastewater management two main practices can be identified. First, the production of wastewater was constrained by the low population density, leading to limited quantities being generated.

The natural environment had the capacity to absorb the limited discharged wastewater capacity without experiencing significant disturbance to the ecology. In the early days scientific knowledge was limited. The understanding of the importance of wastewater management could not be explained with respect to the modern days.

However, as civilizations progressed, the needs of the populace expanded, consequently driving an increase in the production and consumption of goods. With the overshooting demands, the expansion of industrial activities became an inevitable course of action. This expansion, in turn, resulted in a rapid surge in the demand for freshwater resources. Notably, during periods characterized by rapid industrial growth, significant adverse impacts on the environment were reported because of the discharge of untreated wastewater.

Recent advancements in the state of art technologies in the nutrient removal in the wastewater are shifting towards material recovery. This initiative is also linked with the circular economy that has been discussed in the engineering and economic professionals who become the stake holders in the investment and development in the wastewater technologies. The wastewater has now been looked like a source for the material that can become inputs to another process. This is also identified as waste valorization. Anaerobic digestion wastewater treatment process is a classic example of the material and energy recovery process. In this process methane is recovered as the main output which will be used for energy generation. The initial influent can be dewatered where the water or other nutrients can be recovered during subsequent processes. Sludge recovered from the anaerobic digestion process can be used to struvite precipitation, a novel method to remove and recover Phosphorus from wastewater [15]. Sludge (Sewage Sludge) from the post processing in the wastewater treatment plants can be hydrothermally treated (carbonization/liquefaction/gasification) to recover biochar, biogas, biofuels etc. [14]. Recovering the energy from the wastewater treatment systems have taken an innovative turn with the introduction of the electrochemical applications. Natural anodes and cathodes can be set up to make microbial fuel cells in the constructed wetlands. Modified microbial fuel cells can be used to capture the nutrients from the wastewater substrate. For an example macronutrient (N, P, and K) and micronutrients (S, Ca, Mg, and trace metals) required for the plant growth have been recovered from human urine using bio-electrochemical nutrient recovery cell which can do valorisation of nutrient at the same time with supplying the electricity [5].

1.1 Importance of the Nutrient Removal

The introduction of water into wastewater streams leads to the accumulation of nutrients at higher concentrations. These nutrients comprise macronutrients in a significant amount with micronutrients in trace amounts. Notably, nitrogen and phosphorus are key micronutrients and important for the vital roles in animals, plants, and microorganism functions. These nutrient concentrations are significantly increased in wastewater streams, originating from various sources such as human and livestock waste, industrial process discharges, and cleaning agents (soaps/detergents), among others. As a result of the increased nitrogen concentration, eutrophic conditions are developed in natural water bodies, which will eventually be inhabitable for the aquatic creatures and the quality of the water plummeted below the standard levels. Increased nitrogen levels in the drinking water can lead to diseases like infant methemoglobinemia [3].

The quantitative and qualitative determination of these nutrients is crucial for effective wastewater management. Additionally, the presence of micronutrients in wastewater includes metallic compounds such as molybdenum, cobalt, and iron. To get rid of the excessive nutrients that cause many environmental problems, different removal strategies can be employed which are broadly categorized under physical, chemical, and biological processes (Table 1).

Table 1 Different strategies to remove the nutrients from wastewater

Main process category	Examples for the treatment processes
Physical	Membrane filtration (P), stripping (NH_3), ion exchange (inorganic ions), fat removal, screening, sedimentation
Chemical	Chemical precipitation (P), coagulation, flocculation
Biological	Activated sludge process, aerobic and anaerobic nutrient removal (N, P), constructed wetlands, biological aerated filters

2 Conventional Nitrogen Removal Systems

2.1 Identification of the Naturally Occurring Nitrogen Removal Systems

Being an essential macronutrient for cellular formation and biological reactions, different forms of nitrogen are naturally consumed in biological processes. These processes are predominantly driven by microbial communities. Even in the natural environment nitrogen content of the water in a stream or puddle can be increased due to the mixing of animal feces, fallen dead bodies of plants or animals etc. When the waste material is added into the water it will be disintegrated physically by mixing with the water or will be converted from solid matter into soluble ions or molecules in different forms (NH_4^+, NH_3, NO_2, NO_3^-, NO_2^- etc.) (Fig. 1).

Consumption of nitrogenous compounds are essential for any living cell. This natural phenomenon is strategically used in the biological nitrogen removal process to transfer the nitrogen from the water body to cells grown. The nitrogenous compound will be assimilated in the cells which can be subsequently removed from the wastewater. The design of the most wastewater treatment plants (WWTP) are optimized to enhance the growth of nitrogen accumulating organisms to ensure the efficiency of the nitrogen removal. Operation of the WWTP will selectively enhance the proliferation of the microorganisms with higher nitrogen removal capability.

2.2 Nitrification

Nitrification is a sequential process where multiple steps in the operation can be categorized separately according to the chemical and biological conditions. This can be distinct in to two difference process where in the first step ammonia is turned into nitrite and the nitrite is converted in to nitrate subsequently. *Nitrosomonas* and *Nitrosococcus* are examples for the bacterial genera responsible for the ammonia oxidization. These processes are being catalyzed by ammonia monooxygenase and hydroxylamine oxidoreductase enzymes.

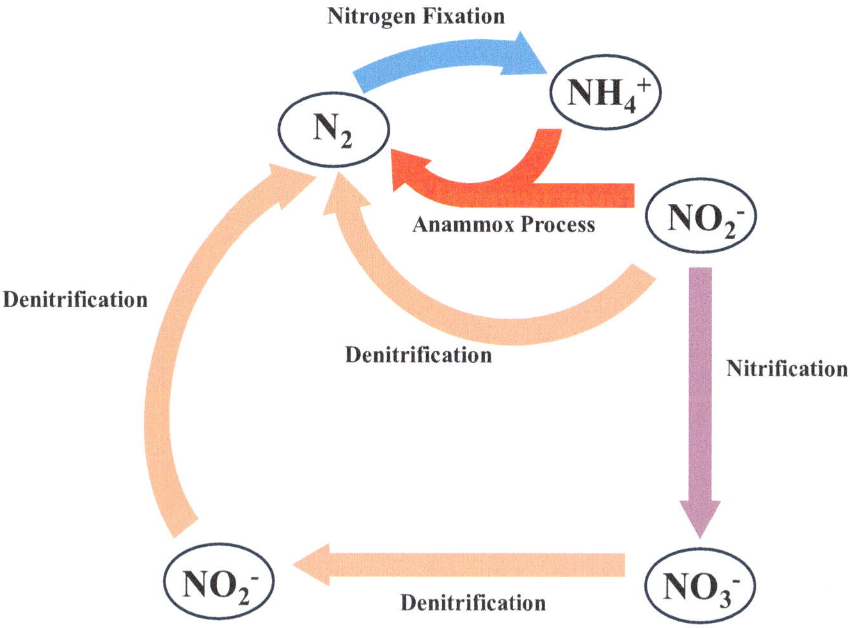

Fig. 1 Conversion of nitrogen species in nitrogen removal processes

There are aerobic ammonia oxidizing bacteria which shows photosynthetic capability however their slow growth rate is a challenge utilization of their nitrification capability in the WWTPs. Role of these bacteria in the nitrogen cycling will help the environmental engineers to optimize their applications in the nitrogen removal process in the industries. Changing operational parameters affect significantly for the nitrogen cycling dynamics which need to be comprehensively identified by the designers of WWTPs.

Nitrification process which is a biologically driven process is applied in the ammonia removal of the wastewater treatment plants. Ammonia oxidizing bacteria (AOB), or archaea (AOA) are used in this process. Both microorganisms use ammonia as an energy source. Ammonia monooxygenase (AMO) enzyme is catalyzing the conversion of ammonia into hydroxyl amine (NH_2OH).

Reaction 1: Ammonia → Hydroxyl amine (NH_2OH).
Reaction 2: Hydroxyl amine → Nitrite (NO^{2-}).

Conversion of hydroxyl amine into nitrite (NO_2^-) is catalyzed by hydroxylamine oxidoreductase (HAO) enzyme. In this nitrite generation step generate a surplus of electrons which can initiate an electron transport chain (ETC). The ETC drives electrons across the cell membrane will synthesize ATP (Adenosine triphosphate) with higher energy which will be subsequently consumed for cellular functions. Fixing of CO_2 as organic carbon will be energized by ATP.

Although there are many microorganisms with the capability of nitrification, *Nitrosomonas* and *Nitrosococcus*. *Nitrosopumilus* are some of well-known AOB genera. *Nitrososphaera* is an example for AOA that have been well studied.

3 Advanced Nitrogen Removal System and Anammox

3.1 Anammox Process

Conventional nitrification process is takes place in aerobic conditions. Anammox process is a recently discovered technology which takes place in the anoxic conditions. The "ANAMMOX" term stands for anaerobic ammonia oxidation. This process is conducted by bacteria which are commonly identified as the anammox bacteria. The anammox bacteria are specialized to use nitrate as the electron acceptor. These bacteria can oxidize ammonium into gaseous nitrogen. Anoxic conditions are known to be maintained under the depleted dissolved oxygen in the media. When the supply of the oxygen is limited microorganisms will use a substitute for oxygen which will serve as the electron accepter. This is important in maintaining the ETC which ensures the energy acquisition of the cell.

Despite the Anammox process is identified during the experiments conducted in wastewater treatment systems, this process is ubiquitous in the nature in the oxygen depleted systems like river and lake sediments. Coastal, marine and estuaries are known to harbor the anammox bacteria. The first anammox bacteria identified is *Brocadia anammoxidans* originated from a wastewater treatment plant. Significant contribution of this bacterial group has been reported in different natural and engineered systems in the removal of ammonia.

4 Phosphorous Removal Systems

4.1 Phosphorous in the Wastewater Systems

Phosphorus is an element critical for the fundamental building blocks (DNA, amino acids, protein, lipids etc.) in any biological system. Biological systems use ATP to absorb energy.

Despite its indispensability, an excessive concentration of phosphorus in water can lead to the undesirable proliferation of algae, resulting in algal blooms. These blooms, in turn, contribute to a depletion of dissolved oxygen (DO) levels in the water. The adverse impact on aquatic biota is inevitable. Furthermore, elevated concentrations of phosphorus can adversely affect individuals who consume the contaminated water. Acute problems arising from exposure to higher phosphorus concentrations include hormonal changes that regulate calcium metabolism and the calcification of non-skeletal tissues, particularly in organs such as the kidneys. Addressing and mitigating these issues are essential for maintaining the ecological balance and safeguarding both aquatic ecosystems and human health.

Influent Phosphorus to the wastewater treatment plants is removed by phytochemical, biological or combination of both these. Although there are many combined processes suggested in the literature, there are very limited mechanisms that are established on the industrial scale. P removal mechanisms can be selective based on the incoming P concentration hence for a variable incoming P loading catering the correct removal mechanism will be challenging.

4.2 Physicochemical P Removal

Removal of the P can be categorized in to physical and chemical technologies. Under the physical methods precipitation, adsorption, absorption, and ion exchange can be given as examples. Addition of the chemical reagents to convert the ionic form of the P species to be convert into a solid form is an example for chemical P removal. Also, variation of the pH will influence significantly for the removal process of P. It is important to highlight that the removal of P cannot be solely done via physical or chemical mechanism along. In the practice removal of P is a hybrid approach combining both physical and chemical methods. P removal treatment is dependent on the level of P in the influent water. The cutoff level generally used is 50 mg/L for the P rich wastewater. Phosphate can be recovered from the P rich wastewater by adding Calcium and Magnesium ions. In this process, P is precipitated into calcium phosphate or magnesium ammonium phosphate (MAP, struvite), crystal form.

4.3 Biological Phosphorous Removal System

Key process used to remove the phosphorous is to enrich the microbial community in the WWTP with microorganisms who can assimilate P in the cells. This is known as Enhanced biological phosphorous removal (EBPR) process [2]. The microorganism involved in this process is known as Polyphosphate Accumulating Organisms (PAO). Phosphate accumulation is a common functional feature observed through a large range of taxonomically diverse bacterial groups.

4.4 Metabolic Mechanism of P Removal by PAO

For the utilization of the PAOs to remove the phosphorus in wastewater treatment systems, it is important to understand the cellular metabolic functions taken place in the PAO cells. A comprehensive understanding of this mechanism is critical for the optimization of the Enhanced Biological Phosphorus Removal (EBPR) process design. EBPR process is a two-step process. Orthophosphates are accumulating in the cells under aerobic conditions in the first step. Then under anaerobic conditions the internally accumulated orthophosphates are released to the outer cell membrane by PAOs. Phosphorus in the wastewater exists in the form of (PO_4^{3-}) but it is stored in the cells as polyphosphates. These polyphosphates can have different molecular structures with linear, cyclic or branched arrangements. This chemical conversion from phosphates to polyphosphates is orchestrated by a series of metabolic pathways in PAOs.

5 Granular Biomass System

5.1 *Activated Sludge Process*

Removal of the organic load in the influent wastewater stream is removed by consuming for the metabolic activities of the microbial consortium. This process in generally conducted inside a bioreactor. Incoming wastewater undergoes an aeration process in the bioreactor providing oxygen required by the bacteria to consume the organic loading. The aeration process is followed by settling stage where the activated sludge is settled at the bottom of the reactor. This is a simple process facilitated by gravity. When the activated sludge is separated from the main water body, the water will be with low concentrated organic loading with respect to the influent water. This treated water will be discharged to the further treatments in the downstream. The activated sludge will contain microorganisms that are associated with organic and inorganic solids such as food particle, fecal matter, grit, sand etc. Microorganisms in the activated sludge can be retained within the bioreactor after the water is discharged. This retained microorganisms can reinoculated the influent wastewater to and grow in the aeration stage consuming the incoming organic load. Hence retention of the activated sludge is important for the function of the bioreactor. This process is a batch process which is operated as a cycle (Fig. 2).

5.2 *Aerobic Granular Sludge*

Aerobic granular sludge is an improved version of conventional activated sludge process. The structure granules have been improved by improvising for the faster settling and enhanced contact area between the granule surface and the wastewater substrate. This structural change is mainly the increased particle diameter and the density which will be significantly improving the aerobic granular process.

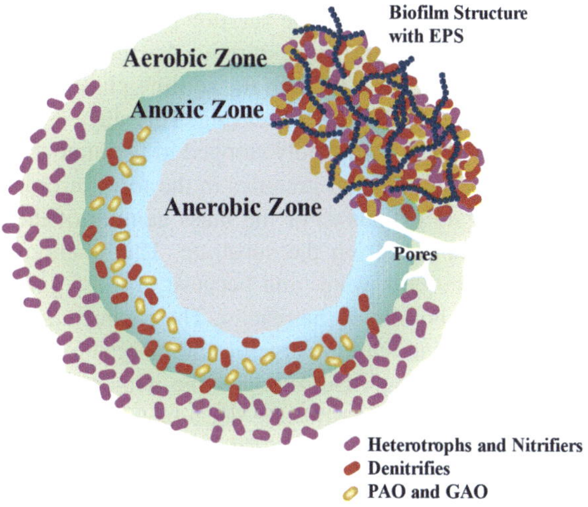

Fig. 2 Microorganism assembly in activated sludge granule structure

This improved granular sludge process is known to be treat higher reactor biomass concentrations (8–15 g/L).

6 Sequencing Batch Reactor (SBR)

The activated sludge process (ASP) is a conventional process which utilizes highly concentrated microbial consortiums to treat wastewater. Sequencing batch reactor (SBR) process is an improved form of ASP with enhanced operational flexibility and process control [13]. The Sequencing Batch Reactor (SBR) technology emerges as one of the most established methods for treating domestic wastewater. The domestic wastewater mainly consists of human excreta and urine, combined with water used for activities such as bathing, laundry, and kitchen usage.

Domestic wastewater is mainly discharged from residential areas such as housing complexes, hotels, apartments, and offices. Fraction of the wastewater generated in industrial settings can also be categorized under domestic wastewater due to human-related discharges. SBR technology is widely used and well understood by the users as well as designers which helped for the development of streamlined infrastructure requirements. This establishment of this technology has significantly helped to reduced spatial footprint, and cost-effectiveness in operational aspects.

Sequencing Batch Reactor (SBR) process can be divided into 5 stages (Fig. 3). The 1st step is Filling (Inflow), where the wastewater is filled into the tank. The time taken for the filling phase depends on the design specifications. For an example requirement of the treatment process such as upstream flowrate of the wastewater and the volume of the tank will decide the filling time. Second step is the reacting or Aeration step, where the active reaction of the SBR is takes place. Once the reactor is filled with wastewater, the aeration step will be carried out by providing oxygen for the bacteria. With the respiration bacteria consume the nutrient matter in the wastewater. After the aeration the settling phase the solid liquid separation will be taken place using the gravity settling. This stage is identified as a clarifications step. Where aeration is completely stopped to facilitate the settling. The microorganisms agglomerate in the solid granules in the form of biofilms, which will subsequently result the gain of mass. For the mass gaining of the cells in the biofilm nutrients will be consumed from the substrate. This phenomenon will reduce the nutrient loading in the bulk water and because of mass gain in the granules, they will be settled at the bottom of the reactor. The clearer liquid free from the solids and with lower concentration of the organic and inorganic nutrient can be downstream for the next treatment stage. Decanting is the withdrawal of the clarified liquid from the top of the bioreactor. After the decantation of the treated liquid, the batch reactor will undergo the idling stage. This stage is also optional based on the design of the wastewater treatment process. This will enhance the settlement of any further settling of any remaining solid particles in the liquid phase. The objective of these steps is to improve the overall efficiency of the process.

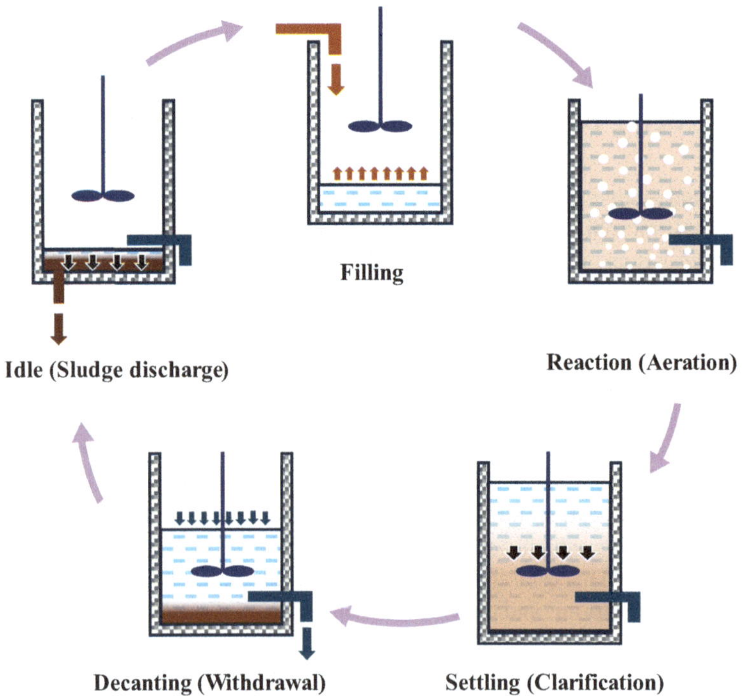

Fig. 3 Stages of sequencing batch reactor (SBR)

7 Integrated Up-Flow Anaerobic Sludge Blanket (UASB) and Short Aerobic System

UASB (Up flow Anaerobic Sludge Blanket) reactors are generally used to treat wastewater with higher organic loading. The advantage of the UASB is improved treatment efficiency with reduced retention time compared to conventional bioreactors used in wastewater treatment systems. Inflow is fed in the bottom-up direction which will move upward facilitating the granules to settle from the substrate to the bottom (Fig. 4). The granules host microorganisms attached to their surfaces in the form of a biofilm. These microorganisms will consume organic loading during this process.

In the UASB reactor, during the cross-directional flow of wastewater substrate and sludge granules, microorganisms efficiently consume the substrate. This anaerobic process, characterized by the absence of aeration, results in anaerobic digestion, generating biogas. Gas bubbles are collected from the top of the reactor using an inverted conical-shaped gas trapping cap. Treated water is then discharged through overflow, flowing into the weir around the circular top of the reactor. The UASB reactor, therefore, provides a competitive and efficient solution for handling higher organic loads in wastewater treatment.

The compact design of the UASB demands less reactor volume and space (low footprint), making them easily installable in congested areas where conventional wastewater treatments cannot be accommodated. The sludge generation in UASB reactors is relatively low, hence the convenience of post-processing material handling

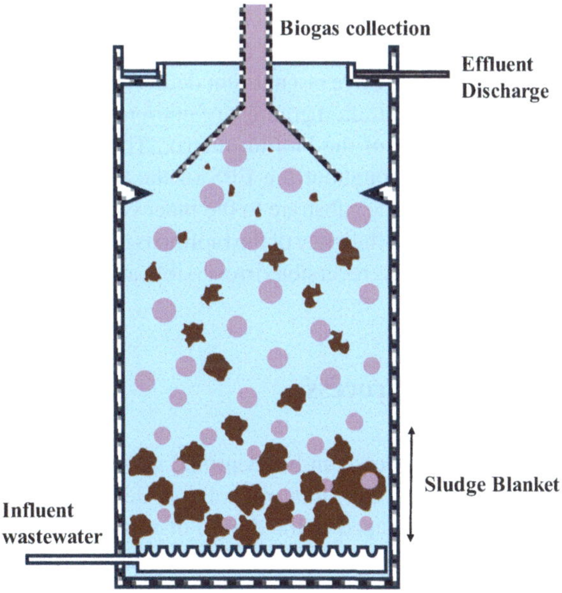

Fig. 4 Up-flow anaerobic sludge blanket (UASB) process

is unmatched. This technology stands out for its biological-driven process, requiring minimal chemical inputs, as the bacteria thrive using incoming nutrients.

8 Biofilm Based Treatment System

Bacterial consortia can be used to remove nutrients from wastewater by absorbing, accumulating, or converting them. The special distribution of bacterial cells in the water body has a significant impact on nutrient removal efficiency. This distribution is mainly dependent on the bacterial lifestyles which can be classified into two main types based on dispersion patterns: planktonic and biofilm. Biofilm-forming bacteria demonstrates free living (planktonic) lifestyle under some conditions (i.e.: nutrient limitation, stress, and maturity).

A biofilm is simply a dense aggregation of single or multiple bacterial cells embedded in a three-dimensional extracellular polymeric substance (EPS) matrix that is attached to a surface [4]. Shifting between planktonic and biofilm lifestyles is important for the bacteria to thrive and disperse throughout ecosystems. In wastewater treatment, nutrient consumption mechanism of biofilms can be used to reduce plant footprint and retention time. Cells attached to a surface (tank walls, floating material) will absorb the nutrient from the waterbody.

Biofilms are ubiquitous in any environment with surfaces. Although biofilms are known to be life-threatening (e.g., dental plaque) and resistant to antibiotics in medicine, they can be used in environmental engineering to remove nutrients from wastewater. Biofilms form on the surfaces of bottom sediments, sludge, and tank walls in both natural and engineered ecosystems.

Bacteria in biofilms do not have a continuous supply of fresh substrate like free-living bacteria, so there is a nutrient deficit away from the solid-water interface of the biofilm. A concentration gradient of nutrients (N, P) is maintained from the substrate to the inner core of the biofilm matrix. The gradient of the nutrient concentration is maintained throughout the EPS of the biofilm. Mass transfer of the nutrient is happening from the substrate to the inner cells embedded in the biofilm matrix. The nutrient removal efficiency of the biofilm is highly dependent on the stages the biofilm is maintained (i.e.: reversible/irreversible attachment, maturation, dispersion).

9 Wetland Process

The wetland process can be identified as one of the oldest strategies for removing contaminants. In fact, this marvel was developed by Mother Nature long before humans were engaged in the design of wastewater treatment plants (WWTP). This process is commonly observed in various environments (i.e.: marshlands, paddy fields, mangroves, peatlands, bogs, and swamps). This mechanism is driven by creating a high demand for incoming nutrients. The animals, plants, and microorganisms living in the wetland ecosystem are working as the sink for the incoming nutrients. The wetland biota consumes organic and inorganic contaminants through a series of complex physical, chemical, and biological transformations. Ultimately, the water released from the wetland exhibits significantly lower contaminant levels.

Streams of wastewater were simply channeled into swamplands during the early stages of getting rid of contaminated wastewater. Although the precise chronology is still unknown, wetlands were initially used as catchment areas for the dumping of wastewater and were frequently thought of as wastelands. The idea that wetlands or bogs already contained unclean water may have led to the preference for wetlands over rivers or water streams for wastewater discharge. This procedure unintentionally made it easier for some wetlands to remove nutrients.

According to historical evidence, wetlands were first designed in the 1950s as a result of a thorough investigation into the intricate physicochemical and biological transformations. Wetlands designed by environmental engineers were simply imitation of treatment processes seen in natural wetlands. The wetland process have been improved significantly during 1990s and 2000s. Combining horizontal and vertical flows to improve treatment effectiveness can he highlighted as an important feature that contributed for the advancement of the artificial wetlands [7, 17].

Environmental scientists and engineers methodically characterized wetland systems as their understanding of natural wetlands deepened. Plant growth, sediment type, soil structure, wastewater flow rate, and nutrient removal effectiveness are some of the key parameters considered during the designing of the wetland. This information paved the way for the creation of artificial wetlands designed to handle wastewater. Removal of heavy metals, color, odor and harmful pollutants was the goal of the creation of artificial wetlands, which will reduce the load for the other unit operations in the full scaled wastewater treatment system.

10 Microalgae

10.1 What Are Microalgae?

The proliferation of algae is typically regarded as inadvertent and problematic, specially in the consideration of the water quality management and recreational characteristics. Nevertheless, the unfavorable proliferation of photosynthetic cells can be purposefully harnessed for the purpose of wastewater remediation. Growth of the algae can be used for a strategy to complete understanding of the mechanisms governing algal-mediated wastewater purification is necessary for the formulation, implementation, and enhancement of strategies pertaining to such remediation processes.

Algae, as a photosynthetic eukaryotic group, exhibit diverse phylogeny and metabolic potentials. They are known to grow in various lacustrine environments and wastewater systems. Microalgae, distinguished by their smaller size, stand apart from other alga types. In contrast to macroalgae, which are visible to the naked eye, microalgae are strictly microscopic and can manifest as either prokaryotic or eukaryotic. The inherent advantage of a higher surface-to-volume ratio in microalgae facilitates enhanced mass transfer through cell membranes, allowing for efficient nutrient uptake in proportion to the substrate volume [1].

10.2 Microalgae Community

Microalgae do not grow in nature as isolated species or strains but as diverse consortia. In these consortia there are multiple species with various metabolic potentials. The outcome that is observed in any given system is the resultant metabolic activity of all the species in the system. Specially the species with higher prevalence in the given ecosystem and greater in numbers are known to be core species which will have a dominant impact on the ecosystem [16]. There are various algal species cohabiting with microorganisms such as bacteria and fungi. As autotrophs, algae establish a robust symbiotic relationship with heterotrophic bacteria. The oxygen produced by the photosynthesis in algea is actively utilized by the bacteria, while carbon dioxide released by heterotrophs is assimilated by algae in the form of bicarbonates. This can be also identified as cohabitation between algae and bacteria. Bacteria release Nitrate and ammonia and byproducts of cellular metabolism. These nutrients are assimilated into algal cells, showcasing a dynamic interdependence in their ecological collaboration [10].

Both Planktonic and attached growth are displayed by algal communities. Every type of algae has a preferred environment, which adds to the diverse range of options for treating wastewater. The diverse range of functions that these algae species possess is essential for wastewater treatment. Micro and macroorganism interactions, in addition to those involving microalgae, all play a part in the effective treatment of wastewater. Microbial ecology classifies these interactions as mutualism and commensalism, for example. It might be difficult to maintain pure cultures, often called axenic cultures, in wastewater treatment systems. Rather, it is advised to use the cooperative efforts of several microbial consortia that operate in mixed cultures [2].

10.3 Selection of the Strain

Many species of microalgae are remarkably efficient at removing nutrients from wastewater. The biological roles of the chosen algae strains have a direct bearing on how well these nutrients are removed from the environment. From an environmental biotechnologist's perspective, the strains are chosen using several important criteria. A high growth rate is the most important factor. This guarantees that enough cells will be generated to properly treat incoming wastewater. Both production and maintenance expenses are important considerations when it comes to large-scale bioreactors. The chosen strain ought to exhibit the capacity to flourish in upgraded reactors while reducing expenses associated with production.

Diverse constituents can add to the treatment facility with incoming wastewater, which can alter the condition of the substrate. Hence, the selected strain must be resilience, where it can withstand the fluctuating environmental conditions. Factors such as pH value, light intensity, and temperature are identified as critical parameters for strain tolerance and success in wastewater treatment [11]. *Chlorella* is one of the most studied microalgae which is nonmotile unicellular with a size of 2–10 μm. *Chlorella vulgaris* algal species have been tested as a model organism to be used in the wastewater treatment processes, it has showed that the removal efficiency of 85% Total Phosphate and 89% of Total Nitrogen [20]. *Chlorella* species have also been demonstrated for higher biosorption capacity which is alternative benefit for the removal of some metal ions up to 100% (Mn/Fe).

10.4 Nitrogen and Phosphate Removal by Microalgae

Algae plays a crucial role in the removal of nitrogen from wastewater, specifically in the forms of nitrates, nitrites, and ammonia. The cellular uptake of nitrite and nitrate ions occurs through co-transportation mechanisms. Within the cellular cytoplasm, nitrate ions undergo reduction to nitrites facilitated by Nitrate Reductase, resulting in the accumulation of nitrites. Subsequently, Nitrite Reductase (*NiR*) enzymes, located in the stroma, further reduce nitrite ions to ammonium ions. Ultimately, these ammonium ions undergo conversion into amino acids, initiating a metabolic pathway that culminates in the synthesis of proteins through anabolism. Algae actively participate in the removal of phosphates, encompassing both organic and inorganic forms. Organic phosphates undergo cleavage by phosphatase enzymes, transforming them into inorganic phosphates (biphosphate anions). The inorganic phosphates within the algal cytoplasm aggregate in the vacuole, manifesting as polyphosphate granules. Meanwhile, cytosolic organic phosphates serve as substrates for nucleotide synthesis within the chloroplast stroma, subsequently undergoing conversion into nucleic acids [19].

10.5 Types of Microalgae Based Wastewater Treatment Technologies

Two distinct microalgae culture systems find application in wastewater treatment plants. The first is the suspended algae culture, wherein planktonic algal cells are immersed in the substrate. The second is the non-suspended or attached growth system, where algae are cultivated on a solid surface, facilitating nutrient removal at the interface. Different algae treatment approaches have been developed to improve the effectiveness of wastewater removal while taking into account the unique features of these two techniques.

As the most popular bioreactor for growing microalgae, photobioreactors provide fine control over temperature, light, nutrients, and mixing conditions that are particular to the type of microalgae growing in the substrate. Different designs of photobioreactors accommodate different growth patterns that correspond with different approaches to water treatment. Paddlewheels are used in high-rate algal ponds, which are designed to handle high flow rates, to gently direct wastewater via a shallow channel and provide uniform solar exposure for cells at all levels. Nutrients are consumed during biomass growth, and the biomass is then removed using a clarifier.

For attached microalgae growth, Algal Turf Scrubber (ATS) technology employs a mesh suspended in the wastewater substrate to nurture algal biofilms. The benthic algae on the mesh can be scraped and removed, allowing for the regrowth of a fresh batch. Membrane Aerated Bacteria-Algae Biofilm Reactor (MABAR) integrates membrane modules supporting attached growth while doubling as air diffusers. This technology has demonstrated compatibility with higher nitrogen removal efficiency [21].

11 Challenges Associated with Nutrient Removal Technologies

The primary challenge in the ANNAMOX process is cultivating the ANNAMOX bacteria under low ammonium concentrations. This can lead to extend startup periods and decrease nitrogen removal rates. The optimal temperature range for ANNAMOX bacteria is 30–40 °C. The changes in the incoming water temperature will disrupt the ideal conditions for growth of ANNAMOX bacteria. Unlike the experiments conducted in controlled conditions, where ANNAMOX bacteria are tested and optimized using synthetic wastewater, actual wastewater carries diverse loads of COD, and the C:N ratio significantly deviates from experimental values. These changes can adversely impact the performance of the ANNAMOX process [18].

When phosphorus concentrations are very low (trace level), the biological phosphorous elimination efficiency decreases. Because bacterial activity is necessary for the phosphorous removal process to occur, it is critical to maintain strict conditions. The overall efficacy of the Enhanced Biological Phosphorus Removal (EBPR) process can be severely hindered by any stress placed on the microorganisms involved [23]. Even though granular biomass technology is a proven system, accelerating the startup process is a major barrier. Because of mixing and aeration in aerobic circumstances, the produced granules are vulnerable to disintegration and external influences. Granule diameter might grow as a result of high organic loads, which lowers the effective surface area available for removing nutrients [6].

Thickness of the biofilm is the key parameter for the biofilm-based nutrient removal. Density of the cells, porosity of the extracellular polysaccharide layer will influence the mass transfer in and out of the biofilm. Over the operation the biofilm thickness will be varying and also the maturity of the biofilm will be changed. This will affect the intact or dispersion behavior of the biofilm. During the operational conditions, it will be very difficult to maintain the thickness of biofilm as a result of varying factors like composition of the wastewater [12].

Despite the advantages of the microalgae technology for nutrient removal in wastewater there are several challenges. The complexity of the treatment mechanism and the high capital cost associated with the treatment method can be identified as key challenges. Selecting the appropriate algal strain proves challenging due to the diverse factors present in incoming wastewater. Most of the microalgal treatment technologies are still operating at the research and development level, translating these advancements into practical applications for existing effluent discharges poses a significant challenge. A notable issue with emerging microalgal technologies lies in their implementation and achieving the expected technoeconomic potentials [8].

12 Conclusions

To maintain biofilm thickness in membrane bioreactors, ongoing experiments in biotechnological applications are exploring the utilization of quorum sensing. Biofilms are recognized for employing quorum sensing to release extracellular polymeric substances (EPS), forming the biofilm matrix. Through external signals, bacteria can be genetically modified using synthetic gene circuits to respond to externally induced signals, including light [9]. Recent advancements in sustainable effluent treatment has pushed the professionals working in the wastewater treatment industry, to emphasis on resource recovery rather than traditional waste treatment methods. This shift is also evident in microalgae-based wastewater treatment systems. The carbon assimilated into microalgal biomass can be harnessed for biodiesel production. Moreover, developments in microalgal wastewater treatment have progressed to the point of recovering raw materials for industrial products such as fatty acids and bioactive components, as well as pharmaceuticals including antibacterial and antiviral compounds, along with anticancer agents [22].

13 Recommendations/Future Prospects

With the development of modern science and engineering knowledge many innovative technologies are introduced to remove the nutrients from the wastewater. Some of these technologies show promising performance against the established classical treatment methods. However, a common challenge observed in the wastewater treatment industry is the reluctance to invest in new technologies. Existing industries should be conducive to new concepts to find the limitations of their designs. Most of the concepts are developed in simulated systems with very limited application in the laboratory or pilot scale. For the establishment of any new concept in the wastewater treatment industry it must be adopted in full scale treatment plant. When it comes to the techno economic feasibilities it is very difficult to predict without running in a real-world environment. To support new concepts and designs stakeholders working with wastewater treatment plants (operators, managers, designers and service providers etc.) should learn and update themselves on the new technologies and try to adopt these novel technologies whenever they can in different scales they manage. These changes will contribute globally to the long run. For example, achieving Sustainable Development Goal 6 (Ensure availability and sustainable management of water and sanitation) can help to contribute to many communities in developing countries and other areas without access to clean water. Also, the sustainable concepts can be practically implemented by adopting the novel technologies in wastewater treatment systems which will ultimately save the limited resources for future generations.

References

1. Abdel-Raouf N, Al-Homaidan AA, Ibraheem IBM (2012) Microalgae and wastewater treatment. Saudi J Biol Sci 19(3):257–275. https://doi.org/10.1016/j.sjbs.2012.04.005
2. Biswas S, Maurya R, Mohanty K (2022) Principles and technology advances in wastewater treatment through microalgae for bioenergy production. In: Biofuels and bioenergy: opportunities and challenges. Elsevier Inc., Amsterdam. https://doi.org/10.1016/B978-0-323-85269-2.00026-5
3. Campbell WAB (1952) Methaemoglobinaemia due to nitrates in well-water. BMJ 2(4780):371–373. https://doi.org/10.1136/bmj.2.4780.371
4. Flemming HC, Wingender J (2010) The biofilm matrix. Nat Rev Microbiol 8(9):623–633. https://doi.org/10.1038/nrmicro2415
5. Freguia S, Logrieco ME, Monetti J, Ledezma P, Virdis B, Tsujimura S (2019) Self-powered bioelectrochemical nutrient recovery for fertilizer generation from human urine. Sustainability (Switzerland) 11(19):1–10. https://doi.org/10.3390/su11195490
6. Guzmán-Fierro V, Arriagada C, Gallardo JJ, Campos V, Roeckel M (2023) Challenges of aerobic granular sludge utilization: fast start-up strategies and cationic pollutant removal. Heliyon 9(2):1–16. https://doi.org/10.1016/j.heliyon.2023.e13503
7. Meyers R (2012) Encyclopedia of sustainability science and technology. https://doi.org/10.1007/978-1-4419-0851-3
8. Li L, Gao K, Yang M, Zheng Q, Zhang M, Deng X (2023) Challenges and potential solutions of microalgae-based systems for wastewater treatment and resource recovery. Front Bioeng Biotechnol 11(June):1–4. https://doi.org/10.3389/fbioe.2023.1210228
9. Mukherjee M, Cao B (2020) Engineering controllable bio films for biotechnological applications. https://doi.org/10.1111/1751-7915.13715
10. Oruganti RK, Katam K, Show PL, Gadhamshetty V, Upadhyayula VKK, Bhattacharyya D (2022) A comprehensive review on the use of algal-bacterial systems for wastewater treatment with emphasis on nutrient and micropollutant removal. Bioengineered 13(4):10412–10453. https://doi.org/10.1080/21655979.2022.2056823
11. Plöhn M, Spain O, Sirin S, Silva M, Escudero-Oñate C, Ferrando-Climent L, Allahverdiyeva Y, Funk C (2021) Wastewater treatment by microalgae. Physiol Plant 173(2):568–578. https://doi.org/10.1111/ppl.13427

12. Saini S, Tewari S, Sharma V (2023) Materials advances biofilm-mediated wastewater treatment: a comprehensive review. Mater Adv 4(6):1415–1443. https://doi.org/10.1039/d2ma00945e
13. Singh A, Srivastava A, Saidulu D, Gupta AK (2022) Advancements of sequencing batch reactor for industrial wastewater treatment: major focus on modifications, critical operational parameters, and future perspectives. J Environ Manage 317(May):115305. https://doi.org/10.1016/j.jenvman.2022.115305
14. Soo A, Kim J, Shon HK (2024) Technologies for the wastewater circular economy—a review. Desalin Water Treat 317(Mar). https://doi.org/10.1016/j.dwt.2024.100205
15. Steimann T, Wollborn D, Röck F, Horstmann R, Schmitt E, Christ JJ, Blank LM, Büchs J (2024) Investigation into struvite precipitation: a commonly encountered problem during fermentations on chemically defined media. Biotechnol Bioeng 121(3):1076–1089. https://doi.org/10.1002/bit.28633
16. Vidanage PW, Ko S-OO, Oh S (2020) Metagenomics uncovers a core SAR11 population in Brackish surface waters of the Baltic Sea. Water, 12(501). https://doi.org/10.3390/w12020501
17. Vymazal J (2019) Constructed wetlands for wastewater treatment. In: Encyclopedia of ecology, 2nd edn, vol 1–4, pp 14–21. https://doi.org/10.1016/B978-0-12-409548-9.11238-2
18. Wang L, Gu W, Liu Y, Liang P, Zhang X, Huang X (2022) Challenges, solutions and prospects of mainstream anammox-based process for municipal wastewater treatment. Sci Total Environ 820:153351. https://doi.org/10.1016/j.scitotenv.2022.153351
19. Whitton R, Ometto F, Pidou M, Jarvis P, Villa R, Jefferson B (2015) Microalgae for municipal wastewater nutrient remediation: mechanisms, reactors and outlook for tertiary treatment. Environ Technol Rev 4(1):133–148. https://doi.org/10.1080/21622515.2015.1105308
20. Yousif YID, Mohamed ES, El-Gendy AS (2022) Using chlorella vulgaris for nutrient removal from hydroponic wastewater: experimental investigation and economic assessment. Water Sci Technol 85(11):3240–3258. https://doi.org/10.2166/wst.2022.157
21. Zhang H, Gong W, Zeng W, Chen R, Lin D, Li G, Liang H (2021) Science of the total environment bacterial-algae bio fi lm enhance MABR adapting a wider COD/N ratios wastewater: performance and mechanism. Sci Total Environ 781:146663. https://doi.org/10.1016/j.scitotenv.2021.146663
22. Zhang S, Zhang L, Xu G, Li F (2022) A review on biodiesel production from microalgae: influencing parameters and recent advanced technologies, 1–20 July. https://doi.org/10.3389/fmicb.2022.970028
23. Zheng Y, Wan Y, Zhang Y, Huang J, Yang Y, Tsang DCW, Wang H, Chen H, Gao B (2023) Recovery of phosphorus from wastewater: a review based on current phosphorous removal technologies. Crit Rev Environ Sci Technol 53(11):1148–1172. https://doi.org/10.1080/10643389.2022.2128194

Enhanced Removal of Nitrogen and Organic Matter from Wastewater in Constructed Wetland Technology by Artificial Aeration

A. K. Ragen, D. Surroop, and P. D. Jeetah

Abstract The main aim of this study was to investigate the enhanced performance of a horizontal subsurface flow constructed wetland (HSSFCW) in terms of nitrogen and organics removal through a proper spatial distribution of aeration to ensure simultaneous aerobic and anaerobic conditions. Two HSSFCWs, one aerated in its front section only and the other one non-aerated (as control), were operated in parallel at a hydraulic retention time (HRT) of 24 h, a hydraulic rate (HLR) of 0.16 m day and a water depth of 0.4 m. Three distinct types of synthetic effluents were prepared and fed in the systems to simulate a wide range of wastewaters. The rate of aeration provided in the aerated (or intensified) bed was also varied to assess its impact on the intensified removal of ammonium-N (NH_4-N) and chemical oxygen demand (COD).The enhanced performance of the aerated bed, when compared to the non-aerated one, in terms of COD, NH_4-N and total nitrogen (TN) removal efficiency were found to be $26.0 \pm 14.7\%$ (N = 68), $61.6 \pm 11.7\%$ (N = 60) and $18.2 \pm 5.5\%$ (N = 60), respectively. Aeration did impact significantly ($p < 0.05$) and positively on the aerated bed performance in terms of COD and NH_4-N removal efficiency. The rate constant of COD and NH_4-N degradation in the aerated bed were 3 times and 10 times higher than in control bed, indicating that aeration indeed enhanced the performance of HSSFCW. This study shows that without aeration, such a natural treatment system may not be capable of effectively treating medium to high strength wastewater. This aeration strategy had overcome its oxygen transfer limitation and will constitute a sustainable solution to treat different kind of wastewater with high organic and nitrogen concentration.

A. K. Ragen (✉) · D. Surroop · P. D. Jeetah
Department of Chemical and Environmental Engineering, Faculty of Engineering, University of Mauritius, Moka, Mauritius
e-mail: ak.ragen@uom.ac.mu

D. Surroop
e-mail: d.surroop@uom.ac.mu

P. D. Jeetah
e-mail: p.jeetah@uom.ac.mu

S. Yadav et al. (eds.), *Emerging Technologies in Biological and Hybrid Wastewater Treatment: Lessons from Developed to Enhancing Practices in Developing Countries*, Earth and Environmental Sciences Library,
https://doi.org/10.1007/978-3-031-74515-7_8

Keywords Horizontal subsurface constructed wetland · Enhanced bed performance · Artificial aeration · Organic and nitrogen

1 Introduction

Despite much progress to improve the horizontal sub surface flow constructed wetland (HSSFCW) performance, it's poor nitrogen removal and low oxygen transfer are two major limitations that need to be further researched [10, 23, 30]. There is thus a need to address oxygen transfer in the bed to intensify its performance in terms of not only nitrogen removal, but also organic matter elimination. The rate of oxygen transfer in conventional HSSFCW, via diffusion from plant roots is 0.02–12 g/m^2 day [18], and through atmospheric diffusion is 5.77–18.45 g/m^2 day [13]. The latter also mentioned that the combined rate of oxygen supply through plant roots and atmospheric diffusion in a HSSFCW is lower than what is generally consumed therein. According to Ilyas and Masih [10], the dissolved oxygen (DO) required to trigger removal of NH_4-N through nitrification process is above 1.50 mg/L. On the other hand, denitrification process, i.e. removal of nitrate-Nitrogen (NO_3-N), as reported by Hao et al. [9], occurs at DO levels lower than 0.50. Ilyas and Masih [10] reported a range of DO of 0.17–1.90 mg/L in HSSFCW, indicating insufficient oxygen to effectively remove ammonium-nitrogen (NH_4-N) via the nitrification process, but enough to ensure removal of NO_3-N. Pascual et al. [18] concluded that HSSFCWs poorly eliminate NH_4-N but remove well NO_3-N. This is because the condition prevailing in the HSSFCW system is mostly anoxic or anaerobic, which inhibits nitrification of NH_4-N, but favours denitrification of NO_3-N.

Various research works showed that NH_4-N elimination efficiencies vary from 9 to 79% in un-aerated HSSFCW [24, 29]. Among the strategies proposed for the intensification of the performance of HSSFCW to overcome its oxygen transfer limitation is artificial aeration (AA) [10], which is being addressed in this study. AA is now being proposed as a solution to generate the ideal aerobic conditions (DO > 1.50 mg/L) for nitrification, and hence improving the performance of the bed to significantly eliminate NH_4-N. AA can be provided in HSSFCW via air pumps [5] or by air blower [7]. NH_4-N removal relies on the DO availability. The effect of AA in HSSFCW has been reviewed by Pascual et al. [18], who reported average NH_4-N removal efficiency as high as 72.3–89.1%. Zapater-Pereyra et al. [27] and Zhang et al. [25] demonstrated a rise in the elimination of NH_4-N by 69% and 90%, respectively. Butterworth et al. [2] studied the effect of AA in a full-scale HSSFCW treating domestic sewage and demonstrated an improved NH_4-N removal from 11% (non-aerated) to 99% (aerated). The consequences of low availability of DO in a conventional bed results in not only poor nitrification, but also a low total nitrogen (TN) elimination. Aerated HSSFCW removed TN in a range of 17–70%, [18]. However, excessive aeration can adversely impact TN (NH_4-N + NO_3-N) removal and lead to accumulation of NO_3-N within the bed through incomplete denitrification [11]. There is thus a need to establish proper spatial aerobic, anoxic and anaerobic zones within the HSSFCW to not only ensure simultaneous nitrification and denitrification [14, 28], but also enhance organic removal. This is achieved via proper distribution of aeration. If aeration is provided along the full length of the bed only NH_4-N will be eliminated and there will be no removal of NO_3-N, since the condition prevailing in their bed will be fully aerobic, inhibiting the process denitrification. Li et al. [11] studied the spatial distribution of oxygen within HSSFCW. They demonstrated that providing AA along the front of their CWs only and creating an anoxic condition in the remaining bed resulted in effective simultaneous elimination of NH_4-N and NO_3-N. They demonstrated that AA in HSSFCW outperformed un-aerated one. The COD, NH_4-N and TN removal efficiencies were 70.3%, 18.4% and 24.6%, respectively in non-aerated bed, and 90.1%, 99.7% and 51.5%, respectively in aerated system. However, they applied unplanted HSSFCW and hence omitted to investigate the effect of plants in removal of organics and nitrogen.

The research element of this study is to demonstrate that provision of artificial aeration (AA) properly distributed at the front section of the HSSFCW will intensify the performance of the HSSFCW in removing not only NH_4-N, but also NO_3-N. Hence, this study is aimed at providing a sustainable solution to eliminate NH_4-N and subsequently NO_3-N within a single bed. Additionally, the efficiency of organic matter removal (chemical oxygen demand—COD) will intensify (or enhance) through aeration.

2 Aim and Objectives

From literature review, it was found that the main limitations in the performance of HSSFCW are the low oxygen availability and poor nitrogen removal (mainly NH_4-N). The aim of this study is to assess its enhanced performance in terms of nitrogen and organics removal through a proper spatial distribution of aeration to ensure simultaneous aerobic and anaerobic conditions. This aeration strategy will overcome its oxygen transfer limitation and will constitute a sustainable solution to treat a wide range of wastewater loaded with both organics and nitrogen. This performance intensification will consequently lead to its increased implementation and its popularity as a sustainable wastewater treatment option, which is low cost during construction and operation.

The specific objectives of this study are:

- To evaluate the beds' performance under varying operational variables such as aeration rate (AR), nitrogen loading rate (NLR) and organic loading rate (OLR). The organic elimination efficiency will be assessed by COD removal. Nitrogen removal efficiency will be assessed in terms of removal of NH_4-N, NO_3-N and TN;
- To assess the progression of nitrogen and COD along the beds, and hence evaluating their kinetics of removal and;

- To evaluate the additional organic and nitrogen removal efficiencies as a result of AA.

3 Methodology

3.1 Experimental Set-Up

Figure 1 illustrates the experimental setup. The experimental set up consisted of two pilot HSSFCWs, operated in parallel, of the same dimensions: 3.0 m long × 0.5 m wide and 0.5 m deep. The length of the inlet and outlet structures were 0.5 m. These structures were filled with 50–100 mm size stones. The planted sections were filled with 6 mm size washed gravel, in which cattails were planted. A feed synthetic wastewater tank of 1000 L was connected via ½-in. PVC pipes to both beds. Each feed inlet pipe was fitted with a ball valve that was used to regulate the flow of wastewater in the beds. Seven perforated pipes, spaced at 0.28 m, 0.53 m, 0.78 m, 1.06 m, 1.37 m, 1.53 m and 1.85 m from the inlet, were provided in each bed for sampling purposes. These distances corresponded to a fractional distance of 0.14, 0.27, 0.39, 0.53, 0.69, 0.77, 0.93, respectively. The fractional distance of the inlet and outlet corresponded to 0 and 1, respectively. The outlet of each bed comprised of an outlet pipe, which regulated the water level below the gravel (or the water depth). One HSSFCW, denoted in this report as the "Aerated bed" was fitted with two air pumps of a total capacity of 9.5 L/min of air. Aeration was provided in the first half portion of the bed, while the remaining portion was not aerated. The AA was provided through two perforated pipes placed at the bottom of the bed at a distance of 0.32 m and 0.70 m from the inlet (or fractional distance of 0.16 and 0.35), respectively. The other bed acted as a control, in which no aeration was provided. The setting up of the systems took around one month and the whole investigation was carried out from September 2020 to December 2021.

3.2 Synthetic Wastewater Preparation

Synthetic influent was prepared by dissolving an appropriate amount of NH_4Cl, KNO_3 and sugar in water to simulate NH_4^+, NO_3-and COD pollution, respectively. 1000 L of synthetic wastewater was prepared at regular frequency to ensure a continuous feed in the systems, including week-ends. Three distinct types of synthetic wastewater (in terms of NH_4-N, NO_3-Nand COD loadings) were prepared and fed in the systems during the whole period of investigation

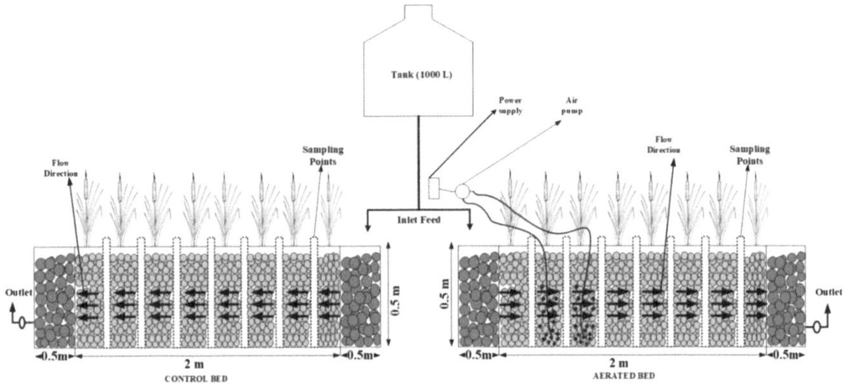

Fig. 1 Schematic representation of the experimental setup

3.3 Operation of the Beds

During the whole period of investigation, unaerated and aerated bed were fed with synthetic wastewater with a wide range of NH_4-N and NO_3-N to imitate a wide range of wastewater such as domestic wastewater, landfill leachate, farm and food-processing plants effluent.

The beds were run at a flow rate of 0.16 m^3/day, corresponding to a hydraulic retention time (HRT) of 1 day and a hydraulic loading rate (HLR) of 0.16 m^3/m^2 day. The depth of water in both beds was maintained at 0.4 m. These operational parameters were maintained throughout this study, during which the concentration of COD, NH_4-N, and NO_3-N of the synthetic influent were varied. There were three distinct types of influent that were investigated in this study. These three synthetic influents were considered as follows:

- The first one was considered as a low organic source of wastewater having a range of COD of 131–359 mg/L and an average of 242 ± 65 mg/L ($n = 19$). Its NH_4-N and NO_3-N level ranged from 9.0 to 32.8 mg/L and 10.0 to 28.8 mg/L with a mean of 20.3 ± 5.8 mg/L ($n = 19$) and 21.8 ± 9.6 mg/L ($n = 19$), respectively. Such a wastewater is considered in this study as domestic wastewater.
- The second synthetic wastewater investigated was considered as an industrial effluent of medium organic strength with COD ranging from 797 to 1323 mg/L and averaging 1108 ± 146 mg/L ($n = 30$). Its NH_4-N concentrated ranged from 37.5 to 98.0 and averaged at 61.8 ± 22.1 ($n = 30$), and its NO_3-N level was 40.6–144.0 mg/L with an average of 67.8 ± 24.1 mg/L ($n = 30$).
- The third synthetic influent can be considered as a high strength industrial effluent (such as farm or food processing effluents or even landfill leachate), which was loaded with COD 2000–3423 mg/L, NH_4-N of 60.0–79.3 mg/L and NO_3-N of 63.1–79.2 mg/L. The average of COD, NH_4-N and NO_3-N were 2648 ± 481 mg/L, 69.3 ± 7.0 mg/L and 73.4 ± 4.6 mg/L, respectively ($n = 19$).

Table 1 Range of operating parameters

Strength of wastewater	OLR (gCOD/ m^2 day)	NH_4-LR (g/ m^2 day)	NO_3-LR (g/ m^2 day)	Air flow (m^3/h)	Run number
Low	21.0–57.4	1.44–5.25	1.60–4.61	0.15	1–9
				0.57	10–19
Medium	127.5–211.7	6.00–15.68	6.50–23.04	0.57	20–41
				0.48	42–49
High	320.0–547.7	9.6–12.7	10.1–12.7	0.48	50–68

Note OLR = organic loading rate = (flow × COD/surface area); NH_4-LR: ammonium loading rate = (flow × NH_4-N/surface area); NO_3-LR = nitrate loading rate = (flow × NO_3-N/surface area); HL = (flow/surface area)

In this study, these three distinct types of synthetic influent are thereafter denoted as low, medium and high strength wastewater. The beds were run under the operating parameters detailed in Table 1.

3.4 Sampling and Analysis

Collection of samples was done at a frequency of twice a week. The inlet and outlet as well as the intermediate seven samples were taken and analysed for NH_4-N, COD, NO_3-N, temperature and DO. The COD was analysed as per the standard APHA (2005) method, while the NO_3-Nand NH_4-N were analysed by the USEPA approved Hach Spectrophotometry. The DO and pH were monitored using a portable DO and pH meter, respectively.

The performance of the bed was evaluated from the analysis of the inlet and outlet of the beds. The progression of NH_4-N, NO_3-Nand COD were monitored from the seven intermediate samples. The results were used to estimate the kinetics of NH_4-N, NO_3-Nand COD removal along the beds. The two kinetic parameters are the (1) order of removal and (2) rate constants of removal (or the k-value) for NH_4-N, NO_3-N, TN and COD and to understand the removal mechanisms. The k-values for the aerated and control were also investigated. The enhanced nitrogen and organic removal efficiencies was assessed by comparing those of the aerated and non aerated bed. In this study, a total of 68 runs out of which 19 for the low strength synthetic wastewater, 30 for the medium strength one and 19 for the high strength wastewater.

3.5 Kinetics Determination

The mean observed values of inlet and outlet COD, NH_4-N, NO_3-N as well as the seven intermediates sampling points (Fig. 1) were used to estimate their kinetics. A plot of these average values against time (corresponding to fractional distance) was generated in the Excel™ software. In the plot, an exponential trend line was inserted. The relationship $C_t = C_0e^{-kt}$ (Eq. 1) as well as the R^2-values were obtained by the Excel™ software. R^2-valuesclose to 1 show that the mean observed values fit well into a 1st order plug flow model or the k-C model.

$$C_t = C_0e^{-kt} \tag{1}$$

where C_t = Concentration at time t, mg/l; C_0 = Inlet concentration, mg/l; k is the rate constant, h^{-1}; t = time, h.

3.6 Statistical Analysis

The following statistical tests at 95% interval confidence (or 5% significance level) were performed in the IBM SPSS Software Version 23.0.

- Paired-Samples T-test to assess difference of the means of the inlet and outlet values. This was performed for both beds separately. The null hypothesis tested was that the means of the inlet will be same as the outlet characteristics will be same, meaning the bed would not treat the wastewater.
- Independent samples T-test to assess whether the mean of the outlet characteristics from the aerated bed were statistically significantly different from those of the non-aerated bed. The null hypothesis tested was that the means of the two outlets were same, indicating that these were treated by the same bed.
- Pearson correlation tests to determine the significant impact the operating parameters would have on the experimental HSSFCWs performance of in terms of NH_4-N. NO_3-Nand COD removal efficiency. The operating parameters (aeration rate, OLR and NLR) were tested. All other operating variables such as HRT, HL and water depth were kept constant.

4 Results and Discussions

4.1 Inlet and Outlet Characteristics

Table 2 details the characteristics of the inlet and outlet of both experimental beds.

As shown in Table 2, the standard deviation values of all parameters of the influents were on the high side. This is was due to the fact that the values comprised of the three types of synthetic influents prepared. The standard deviation values of the outlet of both the un-aerated and aerated beds were much lower than the inlet, indicating more stable performance of the beds.

Table 2 Inlet and outlet characteristics during whole study

Parameters	Influent		Aerated bed (N = 68)		Control (unaerated) bed (N = 68)	
			Effluent		Effluent	
	Range	Mean ± SD	Range	Mean ± SD	Range	Mean ± SD
COD, mg/L	131–3423	1296 ± 960	32–543	190 ± 131	42–1754	648 ± 533
NH_4-N, mg/L	9.0–98.0	50.6 ± 26.3	0.5–10.3	4.0 ± 2.8	6.0–79.0	37.4 ± 20.4
NO_3-N, mg/L	10.0–144.0	54.7 ± 50.6	2.1–52.0	23.1 ± 14.1	1.3–16.5	8.2 ± 4.3
TN (NH_4 + NO_3), mg/L	24.5–240.4	103.6 ± 54.3	4.5–58.9	27.1 ± 15.6	8.9–91.0	45.7 ± 24.7

A Paired-Sample T-test was performed using the IBM SPSS 19.0 to assess the difference of the means of the inlet and outlet parameters at $p = 0.05$. The result showed that $p < 0.05$ at 95% confidence interval for the inlet and outlet COD, NH_4-N, NO_3-N and TN of both aerated and un-aerated beds. The average of inlet characteristics were statistically significantly different from those of the outlets from both beds. Since the null hypothesis was rejected, it was inferred that both the aerated and non-aerated beds did treat the wastewater.

An Independent Samples T test using the IBM SPSS 19.0 was performed at $p = 0.05$ to assess the statistical difference between the averages of outlet characteristics from the aerated and non-aerated beds. It was found that $p < 0.05$ for outlets NH_4-N, NO_3-N TN and COD. The null hypothesis was rejected, indicating that the synthetic influent was indeed treated by two different systems.

4.2 Performance of the CW Aerated and Non-aerated

4.2.1 COD Removal Efficiency

Figure 2 illustrates the COD elimination efficiency achieved in both aerated and control bed. Irrespective of operating conditions, the COD elimination efficiency of the aerated and un-aerated bed ranged from 59.9 to 92.6% and 38.6 to 76.7%, respectively. The mean COD elimination efficiency of the aerated bed and un-aerated bed was 82.5 ± 7.2% and 54.9 ± 9.7% (N = 68), respectively. The enhancement of COD removal due to aeration was found to range from 2.4 to 52.3% and average 27.6 ± 14.4% (n = 68). Figure 2 also indicates that the aerated bed systematically, irrespective of operating conditions, outperformed the non-aerated bed. This evidences that AA can intensify the performance of HSSFCW [11, 17]. Li et al. [11], Zapater-Pereyra et al. [27] and Wu et al. [24] reported lower performance of aerated bed than

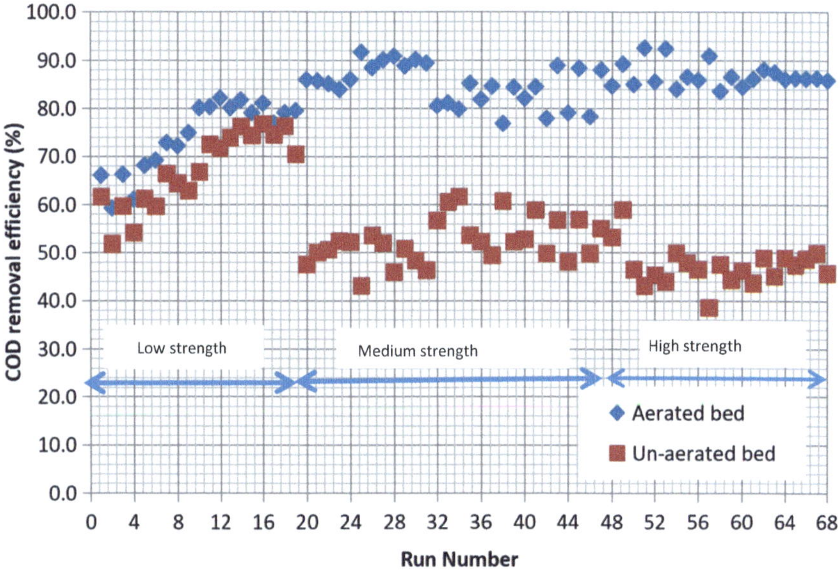

Fig. 2 COD removal efficiency of the aerated and un-aerated bed

in this study. As an indication, Zapater-Pereyra et al. [27] demonstrated an enhancement of COD removal by up to 13% in aerated HFSSCW, while in this study the mean additional COD removal efficiency was 27.6%. When compared to the performance of aerated HSSFCWs studied by Ouellet-Plamondon et al. [17], the COD elimination achieved in this study is lower. Comparable performance was obtained by Zhong et al. [29], who reported 82% elimination of COD by a horizontal bed subjected to continuous AA.

The removal efficiency of the aerated bed was intensified by a mean of 7.1 ± 2.9% (n = 19), 32.4 ± 7.3% (n = 30) and 40.6 ± 4.8% (n = 19) when it was subjected to low, medium and high strength wastewater. Since other operating conditions (HL, water depth and HRT) were kept constant throughout this study, this enhanced performance of the aerated was deemed to have been mainly contributed by aeration. Zhuang et al. [30] reviewed 16 papers on effects of aeration on pollutants removal in constructed wetland and demonstrated that AA did enhance the COD elimination efficiency. The main reason is that aeration speeds up the utilization of organic matter (COD) by heterotrophic microorganisms [30]. This study demonstrated that aeration can contribute in the application of a construction wetland technology to treat medium to high strength industrial wastewater.

The progression of COD removal along the aerated and control bed subjected to the low, medium and high strength synthetic wastewater was found to be a general exponential decline of the COD from the input to the output, however to a non-zero or background BOD. Similar trends are reported in the literature [1, 8, 19].

It was observed that the degradation of COD along the control bed is less pronounced than the aerated, irrespective of the strength of wastewater. Also the performance of the experimental systems operated at higher strength exhibited a much lower performance.

4.2.2 NH_4-N Removal Efficiency

Irrespective of operating conditions applied to the aerated and control bed, the NH_4-N removal efficiency by both beds is illustrated in Fig. 3. It clearly illustrates that the aerated bed significantly outperformed the control one in terms of NH_4-N removal. The respective range of elimination efficiency of NH_4-N by the oxygenated and control bed was 70.2–100% and 9.0–48.3%. Mean efficiency of NH_4-N reduction was 90.4 ± 7.8% (n = 60) with aeration and a paltry 28.8 ± 14.4% (n = 60) without aeration. All similar studies demonstrated that aeration led to higher performance of horizontal constructed wetland than no aeration [2, 6, 11, 16, 17, 24, 25, 27, 29]. Chyan et al. [4] reported that nitrification, sedimentation burial and plant uptake contribute to 54–96%, 0–46% and 7.5–14.5% of NH_4-N removal, respectively. It is widely accepted that the nitrification is the principal route of NH_4-N removal in a constructed wetland [3, 4]. In a conventional HSSFCW, being limited in oxygen supply, poorly NH4-N is poorly removed [10, 30], indicating that nitrification is low or inhibited. Hence, it may be inferred, from this study, that aeration resulted in a higher nitrification rate, indicated by significantly high NH_4-N removal efficiency.

Ilyas and Masih [10] demonstrated that a DO > 1.50 mg/L in the constructed wetland triggers the nitrification process and hence elimination of NH_4-N is ensured. In this study, the aeration provided were as per Table 2. The provision of different air flows in the front portion of the aerated bed (as described in Sect. 4.1 and illustrated in Fig. 1) resulted in a mean DO distribution inside and along the bed as illustrated in Fig. 4.

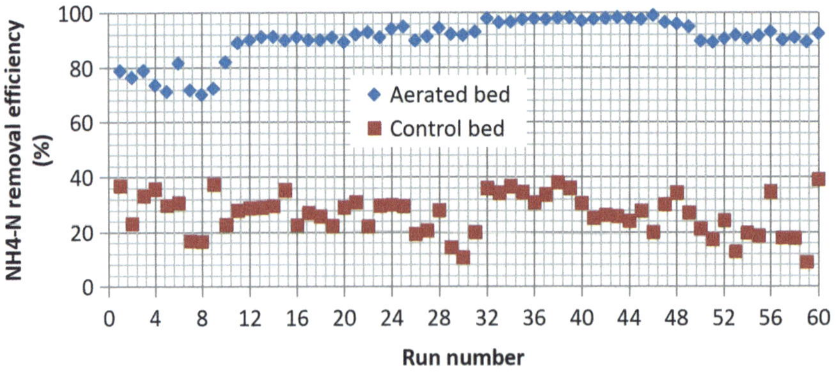

Fig. 3 NH_4-N removal efficiency by the aerated and control bed

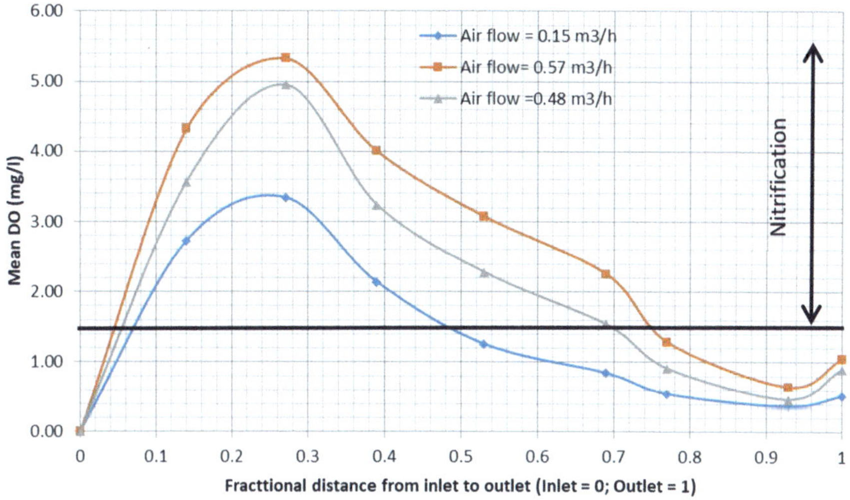

Fig. 4 Mean DO along the bed at three different aeration rates

Figure 4 shows a steady increase in the DO from the inlet to reach a peak value and thereafter, the level of DO decreased steadily. It is seen from Fig. 4, that conditions prevailing in the aerated bed favoured nitrification up to fractional distance of 0.49, 0.69 and 0.72, at aeration rate of 0.15, 0.48 and 0.57 m^3/h, respectively. This shows that as aeration rate increases, there is more lateral diffusion of air along the length of the aerated bed. There was a rapid decline in the NH_4-N in the front section of the aerated bed. Thereafter the rate of removal decreased and steadied thereafter. The majority of the NH_4-N was eliminated in the front section in which aeration was provided. This indicates that the rate of nitrification was higher in the front than rear section of the aerated bed. This is concomitant with the distribution of DO, which was higher in the front than in rear section.

4.2.3 NO_3-N Removal

Figure 5 shows the inlet NO_3-N concentration as well as that of the outlet of bed with and without aeration during the whole period of investigation.

The range and mean of the inlet NO_3-N were 10.0–144.0 mg/L and 54.7 ± 28.4 mg/L (n = 60). The respective outletNO_3-N of the bed with and without aeration ranged from 2.1 to 52.0 mg/L and 1.3 to 16.5 mg/L, respectively. The respective means of the outlet NO_3-N of the bed with and without aeration were 23.1 ± 14.1 mg/L (n = 60) and 8.2 ± 4.3 mg/L (n = 60). Aeration in a horizontal constructed wetland led to higher outlet NO_3-N concentration than non-aerated conventional system. This is because in the non-aerated bed, there was limited DO and hence denitrification (i.e. removal of NO_3-N) was favoured therein.

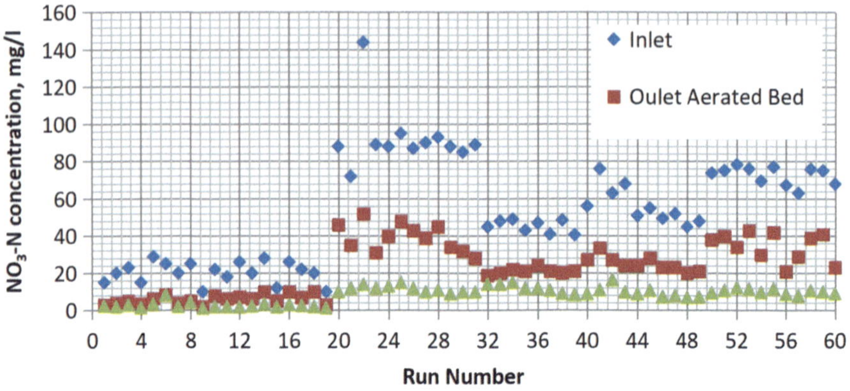

Fig. 5 Inlet as well as outlet NO_3-N from aerated and non-aerated bed

Denitrification process is the principle removal mechanism of NO_3-N removal and is favoured in anoxic conditions of at DO < 0.5 mg/L. As illustrated in Fig. 4, the DO level in the aerated bed ranged from 0.5 to 4.53 mg/L. Such distribution of DO resulted in an increase in the NO_3-N in the front section due to nitrification process (NH_4-N is converted to NO_3-N). Such increase is illustrated in Fig. 6 and thereafter with decreased DO, denitrification (removal of NO_3-N) occurred. A quadratic trend line was inserted to assesses relationship between outlet NO_3-N with time. The relationship was statistically significant ($p < 0.05$) with R^2-value high at 0.8326. Removal of NO_3-N can be expressed as Eq. 2.

$$C_t = 0.2558t^2 + 4.2058t + 62.628 \tag{2}$$

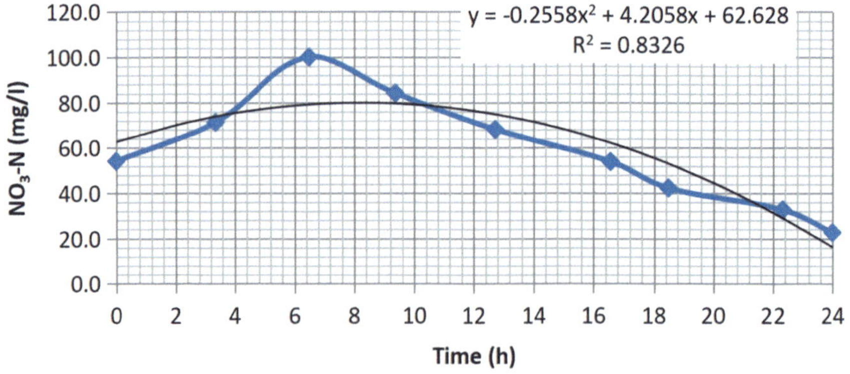

Fig. 6 Progression of the mean NO_3-N (n = 60) along the aerated bed

The removal of NO_3-N was better achieved in constructed wetland without aeration. This is because the mean DO level in the non-aerated (n = 68) was on the low side, ranging from 0.03 to 0.29 mg/l, inside the bed. Hence, denitrification was favoured leading to high removal efficiency of NO_3-N. The respective range of NO_3-N removal efficiency in bed with and without aeration was 43.5–80.0% and 66.4–92.7%. The mean of the aerated and non-aerated bed was 58.8 ± 10.1 and 84.4 ± 5.9, respectively. The standard deviation value of the aerated bed was higher than the conventional one, indicating that the performance of the latter was more stable than the former.

4.3 *Effect of Operational Variables on the Aerated and Non-aerated Bed Performance*

The operating variables investigated in this study include (1) aeration, (2) OLR, (3) NH_4-LR, (4) NO_3-LR and (5) temperature. Other operational parameters such as HRT, HL and water depth are outside the scope of this analysis as these were kept constant. Pearson's correlation tests were performed using the IBM SPSS 19.0 to analyse the effect of various operational variables on the performance of both aerated and non-aerated bed.

4.3.1 Effect of Aeration

Aeration did impact significantly ($p < 0.05$) and positively on the aerated bed performance in terms of elimination of NH_4-N and COD at significance level of 5%. This means that the higher the aeration (or air flow), the greater the removal of COD and NH_4-N. The same trend was demonstrated by Lin et al. [12], Ilyas and Masih [10] and Zhuang et al. [30] when they investigated impact of aeration of constructed wetland performance.

As far as efficiency of removal of NO_3-N is concerned, it has been found that an increase in aeration caused a decrease in the aerated bed performance. This can be attributed to the fact that the more oxygen is supplied to the bed, denitrification is inhibited leading to lower performance of the aerated bed in removing NO_3-N.

No statistical evidence ($p > 0.05$) of the impact of aeration on TN removal efficiency of the aerated bed was found. Zhuang et al. [30] confirmed that there is no consistent evidence that an aerated bed can enhance or inhibit TN removal.

The Pearson's coefficients were 0.744 and 0.640 for NH_4-N and COD removal efficiency, respectively. This means that impact of aeration is more pronounced on NH_4-N than COD removal. Zhuang et al. [30] demonstrated by aeration in a constructed wetlands enhances NH_4-N removal more than COD.

4.3.2 Effects of OLR

In this study, OLR was increased by feeding the beds with synthetic wastewater containing varying (or increasing) inlet COD (or increasing organic strength). OLR had a negative impact on the removal of COD, as indicated by $p < 0.05$ and the negative value of the Pearson's coefficients. It means that at higher COD loads, the performance of the bed deteriorated. However, it was observed that the effect of OLR was more pronounced in the control than the aerated bed, as evidenced by a higher person's coefficient in non-aerated ($r = -0.721$) than for the aerated bed ($r = -0.268$). There was no statistical evidence ($p > 0.05$) of OLR affecting the performance of both the aerated and un-aerated bed in terms of NH_4-N and NO_3-N elimination. However, this study demonstrated that OLR impacted adversely (or negatively) on the removal of TN ($p < 0.5$).

4.3.3 Effects of NH_4-LR

In the aerated bed, there was statistical evidence of the impact of NH_4-LR on the elimination efficiency of TN and NH_4-N ($p < 0.05$). The higher the load of NH_4-N in the aerated bed, the higher the elimination efficiency of NH_4-N, but the lower was the TN elimination efficiency. NH_4-LR affected negatively ($p < 0.05$) the removal of TN in the control bed. There was no evidence of statistical co-relationship ($p > 0.05$) between NH_4-LR and elimination of NO_3-N and NH_4-N in the un-aerated bed.

4.3.4 Effects of NO_3-LR

The effect of increasing NO_3-LR deteriorate the aerated bed performance in terms of NO_3-N, NH_4-N and TN elimination efficiency, since their Pearson's coefficients were all negative and $p < 0.05$. In the control bed, there was no statistical co-relation between NO_3-LR and the elimination of NO_3-N, NH_4-N and TN, since $p > 0.05$.

4.4 *Kinetics of Removal of Pollutants Along the Beds*

4.4.1 Kinetics of COD Removal

Figure 7 illustrates the mean COD values observed along both the aerated and control bed in terms of fractional HRT (or fractional distance) at the 3 strengths of synthetic wastewater, respectively. The fraction HRT at the inlet is 0 h (or fractional distance of 0) and the outlet is 24 h (or fractional distance of 1). Applying Eq. 1, the rate constant of COD removal (k_{COD}) can be deduced from Fig. 7. Table 3 shows the correlation between the mean Ct with fractional distance from the inlet (related to

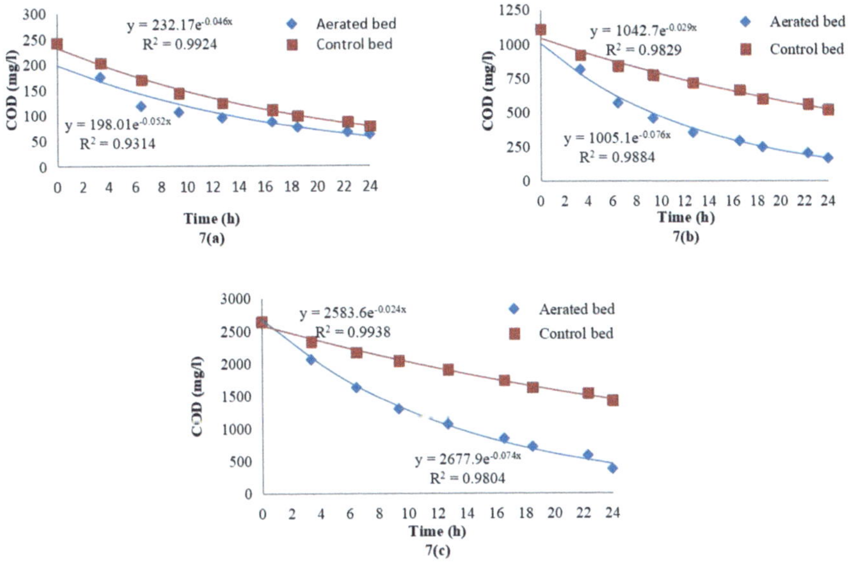

Fig. 7 Progression of mean COD values along the aerated and unaerated bed at low strength (**a**), medium strength (**b**) and high strength (**c**). Time 0 h: inlet and 24 h: outlet

Table 3 Exponential relationship between Ct and time and rate constant of COD degradation in aerated and control bed at 3 wastewater strengths

Strength of wastewater	Aerated bed		Control bed	
	Correlation	K_{COD} (/h)	Correlation	K_{COD} (/h)
Low	$C_t = 198.01e^{-0.052t}$	0.052	$C_t = 232.17e^{-0.046t}$	0.046
Medium	$C_t = 1005.1e^{-0.076t}$	0.076	$C_t = 1042.7e^{-0.029t}$	0.029
Strong	$C_t = 2677.9e^{-0.074t}$	0.074	$C_t = 2583.6e^{-0.024t}$	0.024

HRT) for the K-C model and their respective k_{COD} at different wastewater strengths in both the aerated and control bed.

The k_{COD} value, at the same wastewater strength, was systematically higher in the aerated bed than in the control one. This indicates that aeration benefited in enhancing the rate of COD removal. The rate constant of COD degradation in the aerated bed was 1.13, 2.63 and 3.07 times higher than in the control bed at low, medium and strong wastewater feed, respectively.

4.4.2 Kinetics of NH_4-N Removal

The $k_{NH_4\text{-}N}$ derived from the exponential fit for the aerated and control bed was 0.11/h and 0.01/h, indicating the rate of removal of NH_4-N in an aerated bed was 10 times

Table 4 Enhancement of COD, NH_4-N and TN removal efficiency

	Mean removal efficiency (%)			N
	Aerated bed	Control	Enhancement	
COD	82.5 ± 7.2	56.0 ± 9.8	26.0 ± 14.7	68
NH_4-N	90.5 ± 7.5	26.7 ± 7.3	61.6 ± 11.7	60
TN	74.6 ± 3.7	56.4 ± 4.6	18.2 ± 5.5	60

more than a non-aerated one. Zhuang et al. [30] concluded than aeration benefits more enhanced removal of NH_4-N than COD. This study demonstrates the same observation. Also, the $K_{NH_4\text{-}N}$ values were larger than the K_{COD} values in the aerated bed, evidencing that aeration contributes in removing NH_4-N more than COD.

4.5 Enhancement of Performance Due to Aeration

Table 4 shows the enhancement of the performance of the bed, which is the additional removal efficiency achieved when a conventional bed is aerated.

This study showed that aeration can effectively intensify the HSSFCW performance in terms of its elimination efficiency of TN, NH_4-Nand COD. As per Table 4, the mean enhanced COD, NH_4-N and TN removal efficiency % achieved in this study was 26.0 ± 14.7% (n = 68), 61.6 ± 11.7% (n = 60) and 18.2 ± 5.5% (n = 60).

The main research hypothesis of this study was to demonstrate that supplying AA in the front section of a single horizontal subsurface constructed wetland both NO_3-N and NH_4-N can be eliminated effectively. This has indeed been demonstrated in this study and as illustrated in Fig. 8, which shows the observed mean values of NO_3-N and NH_4-N (N = 60) along the aerated bed. Such performance resulted in an enhanced removal of TN.

5 Conclusions

It can be concluded from this study that a conventional HSSFCW intensified with aeration supply in the front section can effectively treat low, medium and high organic strength wastewater. This aeration strategy had overcome its oxygen transfer limitation and will constitute a sustainable solution to treat a wide range of wastewater loaded with both organics and nitrogen. Without aeration, such a natural treatment system is not capable of treating medium to high strength wastewater. This study has generated a set of rate constant of COD and NH_4-N removal, which can be applied to preliminary design of such intensified HSSFCW.

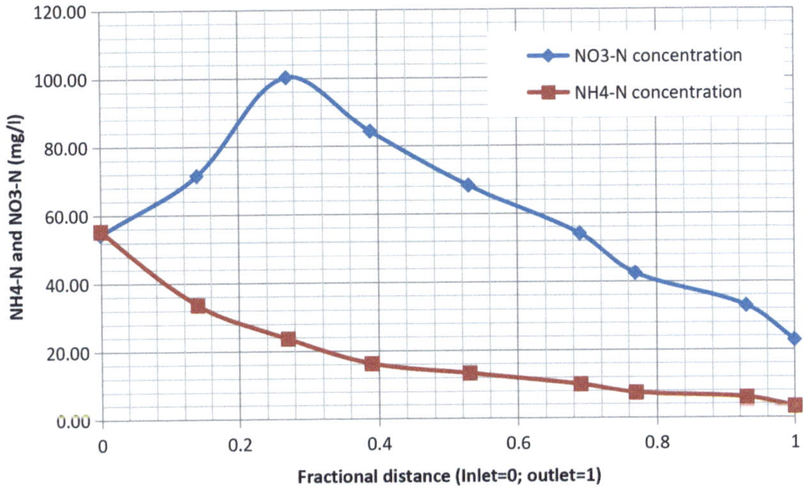

Fig. 8 Observed mean NH_4-N and NO_3-N concentration along the aerated bed (N = 60)

6 Recommendations

The followings are the main recommendations derived from this study.

- In any future intensified HSSFCWs via aeration projects in Mauritius (or other similar tropical regions) treating wastewater loaded with nitrogen and organic matter of similar concentration investigated in this study, the new rate constants obtained therein can earnestly be applied to design full scale systems.
- The aeration should be applied in the front section of the bed to enhance simultaneous elimination of NH_4-N and NO_3-N within a single-stage HSSFCW.
- This study may be applied for all upcoming similar intensified beds projects in industries locally or in countries having same climatic conditions. Moreover, information/data/results generated from this study can be used by the Government authorities (for instance: Wastewater Management Authorities & Ministry of Environment) to lay down or draft conformity limits when tender will be evaluated in this domain during the time the system will be implemented widely in Mauritius.

References

1. Akratos CS, Tsihrintzis VA (2007) Effect of temperature, HRT, vegetation and porous media on removal efficiency of pilotscale horizontal subsurface flow constructed wetlands. J Ecol Eng 29(2):173–191
2. Butterworth E, Dotro G, Jones M, Richards A, Onunkwo P, Narroway Y, Jefferson B (2013) Effect of artificial aeration on tertiary nitrification in a full scale subsurface horizontal flow constructed wetland. Ecol Eng 54:236–244
3. Chen Y, Wen Y, Zhou Q, Vymazal J (2014) Effects of plant biomass on denitrifying genes in subsurface-flow constructed wetlands. Biores Technol 157:341–345
4. Chyan JM, Jhu YX, Chen IM, Shiu RF (2016) Improvement of nitrogen removal by external aeration and intermittent circulation in a subsurface flow constructed wetland of landscape garden ponds. Process Saf Environ Prot 104:587–597
5. Dong H, Qiang Z, Li T, Jin H, Chen W (2012) Effect of artificial aeration on the performance of vertical-flow constructed wetland treating heavily polluted riverwater. J Environ Sci 24:596–601
6. Fan J, Liang S, Zhang B, Zhang J (2012) Enhanced organics and nitrogen removal in batch-operated vertical flow constructed wetlands by combination of intermittent aeration and step feeding strategy. Environ Sci Pollut Res 1–8
7. Foladori P, Ortigara ARC, Ruaben J, Andreottola G (2012) Influence of high organic loads during the summer period on the performance of hybrid constructed wetlands (VSSF þ HSSF) treating domestic wastewater in the Alps region. Water Sci Technol 65(5):890–897
8. Fonder N, Xanthoulis D (2007) Removal processes and their distribution inside a subsurface horizontal flow constructed wetland. In: Borin M, Bacelle S (eds) Proceedings of the international conference on multi functions of wetland systems. P.A.N. s.r.l., Padova, Italy, pp 60–71
9. Hao Z, Ali A, Ren Y, Su J, Zhao W (2022) A mechanistic review on aerobic denitrification for nitrogen removal in water treatment. Sci Total Environ 847:157452. https://doi.org/10.1016/j.scitotenv.2022.157452
10. Ilyas H, Masih I (2017) The performance of the intensified constructed wetland for organic matter and nitrogen removal: a review. J Environ Manage 198:372–383
11. Li F, Lu L, Zheng X, Zhang X (2014) Three stage horizontal subsurface flow constructed wetlands for organics and nitrogen removal: effect of aeration. Ecol Eng 68:90–96
12. Lin JL, Tu YT, Chiang PC, Chen SH, Kao CM (2015) Using aerated gravel-packed contact bed and constructed wetland system for polluted river water purification: a case study in Taiwan. J Hydrol 525:400–408
13. Lu J, Guo Z, Kang Y, Fan J, Zhang J (2020) Recent advances in the enhanced nitrogen removal by oxygen-increasing technology in constructed wetlands. Ecotoxicol Environ Saf 205:111330. https://doi.org/10.1016/j.ecoenv.2020.111330
14. Ma XY, Du YL, Peng WQ, Zhang SH, Liu XB, Wang SY, Yuan SJ, Kolditz O (2021) Modeling the impacts of plants and internal organic carbon on remediation performance in the integrated vertical flow constructed wetland. Water Res 204:117635. https://doi.org/10.1016/j.watres.2021.117635
15. Masi F, Conte G, Martinuzzi N, Pucci B (2002) Winery high organic content wastewaters treated by constructed wetlands in Mediterranean climate. In: Proceedings of the 8th international conference on wetland systems for water pollution control, University of Dar-es-Salaam, Tanzania and IWA, pp 274–282
16. Nivala J, Hoos MB, Cross C, Wallace S, Parkin G (2007) Treatment of landfill leachate using an aerated, horizontal subsurface-flow constructed wetland. Sci Total Environ 380(1–3):19–27
17. Ouellet-Plamondon C, Chazarenc F, Comeau Y, Brisson J (2006) Artificial aeration to increase pollutant removal efficiency of constructed wetlands in cold climate. Ecol Eng 27(3):258–264
18. Pascual A, Alvarez JA, de la Varga D, Arias CA, Van Oirschot D, Kilian R, Soto M (2024) Horizontal flow aerated constructed wetlands for municipal wastewater treatment: the influence of bed depth. Sci Total Environ 908. https://doi.org/10.1016/j.scitotenv.2023.168257

19. Rousseau DP, Vanrolleghem PA, De Pauw N (2004) Model-based design of horizontal subsurface flow constructed treatment wetlands: a review. Water Res 38(6):1484–1493
20. Stefanakis AI, Tsihrintzis V (2009) Effect of outlet water level raising and effluent recirculation on removal efficiency of pilot-scale, horizontal subsurface flow constructed wetlands. Desalination 248:961–976
21. Uggetti E, Garcia J, Lind SE, Martikainen PJ, Ferrer I (2012) Quantification of greenhouse gas emissions from sludge treatment wetlands. Water Res 46(6):1755–1762
22. Von Sperling V, de Paoli AC (2013) First order COD decay coefficient associated with different hydraulic models applied to planted and unplanted horizontal sub surface constructed wetlands. Ecol Eng 57:205–209
23. Wang W, Song X, Li F, Ji X, Hou M (2020) Intensified nitrogen removal in constructed wetlands by novel spray aeration system and different influent COD/N ratio. Bioresour Technol 306:123008. https://doi.org/10.1016/j.biortech.2020.123008
24. Wu H, Fa J, Zhang J, Ngo HH, Guo W, Hu Z, Li J (2016) Optimization of organics and nitrogen removal in intermittently aerated vertical flow constructed wetlands: effects of aeration time and aeration rate. Int Biodeterior Biodegrad 113:139–145
25. Zhang LY, Zhang L, Liu YD, Shen YW, Liu H, Xiong Y (2010) Effect of limited artificial aeration on constructed wetland treatment of domestic wastewater. Desalination 250:915–920
26. Zhang DQ, Jinadasa KBSN, Gersberg RM, Liu Y, Ng WJ, Tan SK (2014) Application of constructed wetlands for wastewater treatment in developing countries—a review of recent developments (2000–2013). J Environ Manage 141:11–131
27. Zapater-Pereyra M, Gashugi E, Rousseau DPL, Alam MR, Bayansa LPNL (2014) Effect of aeration on pollutants removal: biofilm activity and protozoan abundance in conventional and hybrid horizontal subsurface-flow constructed wetlands. Environ Technol 35(16):2086–2094
28. Zheng Y, Zhang N, Zhang R, Li C, Liu Z, Guo Y, Zhang L, Wang Q, Li Y, Salah M, Zheng H, Li F (2024) Improved nitrogen removal by optimizing the distribution of dissolved oxygen in a simulated constructed wetland system with embedding a water-sealed zone for polishing reclaimed water. J Water Process Eng 62:105287. https://doi.org/10.1016/j.jwpe.2024.105287
29. Zhong F, Wu J, Dai Y, Xiang D, Cheng S, Ji H (2015) Performance evaluation of wastewater treatment using horizontal subsurface flow constructed wetlands optimized by micro-aeration and substrate selection. Water Sci Technol 71(9):1317–1324
30. Zhuang L, Yang T, Zhang J, Li X (2019) The configuration, purification effect and mechanism of intensified constructed wetland for wastewater treatment from the aspect of nitrogen removal: a review. Bioresour Technol 293:122086

Harnessing Algae: Advancements in Wastewater Treatment and Resource Recovery

Bavatharny Thevarajah, Sajani Piyatilleke, P. H. V. Nimarshana, Sanja Gunawardena, and Thilini U. Ariyadasa

Abstract Wastewater treatment is an energy- and resource-intensive process due to the necessity to meet the effluent discharge standards. Besides, wastewater contains substantial quantities of valuable resources, mainly carbon, nitrogen, and phosphorous, which could be recovered and recycled for secondary applications. Furthermore, resource recovery from wastewater leads to positive economic and environmental impacts while enhancing the sustainable maintenance of a circular bioeconomy. Among various techniques, algae-based wastewater treatment has gained attention in recent years due to sustainable, low-cost, and highly efficient treatment. In general, the proliferation of algae provides the oxygen needed for the degradation of organic compounds by anaerobic microorganisms, which simultaneously provide the carbon dioxide required for algal photosynthesis. Meanwhile, solar irradiation with elevated oxygen and pH, resulting from algal photosynthesis, enhances the disinfection of pathogens in wastewater. Thus, algae cultivation is being immensely exploited in primary, secondary, and tertiary stages of wastewater treatment plants, which has demonstrated significantly high removal efficiencies of pollutants, nutrients, and pathogens, which consequentially lead to advanced wastewater treatment. In addition, algae biomass generated from wastewater-based cultivation contains high-value metabolites, namely lipids, proteins, carbohydrates, and carotenoids, among others, which are beneficial in biorefining value-added products, such as food/feed, biofuels, biofertilizers, biopolymers, etc. Nevertheless, there are challenges associated with coupling wastewater treatment and algal cultivation including, but not limited to, maximizing biomass yields and enhancing nutrient removal rates, to develop commercially viable full-scale treatment systems. In this context, the current chapter holistically scrutinizes the prospects of integrating algae

B. Thevarajah · S. Piyatilleke · S. Gunawardena · T. U. Ariyadasa (✉)
Department of Chemical and Process Engineering, Faculty of Engineering, University of Moratuwa, Moratuwa 10400, Sri Lanka
e-mail: thilini@uom.lk

P. H. V. Nimarshana
Department of Mechanical Engineering, Faculty of Engineering, University of Moratuwa, Moratuwa 10400, Sri Lanka

S. Yadav et al. (eds.), *Emerging Technologies in Biological and Hybrid Wastewater Treatment: Lessons from Developed to Enhancing Practices in Developing Countries*, Earth and Environmental Sciences Library,
https://doi.org/10.1007/978-3-031-74515-7_9

cultivation in wastewater treatment plants for efficient energy and resource recovery while focusing on the techno-economic feasibility of algae-based technologies.

Keywords Microalgae · Resource recovery · Bioremediation · Value-added production · Wastewater treatment

1 Introduction

Wastewater production is estimated to be 9.86 × 10^5 million liters per day (MLD) globally, with only ~1.87 × 10^5 MLD being adequately collected and treated by conventional wastewater treatment (WWT) methods prior to disposal [1]. A significant quantity of wastewater is released without appropriate treatment, leading to negative environmental and human impacts, such as pollution of freshwater bodies, air pollution due to greenhouse gas (GHG) emission, eutrophication, soil acidification, and soil toxicity by heavy metal leachates, as well as the spread of waterborne diseases [2]. The United Nations' Sustainable Development Goals (SDGs) 2030 comprises 17 goals majorly linked to sustainable and responsible consumption, with well-established waste disposal and water infrastructure in cities [3]. In particular, SDG 6 focuses on "ensure availability and sustainable management of water and sanitation for all", while SDG 11 intends to "make cities and human settlements inclusive, safe, resilient, and sustainable" [4]. Thus, it is of paramount importance to treat wastewater and maintain effluent discharge standards to reduce adverse environmental effects and preserve the ecosystems. Although conventional WWT plants (WWTP) have significantly improved the maintenance and disposal of wastewater, establishing a multi-step process consisting of primary, secondary, and tertiary treatments is often cost- and energy-intensive and requires significant land area [5]. Furthermore, based on the wastewater characteristics, specialized and intensive treatment is required for wastewater carrying toxic materials, namely heavy metals and emerging organic pollutants [6]. Moreover, conventional WWTPs produce sludge in high quantities, up to 160–210 MLD, which is highly concentrated with pollutants, such as heavy metals, pathogens, and organic compounds, and thus, proper sludge disposal is essential for environmental protection and human health [7].

In recent years, microalgae-based bioremediation has gained significant attention as a promising and sustainable alternative to conventional WWT. This approach is considered as an attractive prospect as it offers the dual benefits of effectively treating wastewater while simultaneously generating valuable biomass [8]. For example, a wastewater treatment analysis by Velásquez-Orta et al. [9] indicated that a pilot-scale high-rate algal pond (HRAP) could effectively remove 90% of ammonium (NH_4^+) [9]. Notably, microalgal biomass could be processed to produce value-added products, such as biofuels, animal feed, and biofertilizers [10, 11]. Moreover, as reported by Nobre et al. [12], the integration of microalgae (*Chlorella vulgaris*) cultivation as tertiary treatment at a WWTP in Portugal showcased yearly gross earnings of ~€

619,000, along with an energy production of 10,874 kWh and 7,377 m^3 of treated water per day [12].

On the other hand, treated and polished effluents are useful in irrigation, landscaping, groundwater recharge, restoration of water bodies, and industrial applications, such as cooling and washing, which could significantly reduce freshwater footprints, i.e., 1000–10,000 L kg^{-1} of crops [13, 14]. Significantly, bioremediation of effluent from primary treatment via microalgae cultivation is known as "advanced" treatment due to the remarkable removal efficiency of nutrients to non-detectable levels exceeding 99%, which is challenging to be achieved in the conventional WWT processes [15, 16]. Meanwhile, microalgae cultivation in wastewater, which is a promising alternative for culture media, could reduce the cultivation cost by 35–86% [17]. Also, the freshwater footprint required for microalgae cultivation generally accounts for 240–680 L kg^{-1}, which could be reduced through the utilization of wastewater [8]. Moreover, as reported by Kurniawan et al. [18], microalgae-integrated municipal WWTP in Indonesia would showcase the lowest environmental impact of −0.139 kg CO_2eq MJ^{-1}, positive energy balance of 1.23 MJ m^{-3} wastewater, and lower cumulative energy demand value across various impact categories [18]. Thus, microalgae-based WWT promotes the concept of a circular bioeconomy through nutrient recovery for the generation of a bio-based feedstock [12].

Considering the promising nutrient removal performances, i.e., > 80%, of microalgae-based treatment processes from previous studies, primary and secondary treatment efficiencies could be enhanced with the integration of microalgae cultivation in conventional WWTPs [19]. Despite the inherent advantages, there are challenges associated with the utilization of wastewater as the culture media for microalgae cultivation. For instance, raw wastewater generally contains high amounts of suspended solids which roughly account for 200–500 mg L^{-1} and have a characteristic turbid color hindering the light availability for the photoautotrophic growth of microalgae. Similarly, the presence of toxic pollutants, high organic loads, variations in essential nutrient contents, such as carbon (C), nitrogen (N), and phosphorus (P), and the presence of pathogens could adversely affect microalgae cultivation [20]. For instance, Ahiahonu et al. [21] reported 46.86–96.25% removal of toxic heavy metals including arsenic (As), cadmium (Cd), chromium (Cr), and lead (Pb) during the microalgae cultivation of *Tetradesmus reginae*, *Tetradesmus obliquus* and *Chlorella sorokiniana* in municipal wastewater [21]. Moreover, a quantitative microbial risk assessment for microalgae-mediated wastewater treatment at an existing sewage treatment plant in India revealed that 0.053 disability-adjusted life years per person per year, considering the pathogenic loads of *Escherichia coli O157:H7* and adenovirus [22]. Thus, overcoming the challenges with potential solutions is important in integrating microalgae-based bioremediation in WWTPs. In this context, the present chapter scrutinizes the potential of microalgae cultivation in various wastewater sources and the removal performances of nutrients while assessing biomass productivity and synthesis of bioproducts. Moreover, microalgae cultivation in primary and secondary effluents will be analyzed, along with various bioproducts derived from the harvested biomass from wastewater bioremediation. In addition, challenges associated with microalgae-based WWT will be discussed to identify potential solutions and future perspectives with the aim of real-time integration with WWTPs.

2 Resource Recovery and Bioremediation Potential of Microalgae in Different Wastewater Types

Various industrial activities, such as agriculture, animal husbandry, crop cultivation, textile, paper-pulp, food, tannery, and pharmaceutical, as well as sewage and municipal, produce a variety of nutrient-rich wastewater streams that require remediation prior to disposal, as detailed in Fig. 1. Microalgae-based bioremediation of wastewater occurs through several removal methods namely biosorption, bioaccumulation or biodegradation/biotransformation. Biosorption, a passive action, is the process of entrapping nutrients and contaminants, such as heavy metals, in the extracellular matrix of the microalgae and rapidly absorbed through the cell wall [23]. Processes including chelation/complexation, ion exchange, electrostatic interaction, adsorption, and surface precipitation are involved in the biosorption of nutrients and contaminants [24].

Bioaccumulation is an active process by which the bioadsorbed contaminants are accumulated and assimilated within the microalgal cell [25]. For instance, bioaccumulated heavy metals are reduced or form cellular protein-heavy metal complexes. Biodegradation is a step further in which microalgae transform organic pollutants,

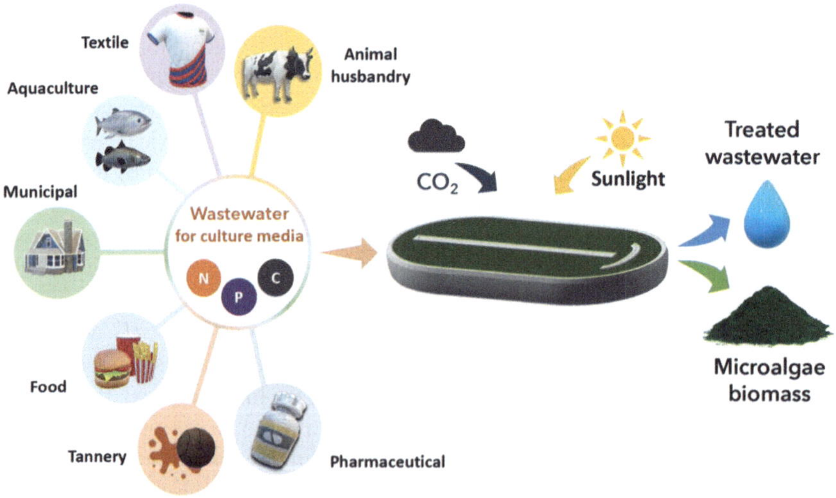

Fig. 1 Microalgae-based bioremediation of different wastewaters

through metabolic action, into intermediate compounds, which may be less toxic [23]. Thus, essential nutrients and other contaminants in wastewater are removed from wastewater by microalgae through biosorption, bioaccumulation, and biodegradation, which would otherwise have been squandered through disposal to the environment [26]. In this context, the following section will detail the potential of diverse microalgal species in the bioremediation of wastewater obtained from varied sources of industrial, municipal, and agricultural operations.

2.1 Industrial Wastewater

Industrial wastewater falls under a wide range, from nutrient-rich food waste streams to pharmaceutical wastes containing emerging organic pollutants and heavy metal-rich wastewater from the processing of leather and textiles. Large quantities of wastewater are generated through industrial processes due to the increase in global population, increase in demand for consumables, and changes to industrial processing practices. For example, an estimated 1–3 L of wastewater is produced per L of milk within dairy industries, whereas tanneries are reported to generate 15–50 tons of toxic wastewater per ton of leather consisting of high heavy metal concentrations [27, 28]. Textile industries have a reported dye effluent discharge of ~280, 000 tons per year worldwide, while the paper pulp industry produces ~2.5 million L of wastewater annually [29, 30].

Numerous studies have delved into the utility of microalgae for the treatment of industrial wastewater. For instance, Gao et al. [31] utilized vinegar wastewater (VWW) obtained from an established industrial plant and assessed the potential to cultivate microalgae *C. vulgaris* over a period of 7 days. VWW contained high initial quantities of chemical oxygen demand (COD) (11,955 mg L^{-1}), total nitrogen (TN) (339.37 mg L^{-1}), and total phosphorous (TP) (131.76 mg L^{-1}) and yielded a biomass concentration of ~2.5 g L^{-1} by day 7. Removal efficiencies were ~60%, ~76%, and ~77% for COD, TN, and TP, respectively. In addition, Gao et al. [31] assessed the utility of winemaking wastewater (WWW) and pickle wastewater (PWW) as culture media for *C. vulgaris* cultivation. Due to the lower initial COD, TN, and TP concentrations of WWW and PWW, *C. vulgaris* demonstrated greater removal efficiencies reaching up to ~100% TP, and ~80% TN in both PWW and WWW (Table 1). However, biomass yield in both PWW and WWW was <50% that of *C. vulgaris* cultivated in VWW. Moreover, *C. vulgaris* cultivation in VWW simultaneously demonstrated highly efficient nutrient removal and biomass production. Furthermore, the nutrient removal efficiencies could be enhanced by optimizing lighting and pH, and utilizing sequential cultivation stages the nutrient removal efficiencies and improving the cultivation process.

Raw wastewater collected from a rice mill's paddy soaking process contained a relatively high COD ($\sim$3,800 mg L^{-1}) and biochemical oxygen demand (BOD) ($\sim$1,500 mg L^{-1}), and adequate quantities of N-NH_3 ($\sim$150 mg L^{-1}) and phosphates (>50 mg L^{-1}) suitable for microalgae cultivation [37]. Umamaheswari and

Table 1 Bioremediation performances and resource recovery potential of microalgae in different wastewater types

Wastewater stream	Initial nutrient concentrations	Microalgal species	Cultivation conditions	Bioremediation performances				References
				COD	N	P	Others[a]	
Vinegar plant wastewater	COD: 11,955 mg L^{-1} TN: 339 mg L^{-1} TP: 131 mg L^{-1}	*Chlorella vulgaris*	Flask Cultivation period: 7 days Working volume: 150 mL Temperature: ~25 °C Light intensity: ~50 μmol m^{-2} s^{-1} L:D of 12:12 h	60%	~76%	~77%	nr	Gao et al. [31]
Winemaking wastewater	COD: 6,495 mg L^{-1} TN: 42 mg L^{-1} TP: 4 mg L^{-1}	*Chlorella vulgaris*	Flask Cultivation period: 7 days Working volume: 150 mL Temperature: ~25 °C Light intensity: ~50 μmol m^{-2} s^{-1} L:D of 12:12 h	~54%	~80%	~100%	nr	
Pickle wastewater	COD: 2,295 mg L^{-1} TN: 51 mg L^{-1} TP: 7 mg L^{-1}	*Chlorella vulgaris*	Flask Cultivation period: 7 days Working volume: 150 mL Temperature: ~25 °C Light intensity: ~50 μmol m^{-2} s^{-1} L:D of 12:12 h	~100%	~80%	~100%	nr	
Dairy wastewater (70%)	COD: 2,581 mg L^{-1} TN: 97 mg L^{-1} TP: 31 mg L^{-1} BOD: 1,953 mg L^{-1}	*Chlorella minutissima*, *Nostoc muscorum* and *Spirulina* sp.	Flask Cultivation period: 12 days Working volume: 300 mL Temperature: 27 °C L:D of 12:12 h 7.5 pH	59%	84%	85%	BOD: 75%	Chandra et al. [32]

(continued)

Table 1 (continued)

Wastewater stream	Initial nutrient concentrations	Microalgal species	Cultivation conditions	Bioremediation performances				References
				COD	N	P	Others[a]	
Pharmaceutical wastewater	COD: 4,704 mg L^{-1} NO_3^-: 72 mg L^{-1} PO_4^{3-}: 99 mg L^{-1}	Unspecified culture from Nacharam Cheruvu lake	Temperature: 25 °C Initial pH: 7.87	97.1%	NO_3^-: 77.6%	PO_4^{3-}: ~100%	nr	Hemalatha and Venkata Mohan [33]
Municipal wastewater containing pharmaceutical compounds	COD: 162 mg L^{-1} Pharmaceutical compounds (μg L^{-1}): CBZ-Diol: 0.98 Carbamazepine: 0.17 Gabapentin: 0.55 Lamotrigine: 0.1 Primidone: <0.1 Clofibric acid: <0.1 Fenofibric acid: 0.23 Bezafibrate: <0.1 Etofibrate: <0.1 Fenofibrate: 0.11 Gemfibrozil: 1.99 Diclofenac: 0.29 Fenoprofen: <0.1 Ibuprofen: 3.18 Indomethacin: <0.1 Ketoprofen: 0.11 Naproxen: 5.91 Paracetamol: 24.78 Diazepam: <0.1 Oxazepam: <0.1 Pentoxifylline: 0.11	Primarily *Parachlorella kessleri*, *Desmodesmus serratus*, *Scenedesmus denticulatus*, and 17 other species	High-rate algal pond HRT: 3 days Flow: 0.2 m^3 day^{-1} Paddlewheel for mixing: 5 rpm	~ 66%	nr	nr	CBZ-Diol: ~ 10% Carbamazepine: ~46% Gabapentin: ~ 83% Lamotrigine: ~ 48% Fenofibric acid: 84% Fenofibrate: 54% Gemfibrozil: 84% Diclofenac: 89% Ibuprofen: 98% Ketoprofen: 58% Naproxen: 83% Paracetamol: 100% Pentoxifylline: 54%	Jiménez-Bambague et al. [34]
Tannery wastewater	COD: 16.54 mg L^{-1} NO_3^-: 5.3 mg L^{-1} NO_2^-: 15.2 mg L^{-1} PO_4^{3-}: 10 mg L^{-1} Cd: 0.033 mg L^{-1} Co: 0.512 mg L^{-1} Pb: 8.72 mg L^{-1}	*Chlorella vulgaris*	Cultivation period: 12 days Working volume: 100 mL Temperature: ~25 °C Light intensity: 2,000–3,000 lux L:D of 12:12 h	nr	NO_3^-: 100% NO_2^-: 80%	PO_4^{3-}: 77%	Cd: 24% Co: 71% Pb: 100%	Arutselvan et al. [35]

(continued)

Table 1 (continued)

Wastewater stream	Initial nutrient concentrations	Microalgal species	Cultivation conditions	Bioremediation performances				References
				COD	N	P	Others[a]	
		Scenedesmus dimorphus	Cultivation period: 12 days Working volume: 100 mL Temperature: ~25 °C Light intensity: 2,000–3,000 lux L:D of 12:12 h	nr	NO_3^-: 100% NO_2^-: 100%	PO_4^{3-}: 85%	Cd: 42% Co: 86% Pb: 100%	
		Chlorella vulgaris and *Scenedesmus dimorphus* consortium	Outdoor pond Cultivation period: 6 days Working volume: 30 L	nr	NO_3^-: 100% NO_2^-: 100%	PO_4^{3-}: ~55%	Cd: 93% Co: 98% Pb: ~ 86%	
Textile wastewater	TDS: 13.24 mg L^{-1} NO_3^-: 8.2 mg L^{-1} NO_2^-: 136.7 mg L^{-1} PO_4^{3-}: 15 mg L^{-1} Cd: 0.128 mg L^{-1} Co: 0.562 mg L^{-1} Pb: 6.04 mg L^{-1}	*Chlorella vulgaris*	Cultivation period: 12 days Working volume: 100 mL Temperature: ~25 °C Light intensity: 2,000–3,000 lux L:D of 12:12 h	nr	NO_3^-: 77% NO_2^-: 76%	PO_4^{3-}: 87%	Cd: 100% Co: 78% Pb: 100%	
		Scenedesmus dimorphus	Cultivation period: 12 days Working volume: 100 mL Temperature: ~25 °C Light intensity: 2,000–3,000 lux L:D of 12:12 h	nr	NO_3^-: 63% NO_2^-: 83%	PO_4^{3-}: 98%	Cd: 100% Co: 94% Pb: 100%	
		Chlorella vulgaris and *Scenedesmus dimorphus* consortium	Outdoor pond Cultivation period: 6 days Working volume: 30 L	nr	NO_3^-: 100% NO_2^-: 100%	PO_4^{3-}: 80%	Cd: 98% Co: 70% Pb: 95%	

(continued)

Table 1 (continued)

Wastewater stream	Initial nutrient concentrations	Microalgal species	Cultivation conditions	Bioremediation performances				References
				COD	N	P	Others[a]	
Piggery wastewater	COD: 6,820 mg L^{-1} $N\text{-}NH_3$: 850 mg L^{-1} P: 16.4 mg L^{-1}	*Chlorella sorokiniana, Coelastrella* sp. and *Acutodesmus nygaardii*	3 stage sequential cultivation process Wastewater sterilized and undiluted Cultivation period: 2 days Temperature: ~25 °C Light intensity: ~750 μmol m^{-2} s^{-1}	92%	90%	100%	Turbidity—99%	Lee et al. [36]
Paddy-soaked wastewater	COD: 3,800 mg L^{-1} BOD: 1,500 mg L^{-1} $N\text{-}NH_3$: 150 mg L^{-1} $P\text{-}PO_4^{3-}$: >50 mg L^{-1}	*Scenedesmus obliquus*	Raceway pond Working volume: 75 L Paddle wheels gentle mixing: 5–10 rpm Light intensity: ~50 μmol m^{-2} s^{-1} L:D of 12:12 h	94%	$N\text{-}NH_3$: 86%	$P\text{-}PO_4^{3-}$: 93%	BOD: ~89%	Umamaheswari and Shanthakumar [37]
Municipal wastewater	COD: 382 mg L^{-1} $N\text{-}NH_3$: 46.83 mg L^{-1} TP: 2.53 mg L^{-1}	*Chlorococcum* sp. and bacteria indigenous to the ww	Flask Cultivation period: 5 days Working volume: 600 mL Temperature: ~25 °C Light intensity: 3,000 lux L:D of 12:12 h	87%	$N\text{-}NH_3$: ~100%	~80%	nr	Yang et al. [38]
Municipal wastewater	COD: 141.41 mg L^{-1} $N\text{-}NH_3$: 16.6 mg L^{-1} TP: 1.65 mg L^{-1} TN: 21.87 mg L^{-1}	*Chlorella vulgaris*	Aeration columnar photobioreactors HRT 24 h Working volume: 1.2 L	~45%	$N\text{-}NH_3$: ~50% TN: ~ 50%	~87%	nr	Li et al. [39]
		Chlorella vulgaris and activated sludge (obtained from an aerobic tank of a wastewater treatment plant)	Aeration columnar photobioreactors HRT 24 h Working volume: 1.2 L	~69%	$N\text{-}NH_3$: ~65% TN: ~62%	~83%		
Starch wastewater	COD: ~12,266 mg L^{-1} TN: ~611 mg L^{-1} TP: ~50 mg L^{-1}	*Chlorella pyrenoidosa* and *Aspergillus oryzae* (microalgae + fungi)	Flask Temperature: 25 °C Continuous illumination light intensity: 30 μmol m^{-2} s^{-1} Algae/fungi ratio at 4000:1	92%	84%	97%	nr	Wang et al. [40]

[a]Other contaminants and pollutants such as heavy metals and pharmaceutical compounds

nr - Not reported; COD—chemical oxygen demand; TN—total nitrogen; TP—total phosphorous; PO_4^{3-}—phosphate; NH_3—ammonia; NO_3^-—nitrate; NO_2^-—nitrite; Cd—cadmium; Co—cobalt; Pb—lead

Shanthakumar [37] utilized this paddy-soaking wastewater to cultivate *S. obliquus* in a raceway pond (75 L) for 7 days, under ~50 μmol m^{-2} s^{-1} (light:dark photoperiod of 12:12 hrs), which showcased N-NH_3 and phosphate removal efficiencies of 86% and 93%, respectively.

Pharmaceuticals and personal care products (PPCP) wastewater contains micropollutants, which could cause significant environmental and health detriments, even in trace quantities. Microalgal species, such as *Chlorella*, *Nannochloris*, *Nannochloropsis*, *Chlamydomonas*, *Haematococcus pluvialis*, and *Scenedesmus obliquus* exhibit potential in bioremediating PPCP from wastewater [41, 42]. For instance, Jiménez-Bambague et al. [34] utilized a microalgal consortium primarily populated by *Parachlorella kessleri*, *Desmodesmus serratus*, *Scenedesmus denticulatus* as well as 17 other species, to remove emerging pollutants found in domestic wastewater obtained from the city of Santiago de Cali, Colombia by cultivating in a pilot-scale HRAP. The pharmaceutical compounds, fenofibric acid, ibuprofen, and paracetamol were removed with efficiencies >70%. The removal efficiency of fenofibrate, gabapentin, lamotrigine, gemfibrozil, ketoprofen, diclofenac, naproxen, and pentoxifylline was within 30–70%. Moreover, Jiménez-Bambague et al. [34] reported that a significant reduction of ecological hazard could be achieved by bioremediation using the selected consortium. However, the hazards posed by gemfibrozil and ibuprofen were still high, and thus, it would require complementary treatment and more detailed research into alternative microalgal species that may possess greater ability for the degradation of gemfibrozil and ibuprofen.

Tannery wastewater, rich in heavy metals and other toxic pollutants, such as phthalates, formaldehydes, chlorophenols, oils, resins, and detergents, can be effectively treated by a variety of microalgal species, including *Spirulina* sp., *C. vulgaris*, *Scenedesmus dimorphus*, *Coelastrella* sp. and *Chlorococcum* sp. [6]. For instance, Rajalakshmi et al. [43] achieved ~93–95% removal of heavy metals and 95% removal of BOD and COD from tannery wastewater utilizing *Chlorella* sp. In a study by Arutselvan et al. [35], raw tannery effluent was utilized in cultivating a consortium consisting of *C. vulgaris* and *S. dimorphus* in an open pond system. The consortium achieved 100% nitrate removal within 3 days, whereas 100% removal of nitrite and 84% removal of phosphate required 6 and 9 days of cultivation, respectively. Moreover, Arutselvan et al. [35] observed 100% removal of iron, magnesium, and heavy metals (Cd, Pb, and cobalt) within 12 days of cultivating the *C. vulgaris* and *S. dimorphus* consortium, ultimately achieving biomass with 28% lipid content suitable for biofuel production. Furthermore, Arutselvan et al. [35] assessed the bioremediation and metabolite synthesis capability of *C. vulgaris* and *S. dimorphus* consortium in raw textile effluent, which exhibited 100% nitrate and nitrite removal within 3 days, and 100% phosphate removal in 6 days. Similar to tannery wastewater, 100% heavy metal removal was achieved in a 12-day cultivation period with biomass lipid content reaching up to 31%. Additionally, Arutselvan et al. [35] reported that the cultivation period required for 100% removal of the heavy metal contaminants and nutrients was less (~3–6 days) when cultivating a consortium as opposed to axenic cultivation (>12 days), emphasizing the potential benefit of utilizing a microalgae consortium over a monoculture. Table 1 presents further details of the industrial wastewater bioremediation of various microalgal species under a diverse cultivation conditions.

2.2 Municipal and Sewage Wastewater

Municipal and sewage wastewater act as major contributors towards the pollution of freshwater bodies, and as reported by UNICEF and World Health Organization [44], approximately 80% is released to the environment, untreated, especially in developing countries. Microalgae offer a cost-effective alternative to municipal WWT and showcase noteworthy results with nutrient and contaminant removal efficiencies reaching >80%, based on numerous studies [45]. For instance, Silambarasan et al. [46] achieved nitrate, N-NH_4^+, phosphate, COD, TOC, and TN removal rates of 84–96%, 95–98%, 89–95%, 78–88%, 81–86% and 85–94%, respectively, in municipal wastewater by cultivating a consortium of *Chlorella* sp. and *Scenedesmus* sp. Notably, increased removal efficiencies through consortium cultivation were observed as opposed to monocultures of *Chlorella* sp. or *Scenedesmus* sp. Biomass concentration of the *Chlorella* sp. and *Scenedesmus* sp. consortium was highest (~1.75 g L^{-1}) at a wastewater dilution of 75%, with protein and lipid contents of 175 μg mL^{-1} and 35% DW, respectively. Table 1 further details the various microalgal species capable of enacting bioremediation in municipal/sewage wastewater (Table 2).

2.3 Animal Husbandry and Agricultural Wastewater

Wastewater produced through various activities, including animal husbandry, aquaculture and crop farming, is a major source of N, C and P, which when released to water bodies can instigate eutrophication. Moreover, the pathogenic content within livestock wastewater can cause diseases, such as diarrhea, vomiting, ulcerative colitis, and fever in human beings [58]. Utilizing microalgae for bioremediation of agricultural wastewater can cause inactivation of pathogens by sunlight and high pH conditions during cultivation, or by the release of high dissolved oxygen and toxins from microalgae [58]. Lee et al. [36] treated undiluted raw piggery wastewater by employing three microalgal species, namely *C. sorokiniana*, *Coelastrella* sp. and *Acutodesmus nygaardii* which were adapted to high ammonia (NH_3). The WWT process was carried out in 3 stages, in which small particles, COD and phosphorus were removed, firstly, by heterotrophic cultivation of *C. sorokiniana* in raw piggery water. It was then followed by mixotrophic cultivation of *Coelastrella* sp., removing NH_3 and COD, and finally the remaining ammonia was removed by photoautotrophic cultivation of *A. nygaardii*. The sequential cultivation process carried out across 6 days was successful in removing 100%, 92% and 90% of P, COD and NH_3, respectively, with a 99% deduction in turbidity of the final wastewater [36]. Lee et al. [36] highlighted that dilution of wastewater, generally utilized

Table 2 Effect of primary and secondary effluents on nutrient removal efficiency and biomass/metabolite yields

Wastewater stream	Initial nutrient concentrations	Microalgal species	Bioremediation performances	Biomass yield	Metabolite yields	References
Primary effluent						
Domestic wastewater (75:25 wastewater:freshwater)	N-NO_3^-: 16.58 mg L^{-1} N-NH_3: 37.64 mg L^{-1} P-PO_4^{3-}: 7.42 mg L^{-1} TOC: 208.15 mg L^{-1} COD: 446.25 mg L^{-1} TN: 61.47 mg L^{-1}	*Chlorella* sp.	N-NO_3^-: ~87% N-NH_3: ~90% P-PO_4^{3-}: ~85% TOC: ~80% COD: ~70% TN: ~82%	1.35 g L^{-1}	Chlorophyll: 21.06 µg L^{-1} Protein: 154 µg L^{-1} Lipids: 29.5% DW	Silambarasan et al. [46]
		Scenedesmus sp.	N-NO_3^-: 90% N-NH_3: ~ 95% P-PO_4^{3-}: ~ 87% TOC: ~ 80% COD: ~ 78% TN: ~ 83%	1.42 g L^{-1}	Chlorophyll: 22.08 µg L^{-1} Protein: 157 µg L^{-1} Lipids: 30% DW	
		Chlorella sp. and *Scenedesmus* sp. consortium	N-NO_3^-: ~ 96% N-NH_3: ~ 96% P-PO_4^{3-}: ~ 95% TOC: ~ 86% COD: ~ 82% TN: ~ 92%	1.78 g L^{-1}	Chlorophyll: 27.03 µgL^{-1} Protein: 175 µgL^{-1} Lipids: 34.8% DW	
Molasses wastewater	COD: ~11,230 mg L^{-1} TN: ~407 mg L^{-1} TP: ~30.4 mg L^{-1}	*Chlorella vulgaris* + *Aspergillus* sp.	COD: 71% TN: 67% TP: 88%	~2.5 g L^{-1}	Protein: 38.6% DW Lipids: 35.2% DW	Yang et al. [47]

(continued)

Table 2 (continued)

Wastewater stream	Initial nutrient concentrations	Microalgal species	Bioremediation performances	Biomass yield	Metabolite yields	References
Septic wastewater	COD: 790 mg L^{-1} N-NH_4^+: 19.3 mg L^{-1} N-NO_3^-: 1.4 mg L^{-1} TP: 5.4 mg L^{-1}	*Chlorella vulgaris* + nitrifying bacteria consortium	COD: ~ 90% N-NH_4^+: ~ 54% N-NO_3^-: increase TP: ~ 77%	nr	nr	Sorenson et al. [48]
Brewery wastewater	Organic carbon: 182.1 mg L^{-1} TN: 14.5 mg L^{-1} TP: 19.49 mg L^{-1}	*Chlorella* sp.	Organic carbon: 100% TN: 100% TP: 100%	~1.2 (OD_{540})	nr	Subramaniyam et al. [49]
Domestic wastewater	COD: 548 mg L^{-1} N-NH_4^+: 51 mg L^{-1} TP: 0.85 mg L^{-1}	*Scenedesmus* sp.	COD: 65% N-NH_4^+: 95% TP: 85%	0.55 g L^{-1}	Polysaccharides: < 5% Lipids: ~ 16% Proteins: 50%	Wang et al. [50]
		Chlorella sp. HL	COD: 25% N-NH_4^+: 85% TP: ~82%	0.38 g L^{-1}	Polysaccharides: ~5% Lipids: 20% Proteins: 50%	
		Raphidocelis subcapitata	COD: 79% N-NH_4^+: 70% TP: 85%	0.45 g L^{-1}	Polysaccharides: <2% Lipids: 25% Proteins: 40%	
		T. obliquus	COD: 70% N-NH_4^+: 95% TP: 80%	0.44 g L^{-1}	Polysaccharides: ~2% Lipids: 25% Proteins: 55%	

(continued)

Table 2 (continued)

Wastewater stream	Initial nutrient concentrations	Microalgal species	Bioremediation performances	Biomass yield	Metabolite yields	References
		Chlorella sp. (A)	COD: 20% N-NH_4^+: 75% TP: 70%	0.42 g L^{-1}	Polysaccharides: ~3% Lipids: 20% Proteins: 42%	
		Chlorella sp. (B)	COD: 36% N-NH_4^+: 65% TP: ~82%	0.31 g L^{-1}	Polysaccharides: ~5% Lipids: 20% Proteins: 35%	
Municipal wastewater	TN: 61.52 mg L^{-1} TP: 2.23 mg L^{-1}	*Chlorella vulgaris*	TN: >90% TP: ~100%	~ 1.15 g L^{-1}	Chlorophyll: 3.8 mg/g_{DW} Carotenoid: 1.7 mg/g_{DW} Lipid: 34.2% DW	Ferro et al. [51]
		Scenedesmus sp.	TN: >90% TP: ~100%	~ 1.25 g L^{-1}	Chlorophyll: 3.4 mg/g_{DW} Carotenoid: 1.6 mg/g_{DW} Lipid: 16.6% DW	
		Desmodesmus sp.	TN: >90% TP: ~100%	~0.9 g L^{-1}	Chlorophyll: 4.1 mgg_{DW} Carotenoid: 1.8 mg/g_{DW} Lipid: 29.8% DW	

(continued)

Table 2 (continued)

Wastewater stream	Initial nutrient concentrations	Microalgal species	Bioremediation performances	Biomass yield	Metabolite yields	References
		Scenedesmus obliquus	TN: >90% TP: ~100%	~1.4 g L^{-1}	Chlorophyll: 4.3 mg/g_{DW} Carotenoid: 2.3 mg/g_{DW} Lipid: 22.5% DW	
Secondary effluent						
Synthetic wastewater	COD: 125 mg L^{-1} TN: 49 mg L^{-1} TP: 8 mg L^{-1}	Natural microalgae consortium	COD: ~66% TN: ~81% TP: 7.73%	nr	nr	Berrio-Restrepo et al. [52]
Reclaimed water	COD: 87 mg L^{-1} TN: 51 mg L^{-1} TP: 2.5 mg L^{-1}	*Desmodesmus* sp.	COD: 51% TN: 89% TP: 95%	1.562 (OD_{680})	Chlorophyll: 5.91 mg L^{-1}	Wang et al. [53]
Municipal wastewater	COD: 95.8 mg L^{-1} BOD: 71.5 mg L^{-1} TSS: 81 mg L^{-1} TN: 21.96 mg L^{-1} TP: 1.28 mg L^{-1}	*Scenedesmus obliquus*	N-NO_3^-: ~91% TP: ~100%	~0.3 g L^{-1}	Chlorophyll: 1.2 mg L^{-1} Lipid: ~0.6 g L^{-1}	Eida et al. [54]
Municipal wastewater	COD: 25.6 mg L^{-1} TN: 12.7 mg L^{-1} TP: 0.6 mg L^{-1}	*Chlorella vulgaris*	nr	1.84 g L^{-1}	Lipid: ~37% DW	Gao et al. [55]
		Scenedesmus obliquus	nr	1.72 g L^{-1}	Lipid: ~30%	
Urban wastewater	TN: 5.15 mg L^{-1} P-PO_4^{3-}: ~0.36 mg L^{-1} TP: 0.16 mg L^{-1}	Mixed microalgae culture	TN: 5.55 mgL^{-1} day^{-1} P-PO_4^{3-}: 0.4 mgL^{-1} day^{-1}	0.037 g L^{-1} day^{-1}	Chlorophyll: 2.46 mg L^{-1}	Solmaz and Işık [56]

(continued)

Table 2 (continued)

Wastewater stream	Initial nutrient concentrations	Microalgal species	Bioremediation performances	Biomass yield	Metabolite yields	References
Municipal wastewater	COD: 59 mg L^{-1} N-NO_3^-: 18 mg L^{-1} N-NH_4^+: 44 mg L^{-1} N-NO_2^-: 0.89 mg L^{-1} TP: 26 mg L^{-1}	*Chlorella vulgaris*	COD: 49.1% N-NO_3^-: 22.2% N-NH_4^+: 70.9% TDP: 11.5%	~2 g L^{-1}	nr	Al Momani and Örmeci [57]
		Neochloris oleoabundans	COD: 42.4% N-NO_3^-: 33.3% N-NH_4^+: 67.5% TDP: 5.67%	~1.5 g L^{-1}	nr	
		Mixed indigenous microalgae from WWTP	COD: 70.3% N-NO_3^-: 66.7% N-NH_4^+: 67.5% TDP: 30.8%	~0.5 g L^{-1}	nr	

nr—not reported; COD—chemical oxygen demand; TN—total nitrogen; TP—total phosphorous; PO_4^{3-}—phosphate; NH_3—ammonia; NO_3^-—nitrate; NO_2^-—nitrite; Cd—cadmium; Co—cobalt; Pb—lead to facilitate higher efficiency of nutrient uptake by the microalgae, was not required for the sequential cultivation process, thus preventing the requirement of additional freshwater footprints in the WWT process.

Leong et al. [59] obtained dairy manure wastewater (DMWW) from a cattle ranch containing COD of 1800–2400 mg L^{-1}, TN of 200–300 mg L^{-1}, and TP of 20–50 mg L^{-1}. The undiluted DMWW was sieved to remove large particles and consequently inoculated with 0.3 g L^{-1} of *C. vulgaris*. Within a cultivation period of 6 days, COD, TN, and TP removal of 24%, 26% and 17%, respectively, were achieved. However, Leong et al. [59] reported that dilution of DMWW to a 1:1 ratio with BG 11 medium increased the removal efficiencies of COD, TN, and TP to 40%, 77% and 81%, respectively. It is evident that dilution offers greater removal efficiency on a lab scale, but the practical applicability in large-scale operation is questionable due to the large requirement of freshwater footprints. Thus, Leong et al. [59] assessed the possibility of two-stage treatment in order to enhance the overall nutrient removal efficiency, with the first stage consisting of indigenous bacterial cultivation over a 7-day period, followed by microalgal inoculation in the second stage. Through the two-stage treatment process ~80% COD, 88% TN and 93% TP removal in undiluted DMWW was observed. Leong et al. [59] attributed this to the synergistic effects of the microalgae-bacteria consortium. Additionally, due to the effectiveness of two-stage cultivation *C. vulgaris* biomass growth was enhanced by 1.5-fold than direct cultivation in raw DMWW.

Similar to Leong et al. [59]'s study, the utilization of various microorganisms alongside microalgae has significantly enhanced the removal efficiencies of nutrients and contaminants from DMWW. Chang et al. [60] observed increased removal efficiency of BOD, COD, TN and TP in DMWW by *C. sorokiniana*, cultivated along with indigenous bacterial species that had been enriched through an incubation process. Monoculture of *C. sorokiniana* in DMWW demonstrated ~80% removal of TN and TP, however, BOD and COD removal was only 65% and 44%, respectively. Utilization of only the indigenous bacterial population was also unsatisfactory despite the BOD and COD removal efficiencies reaching 97% and 72%, respectively, while TN and TP were only ~3% and 29%. Thus, Chang et al. [60] utilized both microalgae and bacteria to effectively increase the overall removal efficiencies to 91%, 66%, 83% and 84% for BOD, COD, TN and TP, respectively.

It is evident that microalgae exhibit a remarkable capacity for efficient nutrient removal from wastewater. Nonetheless, the capacity of bioremediation depends on diverse factors, such as the microalgal species, characteristics of wastewater, and growth parameters. Thus, future studies on specific wastewaters, paired with specific microalgal species under optimized cultivation conditions are required to assess the real-time applicability in conventional WWTPs. Moreover, it should be noted that organic pollutants and nutrients are biotransformed into non-toxic substances through microalgal metabolism. For instance, the conversion of Cr(II) and Cr(VI) into Cr(III) or organic forms by interaction with intracellular organic molecules [23]. The detailed study of hazardous compound biodegradation mechanisms of microalgae is still ambiguous, and thus, there exists a knowledge gap requiring further research. Several studies have assessed the biodegradation of chemicals, such as benzalkonium chloride and sulfamethoxazole, to intermediate transformation products. However, further analysis of the toxicity and understanding of the molecular interactions of intermediate products require further research [61]. Despite the biodegradation and biotransformation potential of microalgae, hazardous substances such as heavy metals and PPCP may not fully degrade, leading to accumulation within the biomass. Thus, strict toxicity assessments may be required to ensure there will not be any toxic effects on humans by usage of microalgal products cultivated in wastewater.

3 Prospects of Integrating Microalgae Cultivation into Conventional Wastewater Treatment Plants

Primary, secondary, and tertiary treatment are the three fundamental operational stages of conventional WWTPs. Primary treatment of wastewater consists of a step to eliminate larger solid materials, followed by a sedimentation process, wherein settleable solids are removed. Upon the sedimentation of large, settleable particles the wastewater can then be directly utilized for microalgae cultivation as primary effluent or can further be forwarded to secondary treatment. Secondary treatment employs a mixed population of heterotrophic bacteria, either indigenous to the wastewater or extraneously inoculated. The bacteria reduce the BOD/COD by utilizing organic matter present in the wastewater for energy and growth. Consequent to a secondary sedimentation process, secondary effluent is obtained and can be forwarded to conventional tertiary treatments, such as coagulation, filtration, reverse osmosis, disinfection, and activated C adsorption of organics in order to further reduce organics, P, N, metals, turbidity, and pathogens. Alternatively, secondary effluents, rich in N and P, have the potential to be utilized for microalgae cultivation (Fig. 2). The following section details the prospects of microalgae cultivation in primary and secondary effluents, noting the variations in biomass yield, metabolite synthesis and nutrient removal efficiencies of each wastewater effluent type.

3.1 Microalgae-based Bioremediation of Primary Effluents from Conventional Wastewater Treatment Plants

Primary effluent wastewater consists of high quantities of COD, N and P, and thus direct release to the environment would create detrimental effects on natural water bodies due to an increase in turbidity, and effects of eutrophication. By the utilization of heterotrophic or mixotrophic cultivation of microalgae, it is possible to accomplish bioremediation of primary effluent WW, achieving COD, N and P levels suitable for discharge to the environment [62]. Tan et al. [63] assessed the bioremediation

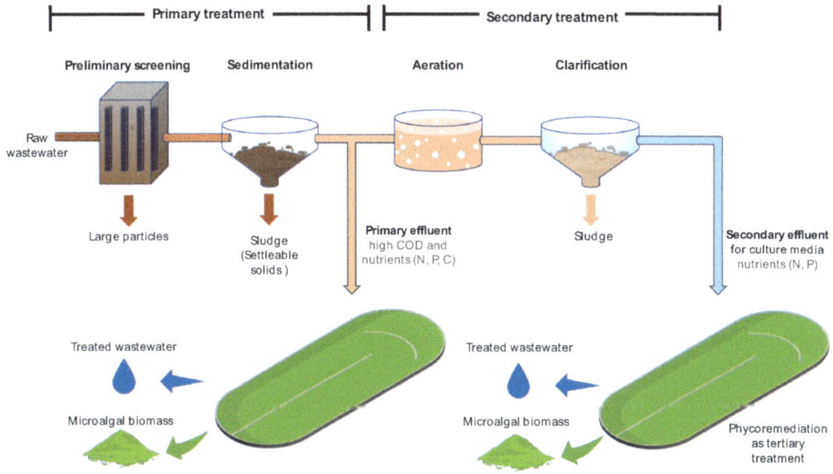

Fig. 2 Microalgae cultivation in primary and secondary effluents of conventional wastewater treatment plants

potential of *Scenedesmus* sp., *Chlorococcum aquaticum*, *Ankistrodesmus augustus*, and *H. pluvialis*, cultivated in non-sterilized undiluted primary domestic wastewater. *Scenedesmus* sp. achieved the highest TN, TP and N-NH_4^+ removal efficiencies of ~100%, and biomass concentration of ~0.6 g L^{-1} within 16 days of cultivation. Moreover, *C. aquaticum* was similarly successful in achieving nutrient removal efficiencies of >90% with a biomass concentration of ~0.7 g L^{-1}. Similarly, *H. pluvialis* achieved the highest biomass concentration of ~0.8 g L^{-1} with TN and N-NH_4^+ removal efficiencies of >80%, however, TP removal efficiency was 68%.

Yukiyo and Hiroyuki [64] assessed the potential of cultivating naturally occurring microalgae (NOM) consortium in primary effluent obtained from an urban sewage WWTP. The microalgae were cultivated in 380 L raceway ponds with a wastewater hydraulic retention time (HRT) of 2 days in summer and autumn, and 4 days in winter and average TN, TP, and COD removal of 43 ± 17%, 60 ± 18%, 74 ± 11%, respectively, were achieved. Furthermore, Yukiyo and Hiroyuki [64] observed that the addition of excess activated sludge from the aeration step of the WWTPs offered better average TN, TP, and COD removal of 63 ± 14%, 84 ± 9.5%, 84 ± 11%, as well as ~2 times higher biomass content. The addition of activated sludge enhanced the bioremediation prospects, as well as enabled easier settling and separation of the microalgal biomass from culture media [64]. Whilst indigenous bacterial consortia, in symbiosis with microalgae, enhance the bioremediation process, certain circumstances may require the sterilization of primary effluent prior to cultivation. It may become necessary to eliminate harmful bacteria which could carry out microalgal predation, leading to lower efficiencies in bioremediation and metabolite synthesis. However, this may lead to higher costs, and can become unfeasible in larger-scale operations. Thus, effectively identifying and maintaining a suitable symbiotic microalgal-bacterial consortium would be beneficial from an economic perspective.

The direct utilization of microalgae-based bioremediation on primary effluents allows for the mitigation of cost, space, and energy required for secondary treatment. Moreover, the sludge formation, generally created during the secondary treatment process, could be eliminated by the microalgae cultivation in primry effluents, and thus reduces the investment required for sludge disposal and/or treatment, i.e., anaerobic digestion systems. Furthermore, utilizing methods, such as two-stage cultivation systems, bacterial symbiosis, and controlling cultivation parameters, could facilitate rectification of the disadvantages associated with microalgal cultivation in primary effluents, while enhancing nutrient recovery and biomass growth.

3.2 *Microalgae-based Bioremediation of Secondary Effluents from Conventional WWT Plants*

Secondary treatment of wastewater within a conventional WWTP facilitates the reduction of COD by bacterial action; however, it oft does not reduce N and P to suit discharge standards. Thus, secondary effluents rich in N and P can be utilized for the cultivation of microalgae whilst accomplishing phycoremediation. In essence, microalgae-based bioremediation of secondary effluents is the act of polishing wastewater, i.e., tertiary treatment [65]. For instance, Wang et al. [53] utilized *Desmodesmus* sp., isolated from Yingxue Lake, for the phycoremediation of secondary effluent from a sewage WWT plant. An initial pH of 9 facilitated the highest COD, TN, TP and $N\text{-}NH_4^+$ removal efficiencies, ~50%, ~90%, 98% and ~100%, respectively, and an initial inoculation of 0.4 (OD_{680}) yielded *Desmodesmus* sp. biomass of ~23% lipid content suitable for biodiesel production.

Biliani and Manariotis [66] employed algal bio-flocs, created using activated sludge, for the remediation of secondary effluent obtained from the WWTP of the University of Patras, Rio. Biomass density of ~7.8 g L^{-1} was observed with nutrient removal efficiencies of 88.4%, 77.5%, 100% and 98.6% for COD, nitrate ($N\text{-}NO_3^-$), ammonia ($N\text{-}NH_3$), and TP, within a low HRT of 1.2 days. Moreover, the comparison between the microalgal cultivation in primary and secondary effluents showcased ~2.7-fold high biomass density and ~50–70% of protein production in secondary effluent [66]. Furthermore, An adequate content of lipids, suitable for biodiesel production, was synthesized at ~15–30% in the secondary effluent. Therefore, a marked difference in the metabolite and biomass content and nutrient removal efficiencies of microalgae cultivated in secondary effluent could be witnessed, compared to primary effluent, mainly due to the initial wastewater characteristics. Hence, suitable microalgae/consortia should be utilized for the optimal bioremediation of different effluent streams.

In this context, it is evident that primary and secondary effluents of WWTPs provide significant opportunities for microalga cultivation and simultaneous bioremediation. Thus, the implementation and integration of microalgae-based bioremediation into WWTPs would be highly beneficial, in contrast to conventional secondary/ tertiary treatment methods. Remarkably, biomassgenerated through microalgae cultivation in wastewater contains notable contents of lipids, protein and carbohydrates that can be processed into various bioproducts, such as biodiesel, biogas, biochar, and biofertilizer.

4 Value-Added Bioproducts Derived from Microalgae-based Wastewater Treatment

4.1 Biofuels

Biomass harvested from the microalgae cultivation integrated to WWTPs could be converted into a diverse spectrum of biofuels, namely biodiesel, bioethanol, biogas, and biohydrogen. As discussed in Sect. 2, microalgal biomass produced during the wastewater bioremediation is enriched with proteins, lipids, and carbohydrates, which could be converted to biofuels through transesterification, anaerobic digestion, fermentation, gasterification, pyrolysis, and hydrothermal liquefaction, among the others. Significantly, up to 70% of microalgae consist of lipids and more than 50% is carbohydrate contents, making the biomass a potential feedstock for biofuel production.

Chlorella and *Scenedesmus* are mainly cultivated in wastewater for biofuel production due to their robustness and adaptability to wastewater characteristics, while *Desmodesmus*, *Nannochloropsis*, and *Tetraselmis* are also utilized during wastewater bioremediation. For instance, Malla et al. [67] cultivated *Chlorella minutissima* in primary treated wastewater from the Indian Agricultural Research Institute in Mayapuri, India. From this study, biomass productivity of 10.1 g L^{-1} and total lipid content of approximately 20.69% were achieved, along with 100% N-NH_4^+ removal, 50% nitrate removal and 60–80% P removal [67]. In another study, *Scenedesmus* sp. was cultivated under mixotrophic mode in primary treated municipal wastewater, which demonstrated 0.069 g $L^{-1}d^{-1}$ of biomass productivity, 22.5% of lipid yield, and 22.04% of fatty acid methyl ester (FAME) content [68]. Notably, the extracted lipids exhibited high-quality biodiesel properties based on the analysis of the fatty acid profile [68]. Similarly, *Chlorella* strains cultivated in unsterilized swine wastewater exhibited biomass growth and carbohydrates content of 3.6 g L^{-1}, and 1.8 g L^{-1}, respectively, along with TN removal efficiency of 92.2%. Fermentation of the obtained biomass showcased bioethanol yield and concentration of 84.2% and 4.2 g L^{-1}, respectively [69].

4.2 Biofertilizers

Microalgal-derived biofertilizers are promising sources of N, P, and K, especially resulting from the biosorption of nutrients during wastewater-based cultivation. Significantly, biofertilizers derived from microalgal biomass releases nutrients at lower rates through biodegradation and possess inherent antimicrobial properties which are beneficial in agriculture. For instance, Das et al. [70] evaluated the microalgae-based treatment of primary effluent collected from Al-Khor WWTP in Qatar with five microalgae namely, *Chlorella* sp., *Dictyosphaerium* sp., *Monoraphidium* sp., *Neochloris* sp., and *Scenedesmus* sp.. Bioremediation of wastewater with *Chlorella* sp. and *Scenedesmus* sp. was effective with concurrent biomass yields of 476 mg L^{-1} and 467 mg L^{-1}, respectively [70]. In accordance with the assessments of N, P and other trace element contents in harvested biomass, the authors determined that it could be a promising alternative for commercial fertilizers [70]. Furthermore, Silambarasan et al. [71] reported that de-oiled biomass obtained from algal consortium of *Chlorella* sp. and *Scenedesmus* sp., combined with inorganic fertilizer demonstrated promising results in *Solanum lycopersicum*. The study showcased 44%, 89%, 95% and 174% enhancements in the plant length, root length, fresh weight and tomato yield, respectively. Thus, it is evident that microalgae harvested from wastewater-based cultivation as a whole or de-oiled biomass is a potential feedstock for the bioproduction of biofertilizers.

4.3 Animal Feed

The nutrient-rich biomass harvested from wastewater-based cultivation could be utilized as animal feed. In a study by Viegas et al. [72], *C. vulgaris* and *S. obliquus* biomass obtained during the microalgal cultivation in raw effluent from brown crab aquaculture composed of 31% and 35% proteins, 6% and 8% lipids, 39% and 30% carbohydrates, respectively [72]. The authors concluded that biomass could be incorporated as a novel and sustainable ingredient in aquaculture feeds, based on the biomass compositions. Similarly, Morais et al. [73] reported that the selenium recovery by *C. vulgaris* and *Scenedesmus* sp. from urban wastewater was up to 43% and 52%, respectively. Furthermore, 1 kg of *C. vulgaris* dry biomass consisted of up to 323 mg of selenium (Se), showcasing a potential supplementation of the Se-rich biomass in the animal feed [73].

Evidently, microalgal biomass derived from wastewater bioremediation is a potential feedstock for biofuels, biofertilizers, and animal feeds. Moreover, the microalgal biomass could be utilized for various other applications, including biochar production and pigment production. For example, *H. pluvialis* cultivation in 7.5% palm oil mill effluent demonstrated a high astaxanthin content of 22.43 mg L^{-1}, simultaneously reducing the freshwater consumption by 50.9% [74]. Nevertheless, there are constraints associated with the application of microalgal biomass obtained from wastewater-based cultivation, mainly due to the presence of toxic pollutants [75]. For instance, high-value pigments derived from wastewater-based microalgae cultivation might contain traces of heavy metals and other hazardous pollutants, which severely affects food- and pharmaceutical-based applications. Hence, studies need to focus on treating wastewater streams to remove hazardous substances prior to microalgae cultivation to ensure the compositional standards suitable for human consumption. Furthermore, techno-economic aspects of bioproduct derivation from biomass harvested from wastewater-based microalgae cultivation need to be evaluated to establish sustainable real-time production processes. In addition, multiproduct biorefineries would facilitate in enhancing the economic viability of microalgae cultivation in wastewater, while maintaining a circular bioeconomy.

5 Challenges and Potential Solutions Associated with Microalgae-based Wastewater Treatment

Numerous challenges are associated with the microalgae-based WWT, hindering the real-world applicability in WWTPs. Hence, it is of paramount importance to identify potential solutions to overcome the challenges, which mainly includes energy-intensive upstream processing and harvesting of microalgae, and wastewater characteristics hindering effective microalgal growth. Wastewater characteristics, namely high organic loads, presence of toxic substances (i.e., heavy metals), variations in nutrient availability, high turbidity and suspended solid contents, and presence of pathogens, have adverse effects on microalgae cultivation. Thus, this section aims to discuss the major challenges associated with microalgae-based wastewater treatment to enhance the performance of resource recovery and biomass synthesis.

The energy intensiveness of upstream processing and biomass harvesting is a major constraint associated with microalgae cultivation. In general, open pond systems are preferred over closed photobioreactors for cost-effective and large-scale microalgae cultivation, due to the simple design, proper aeration, ease of operation, and effective scalability, along with fewer energy requirements [76]. In this context, open pond systems, especially HRAPs, are ideal for wastewater-based microalgae cultivation and integration with conventional WWTPs. Nevertheless, the operating parameters, such as the cultivation period or HRT, of microalgae need to be optimized for the enhancement of biomass productivity while achieving high removal efficiencies. In addition, biomass harvesting is associated with high operation costs as it involves chemicals, materials, and energy-intensive equipment. For instance, centrifugation for biomass harvesting is a high-efficiency method with 80–90% biomass recovery, whereas the energy requirement is intensive, i.e., ~1 MJ kg^{-1} of dry biomass [77]. Similarly, chemical flocculants, i.e., metal salts, such as $Al_2(SO_4)_3$, $Fe_2(SO_4)_3$, $FeCl_3$, $MgCl_2$, $AlCl_3$, and synthetic polyacrylamides polymers, for biomass harvesting have a high risk of contaminating the microalgae biomass with residual chemicals. For example, 205.3 $mg_{Fe}\ g_{dry\ biomass}{}^{-1}$ of iron contamination was showcased during the harvesting of *Dunaliella* sp. with chemical flocculants namely $Al_2(SO_4)_3$ and $FeCl_3$ [78]. Thus, non-toxic bioflocculants, such as cationic starch, chitosan, and bacteria, are promising low-cost alternatives which are highly explored in recent studies.

The wastewater characteristics have severe effects on microalgal biomass productivity and metabolite synthesis. Variations in nutrient availability are a major concern since wastewater is used as an alternative culture medium during wastewater-based microalgae cultivation. Significantly, C:N:P ratio affects the microalgal growth and efficiency of nutrient removal during the bioremediation of wastewater [79]. For instance, the C:N ratio during mixotrophic cultivation of microalgae influences the biomass yield and accumulation of lipids [80]. Low nutrient availability would lead to limited microalgal growth, resulting in lower biomass and targeted metabolite yields. For instance, very low quantities of organic C are present in the secondary treated effluent due to the biodegradation of pollutants during the secondary treatment processes [81]. This could be mitigated by supplementing cost-effective inorganic chemicals and commercial fertilizers, and through the mixing of two or more wastewater types. On the other hand, a high organic load in wastewater would result in lower biomass production by inhibiting microalgal cell growth due to substrate inhibition, along with increasing the probability of biological contamination. Dilution of wastewater prior to microalgal cultivation is a promising solution to overcome the high organic load.

In general, high amounts of suspended solids are present in agricultural and industrial wastewater, which increases the turbidity of the medium. This is one of the major concerns during microalgae-based wastewater bioremediation, as it hinders the availability of light and causes the microalgal cells to agglomerate. Notably, microalgal cultivation in primary effluent from conventional WWT would have a high risk of suspended solids content. As a result, the microalgal growth and metabolite yields are decreased, whereas the nutrient removal performance would be severely affected. Hence, pretreatment techniques, including filtration, gravity settling, coagulation, and flocculation, are required prior to microalgae cultivation to reduce the suspended solids content. Moreover, heterotrophic cultivation using microalgal strains growing well under heterotrophic conditions could be a viable solution to overcome the low light availability [82].

In addition, microalgae cultivation in wastewater is associated with the critical risk of biomass contamination with heavy metals, namely Cd, Mercury (Hg), Pb, Cr, and As which are toxic and carcinogenic to humans. For instance, the maximum contamination limits of Cd, Hg, Pb, Cr, and As are 0.005, 0.002, 0.015, 0.1, and 0.01 mg L^{-1}, respectively. Mostly, wastewater streams from textile, agricultural, and chemical processing industries contain heavy metals. This has negative implications for food- and nutraceutical-based applications of the microalgal biomass and extracts obtained through wastewater bioremediation. Thus, it is crucial to remove heavy metals from wastewater prior to microalgae cultivation, especially to mitigate toxic effects for human consumption. Numerous technologies, including adsorption, coagulation and flocculation, ion exchange, membrane filtration, and precipitation, are commonly used for heavy metal removal from wastewater [17]. Nevertheless, chemical-based heavy metal removal techniques are least referred to due to alterations in the nutritional composition. Meanwhile, most of the physical methods are complex and associated with high operational costs. Thus, studies on the techno-economic feasibility of the heavy removal techniques are essential for the selection of appropriate methods to integrate during the WWT.

The presence of pathogens, such as bacteria, fungi, viruses, and parasites, is another major concern during wastewater-based microalgae cultivation, due to the negative effect on microalgal growth resulting from predation and competition for nutrients [83]. For instance, bacterial species belonging to 36 phyla, 293 families, and 579 genera were identified in activated sludge samples collected from four WWTPs [84]. Hence, it is important to reduce the pathogenic load in wastewater prior to microalgae cultivation. Physical (i.e., membrane filtration, ultra-violet irradiation, and pasteurization) and chemical (i.e., chlorination, ozonation, and alkaline treatment) methods of disinfecting wastewater could be used to minimize the pathogenic load. Nevertheless, the selection of a wastewater sterilization technique highly depends on the process complexity, effect on the nutritional composition of wastewater, removal efficiency, and capital and operational costs. Hence, techno-economic studies are required prior to implementing wastewater disinfection methods for the selection of sustainable and economically feasible techniques for real-time integration into WWT. Moreover, utilizing microalgae strains isolated from the same environment of wastewater collection would facilitate well-acclimatized cultivation due to the robustness of the strains [85]. Thus, it is evident that further state-of-the-art research and developments on the aforementioned techniques to overcome the challenges are essential in achieving real-world applications.

6 Conclusion

Numerous wastewater streams contain high quantities of nutrients, mainly C, N and P, which are essential for microalgal cultivation. Thus, wastewater is a promising alternative nutrient source for the generation of valuable microalgae biomass, while achieving concurrent bioremediation. Notably, high removal efficiencies, i.e., >80%, have been reported from microalga cultivation, thus facilitating it as a primary mode of WWT. Significantly, microalgae-based bioremediation could be an alternative treatment method for secondary WWT in conventional plants to treat organic and inorganic nutrient loads. Similarly, integration of microalgae cultivation after the secondary treatment, i.e., tertiary/advanced treatment, would result in the removal of pollutants and pathogens to non-detectable levels whilst enhancing the effluent water quantity. In addition, microalgae biomass derived from wastewater bioremediation is a promising feedstock in the diversified production of bioproducts within the framework of a circular bioeconomy. Nevertheless, various challenges associated with the utilization of wastewater and integration of microalgae cultivation into conventional WWTPs require future studies, especially based on techno-economic feasibility for real-time applicability.

7 Recommendations

- Comprehensive studies on techno-economic assessments and life cycle analysis are required for the real-time feasibility evaluations of wastewater-integrated microalgae cultivation in HRAPs, to identify the critical processing steps and infrastructure requirements.
- Moreover, the scalability potential and operational challenges must be identified through mathematical modeling and field trials for the real-world implementation of microalgae cultivation into WWTPs.
- Also, future research must focus on the identification of integration strategies for microalgae cultivation with the existing WWTPs.
- Future studies must focus on the employment of microalgae-bacteria consortia as an alternative to the conventional secondary treatment process at WWTPs.
- Exploration studies must aim to isolate and identify microalgal strains from the WWTP sites with superior removal efficiencies for ecosystem conservation.
- Collaborations between researchers, industrial stakeholders, and policymakers need to be established for the progression of the development and adoption of microalgae into WWTPs.

Acknowledgements This work was supported by the National Research Council of Sri Lanka (Grant number: 20–107) and the Senate Research Committee of University of Moratuwa, Sri Lanka (Grant number: SRC/CAP/2020/01).

References

1. Jones ER, van Vliet MTH, Qadir M, Bierkens MFP (2021) Country-level and gridded estimates of wastewater production, collection, treatment and reuse. Earth Syst Sci Data 13:237–254. https://doi.org/10.5194/essd-13-237-2021
2. Premaratne M, Nishshanka GKSH, Anthonio RADP, Liyanaarachchi VC, Thevarajah B, Nimarshana PHV, Malik A, Ariyadasa TU (2022) Resource recovery from waste streams for production of microalgae biomass: a sustainable approach towards high-value biorefineries. Bioresour Technol Rep 18. https://doi.org/10.1016/j.biteb.2022.101070
3. Thevarajah B, Nimarshana PHV, Shehan Sandeepa GD, Boopathy R, Ariyadasa TU (2024a) Upcycling food processing industrial wastes in Spirulina cultivation: a perspective on the advancement of Sustainable Development Goal 12. Trends Food Sci Technol 104537. https://doi.org/10.1016/j.tifs.2024.104537
4. Obaideen K, Shehata N, Sayed ET, Abdelkareem MA, Mahmoud MS, Olabi AG (2022) The role of wastewater treatment in achieving sustainable development goals (SDGs) and sustainability guideline. Energy Nexus 7:100112. https://doi.org/10.1016/j.nexus.2022.100112
5. Chawla P, Gola D, Dalvi V, Sreekrishnan TR, Ariyadasa TU, Malik A (2022) Design and development of mini-photobioreactor system for strategic high throughput selection of optimum microalgae-wastewater combination. Bioresour Technol Rep 17:100967. https://doi.org/10.1016/j.biteb.2022.100967

6. Devi A, Verma M, Saratale GD, Saratale RG, Ferreira LFR, Mulla SI, Bharagava RN (2023) Microalgae: a green eco-friendly agents for bioremediation of tannery wastewater with simultaneous production of value-added products. Chemosphere 139192. https://doi.org/10.1016/j.chemosphere.2023.139192
7. Olejnik D (2024) Evaluation of the heavy metals content in sewage sludge from selected rural and urban wastewater treatment plants in Poland in terms of its suitability for agricultural use. Sustainability 16:5198. https://doi.org/10.3390/su16125198
8. Nishshanka GKSH, Liyanaarachchi VC, Premaratne M, Nimarshana PHV, Ariyadasa TU, Kornaros M (2021) Wastewater-based microalgal biorefineries for the production of astaxanthin and co-products: current status, challenges and future perspectives. Bioresour Technol 342:126018. https://doi.org/10.1016/j.biortech.2021.126018
9. Velásquez-Orta SB, Yáñez-Noguez I, Ramírez IM, Ledesma MTO (2024) Pilot-scale microalgae cultivation and wastewater treatment using high-rate ponds: a meta-analysis. Environ Sci Pollut Res. https://doi.org/10.1007/s11356-024-34000-7
10. Jiao H, He X, Sun J, Elsamahy T, Al-Tohamy R, Kornaros M, Ali SS (2024) A critical review on sustainable biorefinery approaches and strategies for wastewater treatment and production of value-added products. Energy Ecol Environ. https://doi.org/10.1007/s40974-023-00312-6
11. Thevarajah B, Piyatilleke S, Nimarshana PHV, Koushalya S, Malik A, Ariyadasa TU (2024) Exploring effective light spectral conversion techniques for enhanced production of Spirulina-derived blue pigment protein, c-phycocyanin. Bioresour Technol 399:130612. https://doi.org/10.1016/j.biortech.2024.130612
12. Nobre MLF, Tavares D, Fraga C, Oliveira B, Dias M, Mesquita S, Oliveira CM, Pires JCM (2024) Techno-economic analysis of a circular microalgal approach for enhanced wastewater treatment and resource recovery in Northern Portugal. J Clean Prod 434:140389. https://doi.org/10.1016/j.jclepro.2023.140389
13. Nishshanka GKSH, Liyanaarachchi VC, Premaratne M, Ariyadasa TU, Nimarshana PHV (2022) Sustainable cultivation of *Haematococcus pluvialis* and *Chromochloris zofingiensis* for the production of astaxanthin and co-products. Can J Chem Eng. https://doi.org/10.1002/cjce.24317
14. Salgot M, Folch M (2018) Wastewater treatment and water reuse. Curr Opin Environ Sci Health 2:64–74. https://doi.org/10.1016/j.coesh.2018.03.005
15. Badawi AK, Ismail B, Baaloudj O, Abdalla KZ (2022) Advanced wastewater treatment process using algal photo-bioreactor associated with dissolved-air flotation system: a pilot-scale demonstration. J Water Process Eng 46:102565. https://doi.org/10.1016/j.jwpe.2022.102565
16. Hwang J-H, Church J, Lee S-J, Park J, Lee WH (2016) Use of microalgae for advanced wastewater treatment and sustainable bioenergy generation. Environ Eng Sci 33:882–897. https://doi.org/10.1089/ees.2016.0132
17. Liyanaarachchi VC, Premaratne M, Ariyadasa TU, Nimarshana PHV, Malik A (2021) Two-stage cultivation of microalgae for production of high-value compounds and biofuels: a review. Algal Res 57:102353. https://doi.org/10.1016/j.algal.2021.102353
18. Kurniawan KIA, Putra AS, Ishizaki R, Rani DS, Rahmah DM, Al Husna SN, Ahamed T, Noguchi R (2024) Life cycle assessment of integrated microalgae oil production in Bojongsoang Wastewater Treatment Plant, Indonesia. Environ Sci Pollut Res 31:7902–7933. https://doi.org/10.1007/s11356-023-31582-6
19. Mohsenpour SF, Hennige S, Willoughby N, Adeloye A, Gutierrez T (2021) Integrating microalgae into wastewater treatment: a review. Sci Total Environ 752:142168. https://doi.org/10.1016/j.scitotenv.2020.142168
20. Chen C-Y, Kuan S-P, Nagarajan D, Chen J-H, Ariyadasa TU, Chang J-S (2022) A novel two-stage process for the effective treatment of swine wastewater using *Chlorella sorokiniana* AK−1 based algal-bacterial consortium under semi-continuous operation. Bioresour Technol 365:128119. https://doi.org/10.1016/j.biortech.2022.128119
21. Ahiahonu EK, Anku WW, Roopnarain A, Serepa-Dlamini MH, Govender PP (2024) Evaluation of green microalgal strains for simultaneous municipal wastewater treatment and heavy metal bioremediation. J Chem Technol Biotechnol. https://doi.org/10.1002/jctb.7705

22. Bhatt A, Sahu N, Dada AC, Kumar Prajapati S, Arora P (2024) Assessing sustainability of microalgae-based wastewater treatment: environmental considerations and impacts on human health. J Environ Manage 354:120435. https://doi.org/10.1016/j.jenvman.2024.120435
23. Thevarajah B, Nishshanka GKSH, Premaratne M, Wasath WAJ, Nimarshana PHV, Malik A, Ariyadasa TU (2023) Cyanobacterial pigment production in wastewaters treated for heavy metal removal: current status and perspectives. J Environ Chem Eng 11:108999. https://doi.org/10.1016/j.jece.2022.108999
24. Mustafa S, Bhatti HN, Maqbool M, Iqbal M (2021) Microalgae biosorption, bioaccumulation and biodegradation efficiency for the remediation of wastewater and carbon dioxide mitigation: prospects, challenges and opportunities. J Water Process Eng 41:102009. https://doi.org/10.1016/j.jwpe.2021.102009
25. Satya ADM, Cheah WY, Yazdi SK, Cheng Y-S, Khoo KS, Vo D-VN, Bui XD, Vithanage M, Show PL (2023) Progress on microalgae cultivation in wastewater for bioremediation and circular bioeconomy. Environ Res 218:114948. https://doi.org/10.1016/j.envres.2022.114948
26. Premaratne M, Nishshanka G, Liyanaarachchi V, Nimarshana P, Ariyadasa TU (2021) Bioremediation of textile dye wastewater using microalgae: current trends and future perspectives. J Chem Technol Biotechnol 96:3249–3258. https://doi.org/10.1002/jctb.6845
27. Adesra A, Srivastava VK, Varjani S (2021) Valorization of dairy wastes: integrative approaches for value added products. Indian J Microbiol 61:270–278. https://doi.org/10.1007/s12088-021-00943-5
28. Wrzesińska-Jędrusiak E, Czarnecki M, Kazimierski P, Bandrów P, Szufa S (2023) The circular economy in the management of waste from leather processing. Energies (Basel) 16:564. https://doi.org/10.3390/en16010564
29. Toczyłowska-Mamińska R (2017) Limits and perspectives of pulp and paper industry wastewater treatment—a review. Renew Sustain Energy Rev 78:764–772. https://doi.org/10.1016/j.rser.2017.05.021
30. Wang X, Jiang J, Gao W (2022) Reviewing textile wastewater produced by industries: characteristics, environmental impacts, and treatment strategies. Water Sci Technol 85:2076–2096. https://doi.org/10.2166/wst.2022.088
31. Gao Z, Gao K, Li L, Yang M, Cheng P, Zheng Q, Zhang M, Ning R, Liu Q, Ruan R, Deng X (2023) Feasibility of growing microalgae on wastewaters collected from a vinegar plant for, concurrently, nutrient removal and biomass production. Algal Res 74:103188. https://doi.org/10.1016/j.algal.2023.103188
32. Chandra R, Pradhan S, Patel A, Ghosh UK (2021) An approach for dairy wastewater remediation using mixture of microalgae and biodiesel production for sustainable transportation. J Environ Manage 297:113210. https://doi.org/10.1016/j.jenvman.2021.113210
33. Hemalatha M, Venkata Mohan S (2016) Microalgae cultivation as tertiary unit operation for treatment of pharmaceutical wastewater associated with lipid production. Bioresour Technol 215:117–122. https://doi.org/10.1016/j.biortech.2016.04.101
34. Jiménez-Bambague EM, Madera-Parra CA, Ortiz-Escobar AC, Morales-Acosta PA, Peña-Salamanca EJ, Machuca-Martínez F (2020) High-rate algal pond for removal of pharmaceutical compounds from urban domestic wastewater under tropical conditions. Case study: Santiago de Cali, Colombia. Water Sci Technol 82:1031–1043. https://doi.org/10.2166/wst.2020.362
35. Arutselvan C, Narchonai G, Pugazhendhi A, Seenivasan H kumar, LewisOscar F, Thajuddin N (2022) Phycoremediation of textile and tannery industrial effluents using microalgae and their consortium for biodiesel production. J Clean Prod 367:133100. https://doi.org/10.1016/j.jclepro.2022.133100
36. Lee S-A, Lee N, Oh H-M, Ahn C-Y (2021) Stepwise treatment of undiluted raw piggery wastewater, using three microalgal species adapted to high ammonia. Chemosphere 263:127934. https://doi.org/10.1016/j.chemosphere.2020.127934
37. Umamaheswari J, Shanthakumar S (2019) Phycoremediation of paddy-soaked wastewater by indigenous microalgae in open and closed culture system. J Environ Manage 243:435–443. https://doi.org/10.1016/j.jenvman.2019.05.023

38. Yang X, Liu L, Liu X, Xie S, Feng J, Lv J (2023) The responding mechanism of indigenous bacteria in municipal wastewater inoculated with different concentrations of exogenous microalgae. J Environ Manage 345:118547. https://doi.org/10.1016/j.jenvman.2023.118547
39. Li X, Liu J, Tian J, Pan Z, Chen Y, Ming F, Wang R, Wang L, Zhou H, Li J, Tan Z (2023) Co-cultivation of microalgae-activated sludge for municipal wastewater treatment: exploring the performance, microbial co-occurrence patterns, microbiota dynamics and function during the startup stage. Bioresour Technol 374:128733. https://doi.org/10.1016/j.biortech.2023.128733
40. Wang S-K, Yang K-X, Zhu Y-R, Zhu X-Y, Nie D-F, Jiao N, Angelidaki I (2022) One-step co-cultivation and flocculation of microalgae with filamentous fungi to valorize starch wastewater into high-value biomass. Bioresour Technol 361:127625. https://doi.org/10.1016/j.biortech.2022.127625
41. Goswami RK, Agrawal K, Verma P (2022) An exploration of natural synergy using microalgae for the remediation of pharmaceuticals and xenobiotics in wastewater. Algal Res 64:102703. https://doi.org/10.1016/j.algal.2022.102703
42. Hena S, Gutierrez L, Croué J-P (2021) Removal of pharmaceutical and personal care products (PPCPs) from wastewater using microalgae: a review. J Hazard Mater 403:124041. https://doi.org/10.1016/j.jhazmat.2020.124041
43. Rajalakshmi AM, Silambarasan T, Dhandapani R (2021) Small scale photo bioreactor treatment of tannery wastewater, heavy metal biosorption and CO_2 sequestration using microalga *Chlorella* sp.: a biodegradation approach. Appl Water Sci 11:108. https://doi.org/10.1007/s13201-021-01438-w
44. UNICEF, World Health Organization (2020) State of the World's sanitation: an urgent call to transform sanitation for better health, environments, economies and societies, New York
45. Vaz SA, Badenes SM, Pinheiro HM, Martins RC (2023) Recent reports on domestic wastewater treatment using microalgae cultivation: towards a circular economy. Environ Technol Innov 30:103107. https://doi.org/10.1016/j.eti.2023.103107
46. Silambarasan S, Logeswari P, Sivaramakrishnan R, Incharoensakdi A, Cornejo P, Kamaraj B, Chi NTL (2021) Removal of nutrients from domestic wastewater by microalgae coupled to lipid augmentation for biodiesel production and influence of deoiled algal biomass as biofertilizer for *Solanum lycopersicum* cultivation. Chemosphere 268:129323. https://doi.org/10.1016/j.chemosphere.2020.129323
47. Yang L, Li H, Wang Q (2019) A novel one-step method for oil-rich biomass production and harvesting by co-cultivating microalgae with filamentous fungi in molasses wastewater. Bioresour Technol 275:35–43. https://doi.org/10.1016/j.biortech.2018.12.036
48. Sorenson C, Ding L, Zamalloa C, Arsenault A, Debeni Devi N, Hu B (2023) Illuminated septic tank—a microalgae assisted onsite wastewater treatment system. J Environ Chem Eng 11:110768. https://doi.org/10.1016/j.jece.2023.110768
49. Subramaniyam V, Subashchandrabose SR, Ganeshkumar V, Thavamani P, Chen Z, Naidu R, Megharaj M (2016) Cultivation of *Chlorella* on brewery wastewater and nano-particle biosynthesis by its biomass. Bioresour Technol 211:698–703. https://doi.org/10.1016/j.biortech.2016.03.154
50. Wang Q, Wang X, Hong Y, Liu X, Zhao G, Zhang H, Zhai Q (2022) Microalgae cultivation in domestic wastewater for wastewater treatment and high value-added production: species selection and comparison. Biochem Eng J 185:108493. https://doi.org/10.1016/j.bej.2022.108493
51. Ferro L, Gorzsás A, Gentili FG, Funk C (2018) Subarctic microalgal strains treat wastewater and produce biomass at low temperature and short photoperiod. Algal Res 35:160–167. https://doi.org/10.1016/j.algal.2018.08.031
52. Berrio-Restrepo JM, Saldarriaga-Molina JC, Correa-Ochoa MA (2022) A natural microalgae consortium for the biological nutrient removal in a upflow photobioreactor as tertiary wastewater treatment. Bioresour Technol Rep 20:101240. https://doi.org/10.1016/j.biteb.2022.101240
53. Wang P, Shao Y, Geng Y, Mushtaq R, Yang W, Li M, Sun X, Wang H, Chen G (2023) Advanced treatment of secondary effluent from wastewater treatment plant by a newly isolated microalga *Desmodesmus* sp. SNN1. Front Microbiol 14. https://doi.org/10.3389/fmicb.2023.1111468

54. Eida M, Darwesh O, Matter I (2018) Cultivation of oleaginous microalgae *Scenedesmus obliquus* on secondary treated municipal wastewater as growth medium for biodiesel production. J Ecol Eng 19:38–51. https://doi.org/10.12911/22998993/91274
55. Gao F, Cui W, Xu J-P, Li C, Jin W-H, Yang H-L (2019) Lipid accumulation properties of *Chlorella vulgaris* and *Scenedesmus obliquus* in membrane photobioreactor (MPBR) fed with secondary effluent from municipal wastewater treatment plant. Renew Energy 136:671–676. https://doi.org/10.1016/j.renene.2019.01.038
56. Solmaz A, Işık M (2019) Polishing the secondary effluent and biomass production by microalgae submerged membrane photo bioreactor. Sustainable Energy Technol Assess 34:1–8. https://doi.org/10.1016/j.seta.2019.04.002
57. Al Momani FA, Örmeci B (2016) Performance of *Chlorella vulgaris*, *Neochloris oleoabundans*, and mixed indigenous microalgae for treatment of primary effluent, secondary effluent and centrate. Ecol Eng 95:280–289. https://doi.org/10.1016/j.ecoleng.2016.06.038
58. Li S, Qu W, Chang H, Li J, Ho S-H (2022) Microalgae-driven swine wastewater biotreatment: nutrient recovery, key microbial community and current challenges. J Hazard Mater 440:129785. https://doi.org/10.1016/j.jhazmat.2022.129785
59. Leong YK, Su R-H, Varjani S, Lee D-J, Chang J-S (2023) Integrated biohydrogen production and dairy manure wastewater treatment via a microalgae platform. Int J Hydrogen Energy. https://doi.org/10.1016/j.ijhydene.2023.06.308
60. Chang Y-L, Nagarajan D, Chen J-H, Yen Chen C, Wu Y-J, Whang L-M, Lee D-J, Chang J-S (2023) Microalgae-bacteria consortia for the treatment of raw dairy manure wastewater using a novel two-stage process: process optimization and bacterial community analysis. Chem Eng J 473:145388. https://doi.org/10.1016/j.cej.2023.145388
61. Chu Y, Zhang C, Wang R, Chen X, Ren N, Ho S-H (2022) Biotransformation of sulfamethoxazole by microalgae: removal efficiency, pathways, and mechanisms. Water Res 221:118834. https://doi.org/10.1016/j.watres.2022.118834
62. Geremia E, Ripa M, Catone CM, Ulgiati S (2021) A review about microalgae wastewater treatment for bioremediation and biomass production—a new challenge for Europe. Environments 8:136. https://doi.org/10.3390/environments8120136
63. Tan YH, Chai MK, Na JY, Wong LS (2023) Microalgal growth and nutrient removal efficiency in non-sterilised primary domestic wastewater. Sustainability 15:6601. https://doi.org/10.3390/su15086601
64. Yukiyo Y, Hiroyuki S (2021) Natural microalgal cultivation systems using primary effluent and excess sludge. Environ Technol 42:3907–3919. https://doi.org/10.1080/09593330.2020.1753817
65. de Lima Barizão AC, de Oliveira Gomes LE, Brandão LL, Sampaio ICF, de Moura IVL, Gonçalves RF, de Oliveira JP, Cassini ST (2023) Microalgae as tertiary wastewater treatment: energy production, carbon neutrality, and high-value products. Algal Res 72:103113. https://doi.org/10.1016/j.algal.2023.103113
66. Biliani SE, Manariotis ID (2023) Wastewater treatment by high density algal flocs for nutrient removal and biomass production. J Appl Phycol 35:1237–1250. https://doi.org/10.1007/s10811-023-02931-2
67. Malla FA, Khan SA, Rashmi SGK, Gupta N, Abraham G (2015) Phycoremediation potential of *Chlorella minutissima* on primary and tertiary treated wastewater for nutrient removal and biodiesel production. Ecol Eng 75:343–349. https://doi.org/10.1016/j.ecoleng.2014.11.038
68. Devi ND, Sun X, Ding L, Goud VV, Hu B (2022) Mixotrophic growth regime of novel strain *Scenedesmus* sp. DDVG I in municipal wastewater for concomitant bioremediation and valorization of biomass. J Clean Prod 365:132834. https://doi.org/10.1016/j.jclepro.2022.132834
69. Acebu PIG, de Luna MDG, Chen C-Y, Abarca RRM, Chen J-H, Chang J-S (2022) Bioethanol production from Chlorella vulgaris ESP-31 grown in unsterilized swine wastewater. Bioresour Technol 352:127086. https://doi.org/10.1016/j.biortech.2022.127086
70. Das P, Quadir MA, Thaher MI, Alghasal GSHS, Aljabri HMSJ (2019) Microalgal nutrients recycling from the primary effluent of municipal wastewater and use of the produced biomass

as bio-fertilizer. Int J Environ Sci Technol 16:3355–3364. https://doi.org/10.1007/s13762-018-1867-8
71. Silambarasan S, Logeswari P, Sivaramakrishnan R, Incharoensakdi A, Cornejo P, Kamaraj B, Lan Chi NT (2021) Removal of nutrients from domestic wastewater by microalgae coupled to lipid augmentation for biodiesel production and influence of deoiled algal biomass as biofertilizer for *Solanum lycopersicum* cultivation. Chemosphere 268:129323. https://doi.org/10.1016/j.chemosphere.2020.129323
72. Viegas C, Gouveia L, Gonçalves M (2021) Aquaculture wastewater treatment through microalgal. Biomass potential applications on animal feed, agriculture, and energy. J Environ Manage 286. https://doi.org/10.1016/j.jenvman.2021.112187
73. de Morais EG, Murillo AM, Lens PNL, Ferrer I, Uggetti E (2022) Selenium recovery from wastewater by the green microalgae *Chlorella vulgaris* and *Scenedesmus* sp. Sci Total Environ 851:158337. https://doi.org/10.1016/j.scitotenv.2022.158337
74. Fernando JSR, Premaratne M, Dinalankara DMSD, Perera GLNJ, Ariyadasa TU (2021) Cultivation of microalgae in palm oil mill effluent (POME) for astaxanthin production and simultaneous phycoremediation. J Environ Chem Eng 9:105375. https://doi.org/10.1016/j.jece.2021.105375
75. Ariyadasa TU, Thevarajah B, Anthonio RADP, Nimarshana PHV, Wasath WAJ (2023) From present to prosperity: assessing the current status and envisioning opportunities in the industrial-scale cultivation of *Haematococcus pluvialis* for astaxanthin production. Phytochem Rev. https://doi.org/10.1007/s11101-023-09906-8
76. Thevarajah B, Nishshanka GKSH, Premaratne M, Nimarshana PHV, Nagarajan D, Chang J, Ariyadasa TU (2022) Large-scale production of Spirulina-based proteins and c-phycocyanin: a biorefinery approach. Biochem Eng J 185:108541. https://doi.org/10.1016/j.bej.2022.108541
77. Ahmad S, Iqbal K, Kothari R, Singh HM, Sari A, Tyagi VV (2022) A critical overview of upstream cultivation and downstream processing of algae-based biofuels: opportunity, technological barriers and future perspective. J Biotechnol 351:74–98
78. Pirwitz K, Rihko-Struckmann L, Sundmacher K (2015) Comparison of flocculation methods for harvesting Dunaliella. Bioresour Technol 196:145–152. https://doi.org/10.1016/j.biortech.2015.07.032
79. Shashirekha V, Sivakumar M, Seshadri S (2016) Effective C–N–P ratio for growth and nutrient removal efficiency of *Scenedesmus obliquus* in sugar mill effluent. Energy Ecol Environ 1:283–295. https://doi.org/10.1007/s40974-016-0040-9
80. Khanra A, Vasistha S, Kumar P, Rai MP (2020) Role of C/N ratio on microalgae growth in mixotrophy and incorporation of titanium nanoparticles for cell flocculation and lipid enhancement in economical biodiesel application. 3 Biotech 10:331. https://doi.org/10.1007/s13205-020-02323-0
81. Lv J, Feng J, Liu Q, Xie S (2017) Microalgal cultivation in secondary effluent: recent developments and future work. Int J Mol Sci 18:79. https://doi.org/10.3390/ijms18010079
82. da Silva TL, Moniz P, Silva C, Reis A (2021) The role of heterotrophic microalgae in waste conversion to biofuels and bioproducts. Processes 9:1090. https://doi.org/10.3390/pr9071090
83. Thevarajah B, Piyathilleke S, Sahu A, Nimarshana PH V., Malik A, Ariyadasa TU (2024b) Microalgae-based bioproducts and biomaterials towards a sustainable circular bioeconomy. In: Bioeconomy for sustainability. Springer Nature, Singapore, pp 125–162
84. Xu S, Yao J, Ainiwaer M, Hong Y, Zhang Y (2018) Analysis of bacterial community structure of activated sludge from wastewater treatment plants in winter. Biomed Res Int 2018:1–8. https://doi.org/10.1155/2018/8278970
85. Nishshanka GKSH, Thevarajah B, Nimarshana PHV, Prajapati SK, Ariyadasa TU (2023) Real-time integration of microalgae-based bioremediation in conventional wastewater treatment plants: current status and prospects. J Water Process Eng 56:104248. https://doi.org/10.1016/j.jwpe.2023.104248

New Trends in Wastewater: Technologies, Challenges and Opportunity

Treatment of Liquid Waste Loaded with Pharmaceutical Products: A Case of Antibiotics

Amina Khalidi Idrissi, Abdelaziz Madinzi, Latifa Mouhir, Safaa Khattabi Rifi, Roukaya Bouyakhsass, Abdelkader Anouzla, Nor-edine Abriak, and Salah Souabi

Abstract The presence of pharmaceutical compounds (PCs) in aqueous systems poses a significant threat to ecological systems. Additionally, antibiotic contamination poses a risk to microbial communities and human health. If antibiotic-resistant bacteria develop, this could lead to an increase in multi-resistant bacterial infections in the coming days and have serious consequences for human health, ecosystems and the environment. One of the most prospective approaches consists of combining biological and physicochemical methods to treat these effluents. Accordingly, hybrid systems must be fabricated by mixing biological and physicochemical methods. In some cases, these hybrid processes help improve pharmaceutical disposal performance by overcoming each disadvantage of one method over the other. They appear to be a potential means of reducing the impact of pharmaceutical emissions on the environment, ecosystems, wastewater treatment plants and public health. This is about better and more sustainable management of this complex environmental issue.

Keywords Pharmaceutical wastewater · Antibiotics · Physicochemical · Biological treatment

A. Khalidi Idrissi · A. Madinzi · L. Mouhir · S. K. Rifi · R. Bouyakhsass · A. Anouzla · S. Souabi (✉)
Laboratory of Process Engineering and Environment, Faculty of Science and Technology, University Hassan II, BP 146, 28806 Mohammedia, Morocco
e-mail: salah.souabi@gmail.comssss

A. Madinzi
e-mail: abdelaziz.madinzi@fstm.ac.ma

N. Abriak
2IMT Lille Douai, GCE, 764 Bd Lahure, BP 10838, 59508 Douai, France
e-mail: nor-edine.abriak@imt-lille-douai.fr

S. Yadav et al. (eds.), *Emerging Technologies in Biological and Hybrid Wastewater Treatment: Lessons from Developed to Enhancing Practices in Developing Countries*, Earth and Environmental Sciences Library,
https://doi.org/10.1007/978-3-031-74515-7_10

1 Introduction

The growth of the pharmaceutical industry has generated a large amount of liquid waste that must be appropriately disposed of globally [43, 5]. In-depth research has enabled scientists to identify substances harmful to the environment, including drugs, cosmetics, endocrine disruptors, and antibiotics. These toxic pollutants threaten human health, marine life, and ecosystems [83].

Conventional wastewater treatment technologies cannot eliminate trace amounts of these newly discovered pharmaceutical contaminants [21]. This inefficiency results in the leaching of pollutants into rivers, streams, and groundwater. Therefore, to achieve effective removal, wastewater treatment plants need to employ advanced treatment methods such as electrocoagulation [31], oxidation treatment [71], reverse osmosis [53], nanofiltration [61], ultrafiltration [13], membrane bioreactors [68], and hybrid technologies [51].

Ahmed et al. [8] emphasized that to remove micropollutants from pharmaceutical wastewater more effectively, it is essential to adopt modern wastewater treatment methods. In particular, denitrification using biofilm reactors has demonstrated micropollutant removal efficiencies ranging from 15 to 42%.

Dutta and Bhattacharjee [25] argue that the advantages and disadvantages of biological and physicochemical treatments are different. For example, the microbial communities used in biotherapy can become toxic because biotherapy can often not properly mineralize the drug [52]. However, complex treatments can be costly and energy-intensive [58]. The use of chemicals and high pH in advanced oxidation processes can put additional financial pressure on treatment processes, noted Yu et al. [88].

Eniola et al. [26] focused on the integration of biological and physicochemical treatments to develop cost-effective wastewater treatment systems. Furthermore, Jhunjhunwala et al. [43] employed a combined approach that involved three-dimensional electrolysis and ozonation, which yielded promising outcomes. Specifically, within a span of five days, they achieved a remarkable 64% decrease in total organic carbon, a 70% decrease in chemical oxygen demand, and an impressive 97% reduction in biochemical oxygen demand.

2 Problems of Liquid Discharges Loaded with Antibiotics

Liquid discharges containing antibiotics can contribute to many significant issues in multiple domains, including the environment, human health, ecosystems, and wastewater treatment plants [65]. These problems stem from the widespread use of antibiotics in medical practices and agriculture, which releases these potent compounds into the environment via sewage and other channels. Environmentally, antibiotics contaminate natural water bodies such as rivers and lakes, disrupting aquatic ecosystems. They can also seep into the soil by various means, potentially endangering soil organisms and contaminating crops [62].

One of the most pressing concerns is the development of antibiotic-resistant bacteria in the environment, resulting from the persistent presence of antibiotic residues. This alarming development poses a formidable threat to public health, making it increasingly difficult to combat common diseases [64]. Beyond that, exposure to water or food tainted with antibiotic remnants can lead to antibiotic-resistant infections in humans, complicating treatment strategies [67]. In some cases, sensitive individuals may also experience allergic reactions due to the presence of antibiotic residues in drinking water or food, adding to the complexity of the problem. Moreover, certain antibiotics can disrupt the endocrine system, potentially causing hormonal imbalances in humans [1].

The ramifications extend further into ecosystems, where antibiotic contamination can disrupt microbial communities in both soil and water. This disruption can significantly affect nutrient cycling and the overall functioning of these ecosystems [46]. Aquatic organisms may suffer reduced reproductive success when exposed to antibiotics, leading to population declines. At the same time, some species may display greater tolerance to antibiotic exposure, leading to changes in ecosystem composition [10].

Many wastewater treatment plants are ill-equipped to effectively remove antibiotics from wastewater, allowing these compounds to be released into the environment [11]. Upgrading such facilities to solve this problem can be prohibitively expensive, impacting consumers. Furthermore, managing wastewater containing antibiotics can be operationally demanding, often requiring advanced treatment methods [81].

To mitigate these large-scale problems, it is imperative to adopt comprehensive strategies. These strategies should encompass improved wastewater treatment technologies, responsible antibiotic use in healthcare and agriculture, and increased awareness of antibiotic pollution's environmental and health risks [26]. Ongoing monitoring and research efforts are vital to assess the problem and to develop and implement effective mitigation measures [58].

3 Activated Sludge Treatment of Pharmaceutical Wastewater

The traditional biological activated sludge process is widely applied to purify wastewater the healthcare industry generates [66]. Investment costs and operational requirements are lower than for more sophisticated treatment techniques. Pollutants are transformed into digested gases and sludge without harming the environment [37]. In contrast, activated sludge plants are associated with operational problems such as excessive energy requirements and large quantities of sludge [15].

The activated sludge process with continuous feeding cannot efficiently remove the targeted drugs [77]. Using an aeration chamber reactor, a study of pharmaceutical wastewater removal, with calculated conventional pollutants removal rates ranging from 94.4% to 9.5%, was conducted by Kołecka et al. [48].

Activated sludge treatment gave a mineralization of 58.1%, with the degree of mineralization of the metronidazole being the highest under the combination of biostimulation and biomagnification. According to Aboudalle et al. [2], the mineralization rate was 96.1%.

To increase drug removal, Alfonso-Muniozguren et al. [12] proposed a benchtop reactor inoculated with activated sludge and fed with a synthetic medium to treat pharmaceuticals such as paracetamol and ibuprofen. According to Alfonso-Muniozguren et al. [12], removal efficiencies ranged from 99.1 to over 99.5% and were independent of the initial ibuprofen concentration. Paracetamol elimination rates ranged from 93.3 to 98.8%, decreasing with increasing initial dose.

4 Pharmaceutical Wastewater Treatment by Membrane Bioreactor

The membrane bioreactor has found wide application as an innovative approach to pharmaceutical wastewater treatment, as explained by Hosseinpour et al. [36]. This type of bioreactor offers many advantages, such as reduced sludge production, enhanced wastewater quality, superior removal of organic substances, longer sludge retention time, and increased resistance to toxic elements [9]. Nevertheless, Campo et al. [18] have highlighted a persistent issue related to membrane clogging, which reduces the porosity. This significantly affects membrane performance and service life and increases operating costs [23].

According to the analysis conducted by Guo et al. [32], Anaerobic Membrane Bio-Reactors (AnMBRs) can manage complex waste streams, generate energy, and produce nutrient-rich wastewater suitable for irrigation. In addition, Xu et al. [87] studied antibiotic removal using a sequential membrane bioreactor for widely used animal antibiotics, including tetracyclines, sulfonamides, and fluoroquinolones. Results indicated that the SMBR effectively removed 90% of sulfonamides and 90% of tetracyclines, with a slightly lower removal rate (70%) for fluoroquinolones.

5 Anaerobic Treatment for Pharmaceutical Wastewater

Anaerobic treatment is an environmentally friendly technology that can potentially recover methane energy [38]. The central wastewater anaerobic treatment systems include anaerobic batch sequencing reactors, membrane bioreactors [22], sludge bed flow [49], and anaerobic digesters [78].

It is commonly acknowledged that biological treatment techniques are the most cost-effective and successful means of extracting nitrogen from wastewater. This is because of their low production of byproducts, excellent efficiency, and affordability [89].

Sludge reduction through anaerobic digestion has become a viable and sustainable method. However, this produces concentrated wastewater that requires further treatment because it contains many nutrients and organic matter. In addition, biogas (the gaseous byproduct of anaerobic digestion systems) needs to be improved to increase its biomethane content and reduce carbon dioxide emissions, making it a more renewable energy source. This was the aspect that the research focused on [75].

Long-lived substances such as amitriptyline, clozapine, sulfamethoxazole, and detergents can be degraded in anaerobic systems with removal rates exceeding 80% [34]. Anaerobic treatment is limited by low mineralization efficiency, low degradation efficiency, and weak organic waste.

A hybrid treatment system for wastewater treatment was proposed to improve biological disposal. The results showed a 95% removal of pollutants [79]. Furthermore, to enhance the treatment of acetopyrimidine-laden wastewater, Wang et al. [86] studied a unique anaerobic ascending sludge using bioelectrocatalysis technology. The technique positively affected acetopyrimidine degradation, with higher removal and total carbon rates (96.3% and 92.9%, respectively). This combination provides a more suitable route and improves environmental performance.

6 Coagulation-Flocculation Treatment of Pharmaceutical Waste Liquids

The main objective of the coagulation-flocculation process is to remove colloidal pollutants and turbidity from wastewater. This goal is achieved by introducing flocculants, which are typically powerful chemicals that change the physical properties of wastewater [39]. After the coagulation stage, flocculants (usually polymers) are added to promote the formation of larger, denser flocs, thereby improving the separation process [35]. Mohamed Noor et al. [59] it emphasized the critical importance of a reasonable selection of coagulants and flocculants to maximize the efficiency of the treatment process.

Remove pollutants from wastewater from a pharmaceutical wastewater treatment plant using coagulation and flocculation. The treatment uses coagulants, specifically copper sulfate, ferric chloride, and a specific ratio ($CuSO_4 + FeCl_3$). A novel composite design approach was used to optimize the treatment process. The treatment efficiency was observed: 83.92% for $FeCl_3$, 88.06% for $CuSO_4$, and the highest efficiency of 92.68% for the combination of $CuSO_4$ and $FeCl_3$ [82].

The coagulation-flocculation process appears to be a practical solution for treating pharmaceutical waste. On the other hand, the chemical coagulation process also has disadvantages. To overcome these limitations, researchers have investigated the potential of natural coagulants. These natural alternatives have attracted attention due to their biodegradability, safety, and wide availability. According to a study comparing treatment methods [70], coagulation-flocculation methods have several advantages. These include ease of processing, cost-effectiveness, improved biodegradability, effective bacterial inactivation, promoted sludge stability, and ease of dewatering. On the other hand, the process also has some disadvantages, including high sludge management costs, financial impacts, long contact times, and the production of potentially harmful end products. Another study [50] found that the process efficiency of removing all dissolved solids from pharmaceutical wastewater by chemical coagulation was only 16%.

In addition, it is essential to emphasize that electrocoagulation treatment can effectively eliminate harmful microorganisms. Zaied et al. showed that the removal efficiency of pharmaceutical impurities can be improved by using different anode materials such as aluminum, stainless steel, and iron. In 2020, the chemical oxygen demand (COD) was reduced by 92.3%, and the total dissolved solids (TDS) was decreased by 91.5%. Furthermore, it is noteworthy that electro-flocculation requires fewer flocculants than chemical flocculation, making it a cost-effective method [70].

7 Membrane Filtration for Pharmaceutical Wastewater Treatment

One current approach that can effectively reduce pollutants to the desired level is membrane technology. Nanofiltration, microfiltration, reverse osmosis, ultrafiltration, and membrane bioreactors are widely used membrane processes, depending on characteristics such as size or charge. Membrane technology has several advantages over traditional treatment methods [57].

According to Mallakpour and Azadi [56], membrane separation is considered energy-efficient. The technique's advantages—low operating pressure, cost-effectiveness, high permeate flow rate, and fixed cost—have attracted significant attention in various pressure-controlled membrane applications.

The connection between membrane filtration and peroxide-ozonation-adsorption (POA) processes is increasingly close [6]. Membrane filtration can trap hazardous intermediates produced during POA treatment, while POA can effectively combat membrane fouling and organic residues in the concentrate [45]. One study looked at removing antibiotics from wastewater by combining nanofiltration with an ozone-based oxidation process. The results showed an average removal efficiency of up to 87%, a partial removal of dissolved organic carbon (DOC) of 40%, and a significant increase in the ratio of chemical oxygen demand to biochemical oxygen demand.

8 Advanced Oxidation Treatment of Pharmaceutical Liquid Waste

Ma et al. [55] considered that advanced oxidation processes (AOPs) represent a highly competitive technology for eliminating pathogenic microorganisms and emerging pharmaceutical contaminants. Although advanced oxidation, especially peroxide-ozonation-adsorption (POA), requires energy and chemicals, it is worth noting that the primary goal of advanced oxidation is the complete decomposition and mineralization of persistent pollutants [69]. AOP is a promising option for the effective treatment of persistent substances, especially pharmaceuticals, due to its excellent degradation efficiency, ability to eliminate toxic pollutants, general feasibility, and wide acceptance in the scientific community [60].

Satizabal-Gómez et al. [80] introduced a novel system combining hydrogen peroxide and ultraviolet C radiation, providing one of the most effective methods to generate hydroxyl radicals to oxidize pollutants. In addition, advanced Fenton processes (e.g., electro-Fenton and photo-Fenton) have been widely studied to reduce emerging contaminants in industrial wastewater [17]. In the photo-Fenton reaction, the traditional Fenton system is exposed to UV light to enhance the photoreduction of iron species (Fe_3^+) to dissolved iron species (Fe_2^+). In the case of the electro-Fenton reaction, both Fenton reagents can be generated electrochemically. However, it is essential to remember that this approach still requires a lot of capital for practical application [28].

9 Pharmaceutical Wastewater Treatment by Adsorption

Adsorption has been recognized as a promising technology for treating pharmaceutical wastewater [74]. Zhu et al. [91] reported its potential due to its user-friendly nature, flexibility, minimal generation of non-hazardous byproducts, and low energy requirements, as highlighted. Kabir et al. [44] find adsorption attractive, mainly when a cost-effective adsorbent doesn't require additional pre-treatment steps before use.

In the environmental context, Xu et al. [87] have demonstrated that adsorption outperforms other techniques regarding simplicity, design, acquisition cost, and resistance to toxic pollutants. Various factors influence adsorption efficiency, including adsorbent interactions, surface area, adsorbent-to-adsorbate ratio, particle size, temperature, pH, and contact time [63].

Despite the widespread use of activated carbon in wastewater treatment, its high production cost limits its widespread application [30]. Many researchers, including [42], have explored alternative carbon sources to produce cost-effective activated carbon. Raw agricultural solid waste, including leaves, seeds, and forest residue, has been used as adsorbents [27]. Due to their physicochemical properties, affordability, and eco-friendliness, these materials can potentially be used as adsorbents

[4]. Additionally, biochar, a highly porous carbonaceous material, shows promise as an eco-friendly option for efficient drug removal, particularly paracetamol [40].

Adsorption using powdered activated carbon has demonstrated its effectiveness in pharmaceutical wastewater treatment, being considered cost-effective compared to other wastewater treatment technologies, as Viegas et al. [84] recommended. However, it's worth noting that adsorption on powdered activated carbon has limitations, including relatively weaker adsorption of polar compounds and a slower reaction rate [85].

10 Pharmaceutical Wastewater Treatment by Flotation

Flotation offers an alternative to traditional wastewater treatment methods [47]. According to Prabu et al. [73], flotation is a process for segregating solid or liquid contaminants from wastewater by introducing tiny gas particles. Bhatia et al. [16] stress that flotation is often considered one of the most promising technologies due to its notable efficiency, low sludge production, and ease of use. In summary, flotation represents a cost-effective and efficient treatment approach for removing harmful substances from pharmaceutical wastewater [72].

The flotation process includes electro-flotation, induced air flotation, and dissolved air flotation [73]. Dissolved air flotation (DAF) has considerable potential for wastewater treatment. Generating large quantities of DCP (Dissolved Carbon Particles) sludge represents a significant advance in clean energy production, enabling this waste to be used in biogas production [3].

Zhang et al. [90] conducted studies on removing organic contaminants from benzene hydroxamic acid (BHA) using a simple sedimentation flotation process. Their results demonstrate BHA removal rates exceeding 94%, COD removal rates of 83.91%, and purified water with COD and turbidity levels below 25 mg/L and 0.1 NTU, respectively, meeting international standards, as observed by Han et al. [33]. Additionally, Ding et al. [24], showed the effectiveness of flotation in treating industrial wastewater containing BHA.

Flotation requires mixing techniques to enhance the removal of pharmaceutical residues [54]. Electrocoagulation techniques and electro-flotation have demonstrated seamless integration with advanced electrochemical oxidation to create a hybrid pharmaceutical wastewater purifying system, achieving complete COD removal and treating soluble coffee effluent, resulting in a COD removal rate of 74% [76].

Moreover, Ansari et al. [14] studied a combined electrocoagulation/flotation (ECF) peroxide/acetone process. In the ECF process, the maximum removal efficiencies for COD, color, TDS, and TSS were 64%, 60%, 61%, and 50%, respectively. However, after the ECF and peroxide processes, the maximum COD and color removal efficiencies increased considerably to 99.2%, 99.0%, 99.0%, and 99.1%, respectively. This remarkable performance and cost-effectiveness underline the suitability of this integrated system for advanced wastewater treatment in the pharmaceutical industry.

11 Hybrid Treatment of Pharmaceutical Wastewater

Innovative approaches are needed to ensure sustainable development and meet the challenges of emerging wastewater pollutants. Pharmaceutical pollutants (PP) represent a class of toxic and aggressive substances that must be removed from wastewater before it's released into the environment. However, traditional treatment methods have several limitations and show limited effectiveness in PP removal. Consequently, new and advanced treatment systems have been introduced [26].

In recent decades, conventional and advanced treatment processes have received considerable attention [21]. However, these processes are energy-intensive and require significant resources for operation and maintenance, making them costly. For this reason, hybrid systems have been investigated as alternatives, drawing on the strengths of different techniques to remove PP effectively [7].

Hybrid technology combines two treatment systems to enhance the efficiency of PP removal from water. It's essential to note that integrating two or more technologies, a core aspect of hybrid technology, is expected to remain a central theme in the ongoing development of improved technologies to remove a complete range of pharmaceuticals in a single operational cycle [26].

The literature review reveals that advanced oxidation can break down persistent pharmaceuticals. However, if these oxidation processes are not carefully monitored and operated, they can introduce toxic intermediates or byproducts. Physical treatments like carbon adsorption and membrane filtration act as a protective barrier to prevent original compounds and toxic intermediates from entering treated wastewater. However, these methods involve phase-change technologies that move contaminants from one phase to another. As a result, additional treatment and appropriate disposal management are required for the remaining water and sorbent. Combining various treatment processes to concurrently retain and break down parent compounds and process intermediates is an optimal approach. Hybrid technologies combine the advantages of these methods and enable maximum contaminant removal [29].

Furthermore, advanced oxidation processes are commonly used for water purification but are accompanied by high costs, high electricity consumption, and the need to use expensive materials. Integrating these adsorption methods can reduce operational expenses and improve process efficiency. Therefore, a hybrid approach combining oxidation and adsorption is a promising alternative for purifying pharmaceutical-contaminated water [51]. A new high-strength pharmaceutical wastewater treatment method has been proposed by Zhang et al. [89]. This approach has minimal energy consumption and operating costs while maintaining commendable efficiency. It revolves around integrating partial denitrification and anaerobic ammonia oxidation (PD/A) within a sequencing biofilm batch reactor (SBBR). The results indicate a substantial removal rate for pharmaceuticals, specifically for compounds like bismuth nitrate and potassium citrate, reaching an impressive 81.2% removal rate at the total nitrogen level. Using ionizing radiation, such as gamma irradiation, appears to be an extremely cost-effective, robust, and ecologically friendly solution for wastewater treatment. Additionally, gamma irradiation in wastewater treatment presents numerous advantages, including the absence of additional chemicals, sludge generation, exceptional penetration capacity in various water compositions, and insensitivity to solid and insoluble elements in wastewater [19].

A hybrid strategy was conceived in light of the economic challenges linked to single treatment systems and the effectiveness of biological treatment as an environmentally preferred option for wastewater treatment. This strategy combines coagulation treatment, biological treatment, and gamma irradiation, both individually and sequentially, to determine the most suitable treatment sequence for effectively purifying actual pharmaceutical wastewater. The stepwise approach involving coagulation, gamma irradiation, and biological treatment considers persistent pharmaceutical wastewater's combined breakdown and detoxification. It results in an overall removal efficiency of 92.7% ± 2.3% and 90.2% ± 2.9% of COD from low-organic (LSW) and high-organic (HSW) pharmaceutical wastewater, respectively [20].

The research is dedicated to exploring the treatment of current veterinary pharmaceutical wastewater using a hybrid reactor that combines synergistic catalytic ozonation with an iron-loaded zeolite A (Fe-Z) catalyst and an electro-flocculation process. The outcomes demonstrated that under optimal conditions, including a voltage of 5 V, a pH of 7, an ozone dose of 0.4 mg/min, and a catalyst dose of 1.5 g/L, COD removal efficiency reached 85.12%. At the same time, turbidity in pharmaceutical wastewater decreased to 78 NTU. Remarkably, enrofloxacin was eliminated, and amoxicillin recorded a substantial 90%removal rate. This underlines the effectiveness of the synergistic approach in the hybrid reactor for the efficient treatment of veterinary pharmaceutical wastewater, making it a viable alternative [41].

In parallel, a new approach was adopted by Masood et al. [58], encompassing electro-flocculation (EF), catalytic ozonation (CO) with the use of a novel Ni-Co-Zeolite 5A° catalyst, and subsequent ceramic membrane filtration (CMF) for pharmaceutical wastewater treatment. Notably, including CMF after EF + CO significantly increased pollutant removal throughout the process. The outcomes showcased that the combined EF + CO process achieved removal efficiencies of 33% for EFC, 93% for TT, 78% for COD, 45% for BOD, and 60% for other contaminants. Consequently, this inventive hybrid treatment process is promising for large-scale industrial applications in treating veterinary antibiotics [58].

12 Conclusions

In conclusion, liquid pharmaceuticals in the aquatic environment pose a significant environmental risk to aquatic life and human health. These compounds are often persistent and bioaccumulative, can disrupt hormonal balance and have other harmful effects on organisms. To solve this problem, pharmaceutical wastewater treatment methods need to be introduced. The most effective method is a hybrid approach, which combines physical and biochemical methods by integrating the advantages of each technology and mitigating their disadvantages, this approach ensures more complete degradation of pharmaceutical contaminants and significantly reduces their impact on the environment.

13 Recommendations

The treatment of pharmaceutical effluents is essential to protect the environment and public health. Phytoremediation is a technique that can be used to treat this type of effluent. Regular analysis of effluents is necessary to verify treatment effectiveness. Researchers need to develop advanced sensor technologies for real-time monitoring of contaminant concentrations. Finally, training and awareness-raising of personnel on the importance of effluent management is important.

References

1. Abia Akebe LK, Sibanda T, Selvarajan R et al (2022) Environmental reservoirs of antibiotic resistance determinants: a ticking time bomb for the future emergence of super-bugs of environmental and public health importance. Front Environ Sci 10:941847
2. Aboudalle A, Djelal H, Domergue L et al (2021) A novel system coupling an electro-Fenton process and an advanced biological process to remove a pharmaceutical compound, metronidazole. J Hazard Mater 415:125705
3. Abuhasel K, Kchaou M, Alquraish M et al (2021) Oily wastewater treatment: overview of conventional and modern methods, challenges, and future opportunities. Water 13:980
4. Adegoke KA, Olagunju AO, Alagbada TC et al (2022) Adsorptive removal of endocrine-disrupting chemicals from aqueous solutions: a review. Water Air Soil Pollut 233:38
5. Adeoye JB, Tan YH, Lau SY et al (2024) Advanced oxidation and biological integrated processes for pharmaceutical wastewater treatment: a review. J Environ Manage 353:120170
6. Ahmad A, Priyadarshini M, Das S, Ghangrekar MM (2022) Electrocoagulation as an efficacious technology for the treatment of wastewater containing active pharmaceutical compounds: a review. Sep Sci Technol 57:1234–1256
7. Ahmed SF, Mofijur M, Nuzhat S et al (2021) Recent developments in physical, biological, chemical, and hybrid treatment techniques for removing emerging contaminants from wastewater. J Hazard Mater 416:125912
8. Ahmed Y, Zhong J, Wang Z et al (2022) Simultaneous removal of antibiotic resistant bacteria, antibiotic resistance genes, and micropollutants by FeS_2@ GO-based heterogeneous photo-Fenton process. Environ Sci Technol 56:15156–15166
9. Al-Asheh S, Bagheri M, Aidan A (2021) Membrane bioreactor for wastewater treatment: a review. Case Stud Chem Environ Eng 4:100109
10. Al-Hazmi HE, Łuczak J, Habibzadeh S et al (2023) Polysaccharide nanocomposites in water treatment: a review. Chemosphere 140578
11. Alam G, Ihsanullah I, Naushad M, Sillanpää M (2022) Applications of artificial intelligence in water treatment for optimization and automation of adsorption processes: recent advances and prospects. Chem Eng J 427:130011
12. Alfonso-Muniozguren P, Serna-Galvis EA, Bussemaker M et al (2021) A review on pharmaceuticals removal from waters by single and combined biological, membrane filtration and ultrasound systems. Ultrason Sonochem 76:105656

13. Ali SS, Sabry R, Ali HM et al (2023) Integrated system of anoxic/activated sludge and ultrafiltration membrane for zero liquid discharge of pharmaceutical industrial wastewater treatment. J Environ Chem Eng 11:109068
14. Ansari A, Khoram MM, Nematollahi D et al (2021) A comprehensive study on opium pharmaceutical wastewater treatment in laboratory and semi-industrial scales. J Water Process Eng 44:102353
15. Barati B, Zafar FF, Wang S (2022) Different waste management methods, applications, and limitations. In: Waste-to-energy: recent developments and future perspectives towards circular economy. Springer, pp 21–58
16. Bhatia SK, Mehariya S, Bhatia RK et al (2021) Wastewater based microalgal biorefinery for bioenergy production: progress and challenges. Sci Total Environ 751:141599
17. Brillas E (2020) A review on the photoelectro-Fenton process as efficient electrochemical advanced oxidation for wastewater remediation. Treatment with UV light, sunlight, and coupling with conventional and other photo-assisted advanced technologies. Chemosphere 250:126198
18. Campo R, Lubello C, Lotti T, Di Bella G (2021) Aerobic granular sludge–membrane bioReactor (AGS–MBR) as a novel configuration for wastewater treatment and fouling mitigation: a mini-review. Membranes (Basel) 11:261
19. Changotra R, Guin JP, Dhir A, Varshney L (2018) Decomposition of antibiotic ornidazole by gamma irradiation in aqueous solution: kinetics and its removal mechanism. Environ Sci Pollut Res 25:32591–32602
20. Changotra R, Rajput H, Guin JP et al (2019) Hybrid coagulation, gamma irradiation and biological treatment of real pharmaceutical wastewater. Chem Eng J 370:595–605
21. Chen L, Cheng P, Ye L et al (2020) Biological performance and fouling mitigation in the biochar-amended anaerobic membrane bioreactor (AnMBR) treating pharmaceutical wastewater. Bioresour Technol 302:122805
22. Chen Y, Shen W, Wang B et al (2020) Occurrence and fate of antibiotics, antimicrobial resistance determinants and potential human pathogens in a wastewater treatment plant and their effects on receiving waters in Nanjing, China. Ecotoxicol Environ Saf 206:111371
23. Cheng L, Liu G, Zhao J, Jin W (2021) Two-dimensional-material membranes: manipulating the transport pathway for molecular separation. Accounts Mater Res 2:114–128
24. Ding Z, Li J, Bi Y et al (2021) The adsorption mechanism of synergic reagents and its effect on apatite flotation in oleamide-sodium dodecyl benzene sulfonate (SDBS) system. Miner Eng 170:107070
25. Dutta S, Bhattacharjee J (2022) A comparative study between physicochemical and biological methods for effective removal of textile dye from wastewater. In: Development in wastewater treatment research and processes. Elsevier, pp 1–21
26. Eniola JO, Kumar R, Barakat MA, Rashid J (2022) A review on conventional and advanced hybrid technologies for pharmaceutical wastewater treatment. J Clean Prod 356:131826
27. Faiad A, Alsmari M, Ahmed MMZ et al (2022) Date palm tree waste recycling: treatment and processing for potential engineering applications. Sustainability 14:1134
28. Garrido-Cardenas JA, Esteban-García B, Agüera A et al (2020) Wastewater treatment by advanced oxidation process and their worldwide research trends. Int J Environ Res Public Health 17:170
29. Ghazal H, Koumaki E, Hoslett J et al (2022) Insights into current physical, chemical and hybrid technologies used for the treatment of wastewater contaminated with pharmaceuticals. J Clean Prod 361:132079
30. Guellati A, Maachi R, Chaabane T et al (2022) Aluminum dispersed bamboo activated carbon production for effective removal of Ciprofloxacin hydrochloride antibiotics: optimization and mechanism study. J Environ Manage 301:113765
31. Guerreiro Crizel M, Barreto TM, Fiori MA et al (2024) Electrocoagulation process as a consolidated technology in the treatment of industrial effluents and as a promising process in the treatment of effluents generated by car washes: a brief review. ACS ES&T Water 4:1978–2004

32. Guo W, Cheng D, Ngo HH et al (2020) Anaerobic membrane bioreactors for antibiotic wastewater treatment. In: Current developments in biotechnology and bioengineering. Elsevier, pp 219–239
33. Han G, Wen S, Wang H, Feng Q (2021) Surface sulfidization mechanism of cuprite and its response to xanthate adsorption and flotation performance. Miner Eng 169:106982
34. Harb M, Lou E, Smith AL, Stadler LB (2019) Perspectives on the fate of micropollutants in mainstream anaerobic wastewater treatment. Curr Opin Biotechnol 57:94–100
35. Ho Y-C, Chua S-C, Chong F-K (2020) Coagulation-flocculation technology in water and wastewater treatment. In: Handbook of research on resource management for pollution and waste treatment. IGI Global, pp 432–457
36. Hosseinpour S, Azimian-Kivi M, Jafarzadeh Y, Yegani R (2021) Pharmaceutical wastewater treatment using polypropylene membranes incorporated with carboxylated and PEG-grafted nanodiamond in membrane bioreactor (MBR). Water Environ J 35:1249–1259
37. Hu D, Min H, Wang H et al (2020) Performance of an up-flow anaerobic bio-electrochemical system (UBES) for treating sulfamethoxazole (SMX) antibiotic wastewater. Bioresour Technol 305:123070
38. Huang B, Wang H-C, Cui D et al (2018) Treatment of pharmaceutical wastewater containing β-lactams antibiotics by a pilot-scale anaerobic membrane bioreactor (AnMBR). Chem Eng J 341:238–247
39. Huang X, Wan Y, Shi B, Shi J (2020) Effects of powdered activated carbon on the coagulation-flocculation process in humic acid and humic acid-kaolin water treatment. Chemosphere 238:124637
40. Ihsanullah I, Khan MT, Zubair M et al (2022) Removal of pharmaceuticals from water using sewage sludge-derived biochar: a review. Chemosphere 289:133196
41. Ikhlaq A, Javed F, Akram A et al (2020) Synergic catalytic ozonation and electroflocculation process for the treatment of veterinary pharmaceutical wastewater in a hybrid reactor. J Water Process Eng 38:101597
42. Januário EFD, Fachina YJ, Wernke G et al (2022) Application of activated carbon functionalized with graphene oxide for efficient removal of COVID-19 treatment-related pharmaceuticals from water. Chemosphere 289:133213
43. Jhunjhunwala A, Pathak U, Sarkar KK et al (2021) Removal of levosulpiride from pharmaceutical wastewater using an advanced integrated treatment strategy comprising physical, chemical, and biological treatment. Environ Prog Sustain Energy 40:e13482
44. Kabir MM, Alam F, Akter MM et al (2022) Highly effective water hyacinth (Eichhornia crassipes) waste-based functionalized sustainable green adsorbents for antibiotic remediation from wastewater. Chemosphere 304:135293
45. Kaya SI, Gumus E, Cetinkaya A et al (2022) Trends in on-site removal, treatment, and sensitive assay of common pharmaceuticals in surface waters. TrAC Trends Anal Chem 149:116556
46. Khalidi-Idrissi A, Souabi S, Madinzi A et al (2024) Recent advances in the treatment of wastewater contaminated with pharmaceutical pollutants: a critical review. Euro-Mediterranean J Environ Integr 9:23–47
47. Kim J-P, Jin DR, Lee W et al (2020) Occurrence and removal of veterinary antibiotics in livestock wastewater treatment plants, South Korea. Processes 8:720
48. Kołecka K, Gajewska M, Cytawa S et al (2020) Is sequential batch reactor an efficient technology to protect recipient against non-steroidal anti-inflammatory drugs and paracetamol in treated wastewater? Bioresour Technol 318:124068
49. Kong L, Shi X (2022) Effect of antibiotic mixtures on the characteristics of soluble microbial products and microbial communities in upflow anaerobic sludge blanket. Chemosphere 292:133531
50. Lalwani J, Gupta A, Thatikonda S, Subrahmanyam C (2020) An industrial insight on treatment strategies of the pharmaceutical industry effluent with varying qualitative characteristics. J Environ Chem Eng 8:104190
51. Lapchuk I, Shyichuk A, Tatarchuk T (2023) Application of hybrid advanced oxidation and adsorption processes for pharmaceutical wastewater treatment. In: The treatment of pharmaceutical wastewater. Elsevier, pp 247–275

52. Limketkai BN (2022) Biologic treatment withdrawal: two sides of a coin. NEJM Evid 1:EVIDe2200152
53. Liu C, Guo Y, Zhou Y et al (2022) High-hydrophilic and antifouling reverse osmosis membrane prepared based an unconventional radiation method for pharmaceutical plant effluent treatment. Sep Purif Technol 280:119838
54. López-Guzmán M, Flores-Hidalgo MA, Reynoso-Cuevas L (2021) Electrocoagulation process: an approach to continuous processes, reactors design, pharmaceuticals removal, and hybrid systems—a review. Processes 9:1831
55. Ma D, Yi H, Lai C et al (2021) Critical review of advanced oxidation processes in organic wastewater treatment. Chemosphere 275:130104
56. Mallakpour S, Azadi E (2022) Nanofiltration membranes for food and pharmaceutical industries. Emergent Mater 5:1329–1343
57. Mansor E, Abdallah H, Shaban AM (2024) The role of membrane filtration in wastewater treatment. Environ Qual Manag
58. Masood Z, Ikhlaq A, Rizvi OS et al (2023) A novel hybrid treatment for pharmaceutical wastewater implying electroflocculation, catalytic ozonation with Ni-Co Zeolite 5A° catalyst followed by ceramic membrane filtration. J Water Process Eng 51:103423
59. Mohamed Noor MH, Wong S, Ngadi N et al (2022) Assessing the effectiveness of magnetic nanoparticles coagulation/flocculation in water treatment: a systematic literature review. Int J Environ Sci Technol 19:6935–6956
60. Monteoliva-García A, Martín-Pascual J, Muñío M del M, Poyatos JM (2020) Effects of carrier addition on water quality and pharmaceutical removal capacity of a membrane bioreactor–advanced oxidation process combined treatment. Sci Total Environ 708:135104
61. Moradi G, Rahimi M, Zinadini S et al (2022) Natural deep eutectic solvent modified nanofiltration membranes with superior antifouling properties for pharmaceutical wastewater treatment. Chem Eng J 448:137704
62. Nassri I, Souabi S (2023) Occurrence, pollution sources, and mitigation prospects of Antibiotics, anti-inflammatories, and endocrine disruptors in the aquatic environment. Environ Nanotechnol Monit Manag 100878
63. Natarajan R, Saikia K, Ponnusamy SK et al (2022) Understanding the factors affecting adsorption of pharmaceuticals on different adsorbents—a critical literature update. Chemosphere 287:131958
64. Nava AR, Daneshian L, Sarma H (2022) Antibiotic resistant genes in the environment-exploring surveillance methods and sustainable remediation strategies of antibiotics and ARGs. Environ Res 215:114212
65. Ng K, Alygizakis NA, Thomaidis NS, Slobodnik J (2023) Wide-scope target and suspect screening of antibiotics in effluent wastewater from wastewater treatment plants in Europe. Antibiotics 12:100
66. Nguyen PY, Carvalho G, Reis MAM, Oehmen A (2021) A review of the biotransformations of priority pharmaceuticals in biological wastewater treatment processes. Water Res 188:116446
67. Noman EA, Mohamed RMSR, Al-Gheethi AA et al (2022) Antibiotics and antibiotic-resistant bacteria in greywater: challenges of the current treatment situation and predictions of future scenario. Environ Res 212:113380
68. Oberoi AS, Surendra KC, Wu D et al (2022) Anaerobic membrane bioreactors for pharmaceutical-laden wastewater treatment: a critical review. Bioresour Technol 361:127667
69. Otter P, Mette K, Wesch R et al (2020) Oxidation of selected trace organic compounds through the combination of inline electro-chlorination with UV radiation (UV/ECl2) as alternative AOP for decentralized drinking water treatment. Water 12:3275
70. Padmaja K, Cherukuri J, Reddy MA (2020) A comparative study of the efficiency of chemical coagulation and electrocoagulation methods in the treatment of pharmaceutical effluent. J Water Process Eng 34:101153
71. Pandis PK, Kalogirou C, Kanellou E et al (2022) Key points of advanced oxidation processes (AOPs) for wastewater, organic pollutants and pharmaceutical waste treatment: a mini review. ChemEngineering 6:8

72. Pooja G, Kumar PS, Indraganti S (2022) Recent advancements in the removal/recovery of toxic metals from aquatic system using flotation techniques. Chemosphere 287:132231
73. Prabu SL, Suriyaprakash TNK, Kandasamy R, Rathinasabapathy T (2020) Effective Waste water treatment and its management. In: Waste management: concepts, methodologies, tools, and applications. IGI Global, pp 49–72
74. Priya AK, Gnanasekaran L, Rajendran S et al (2022) Occurrences and removal of pharmaceutical and personal care products from aquatic systems using advanced treatment—a review. Environ Res 204:112298
75. Qiu G, Chen H, Raghavan DSS, Ting Y-P (2021) Removal behaviors of antibiotics in a hybrid microfiltration-forward osmotic membrane bioreactor for real municipal wastewater treatment. Chem Eng J 417:129146
76. Raj R, Tripathi A, Das S, Ghangrekar MM (2021) Removal of caffeine from wastewater using electrochemical advanced oxidation process: a mini review. Case Stud Chem Environ Eng 4:100129
77. Rios-Miguel AB, Jetten MSM, Welte CU (2021) Effect of concentration and hydraulic reaction time on the removal of pharmaceutical compounds in a membrane bioreactor inoculated with activated sludge. Microb Biotechnol 14:1707–1721
78. Rorke DCS, Lekha P, Kana GEB, Sithole BB (2022) Effect of pharmaceutical wastewater as nitrogen source on the optimization of simultaneous saccharification and fermentation hydrogen production from paper mill sludge. Sustain Chem Pharm 25:100619
79. Sakurai KSI, Pompei CME, Tomita IN et al (2021) Hybrid constructed wetlands as post-treatment of blackwater: an assessment of the removal of antibiotics. J Environ Manage 278:111552
80. Satizabal-Gómez V, Collazos-Botero MA, Serna-Galvis EA et al (2021) Effect of the presence of inorganic ions and operational parameters on free cyanide degradation by ultraviolet C activation of persulfate in synthetic mining wastewater. Miner Eng 170:107031
81. Singh G, Haileselassie Y, Briscoe L et al (2022) The effect of gastric acid suppression on probiotic colonization in a double blinded randomized clinical trial. Clin Nutr ESPEN 47:70–77
82. Tahraoui H, Belhadj A-E, Triki Z et al (2023) Mixed coagulant-flocculant optimization for pharmaceutical effluent pretreatment using response surface methodology and Gaussian process regression. Process Saf Environ Prot 169:909–927
83. Tiwari B, Drogui P, Tyagi RD (2020) Removal of emerging micro-pollutants from pharmaceutical industry wastewater. In: Current developments in biotechnology and bioengineering. Elsevier, pp 457–480
84. Viegas RMC, Mestre AS, Mesquita E et al (2020) Assessing the applicability of a new carob waste-derived powdered activated carbon to control pharmaceutical compounds in wastewater treatment. Sci Total Environ 743:140791
85. Wang C, Liu J, Li Q et al (2022) A review of the effects of antibiotics on the anaerobic digestion of swine waste. Curr Opin Environ Sci Heal 25:100312
86. Wang Y, Kang Z, Zhang L, Zhu Z (2022) Elucidating the interactions between a [NiFe]-hydrogenase and carbon electrodes for enhanced bioelectrocatalysis. ACS Catal 12:1415–1427
87. Xu X, Deng F, Shao P et al (2022) Internal electric field driving separation and migration of charge carriers via Z-scheme path in $AgIn_5S_8$/ZnO heterojunction for efficient decontamination of pharmaceutical pollutants. Chem Eng J 428:132096
88. Yu S-Y, Xie Z-H, Wu X et al (2024) Review of advanced oxidation processes for treating hospital sewage to achieve decontamination and disinfection. Chinese Chem Lett 35:108714
89. Zhang J, Peng Y, Li X, Du R (2022) Feasibility of partial-denitrification/anammox for pharmaceutical wastewater treatment in a hybrid biofilm reactor. Water Res 208:117856
90. Zhang L, Jiao X, Wu S et al (2021) Study on optimization of tungsten ore flotation wastewater treatment by response surface method (RSM). Minerals 11:184
91. Zhu X, He M, Sun Y et al (2022) Insights into the adsorption of pharmaceuticals and personal care products (PPCPs) on biochar and activated carbon with the aid of machine learning. J Hazard Mater 423:127060

Enhancing Olive Oil Mill Wastewater Treatment via Coagulation Process Optimization: Leveraging *Moringa oleifera* as a Natural Coagulant, with CCD-RSM Integration for Improved Efficiency

Safaa Khattabi Rifi, Roukaya Bouyakhsass, Khadija Zahidi, Abdelaziz Madinzi, Ilham Nassri, Abdelkader Anouzla, Amina Khalidi Idrissi, and Salah Souabi

Abstract An optimization study using composite central designs, responce surface methodology, and Analysis of Variance (ANOVA) was conducted to treat olive oil mill wastewater using *Moringa oleifera.* This study uses the natural coagulant *Moringa oleifera,* to explore the coagulation-based treatment of OMW. Process optimization is accomplished by combining the response surface methodology (RSM) with central composite design (CCD). This process examines at four factors: pH, agitation speed, duration, and concentration of *Moringa oléifera,* in order to determine the relationship between Turbidity, COD, and polyphenols. The models' significance is verified by the ANOVA. The study's highly adjusted coefficient of determination (R2Adj = 86.73%, R2Adj = 90.39%, and R2Adj = 93.33% for turbidity, COD, and polyphenols, respectively) indicates a strong model fit. Moreover, the following conditions are shown to be optimal for achieving effective OMW purification with *Moringa oleifera*: *Moringa oleifera* concentration [Mo] = 65 g/L, pH = 6, agitation time (St) = 25 min, and speed (Ss) = 35 rpm. Turbidity removal efficiency is 96%, COD removal efficiency is 88%, and polyphenol removal efficiency is 86%

S. Khattabi Rifi · R. Bouyakhsass · K. Zahidi · A. Madinzi · A. Anouzla · A. K. Idrissi · S. Souabi (✉)
Laboratory of Process Engineering and Environment, Faculty of Science and Technology, University Hassan II, Mohammedia, Morocco
e-mail: salah.souabi@gmail.com

A. Madinzi
e-mail: abdelaziz.madinzi@fstm.ac.ma

I. Nassri
Géo-Biodiversity and Natural Patrimony Laboratory, Scientific Institute, Mohammed V University, Rabat, Morocco

S. Yadav et al. (eds.), *Emerging Technologies in Biological and Hybrid Wastewater Treatment: Lessons from Developed to Enhancing Practices in Developing Countries*, Earth and Environmental Sciences Library,
https://doi.org/10.1007/978-3-031-74515-7_11

at these optimal conditions. To sum up, this method provides a practical way for addressing the environmental issues that OMW raises.

Keywords Olive mill wastewater · RSM · *Moringa oleifera* · CCD

1 Introduction

The production of olive oil produces a significant amount of wastewater annually, about 5.4 × 106 m^3 globally. This wastewater is produced by a multitude of processes, including the addition of water at different points during the extraction process and the irrigation of fruit, olives, and olive oil. The composition of wastewater (OMW) varies according to the type of extraction process used, whether a press or a two- or three-phase process. Disposal of OMW by infiltration into the ground or evaporation onto rafts can pose a serious environmental problem, mainly due to its high organic load. OMW from continuous two- and three-phase centrifugation procedures can have up to 12.2 g O2/L and 220 g O2/L of COD and 100 g O2/L of BOD5, respectively [9]. The main challenge in managing wastewater from olive oil extraction is coming up with an affordable and sustainable solution. In the past, wastewater was dumped into surrounding rivers and streams, which had a negative impact on aquatic ecosystems and watercourses. The release of wastewater into undeveloped areas lowers the amounts of dissolved oxygen, which is detrimental to aquatic life and the ecosystem [17]. Different technologies have been applied to the treatment of traditional OMW, focusing mainly on the removal of phenolic compounds and COD, and on increasing their biodegradability. Biological processes are generally not applied directly to the effluent due to its high toxicity, requiring pre-treatment to remove polyphenols or multiple dilutions, implying higher costs and greater complexity [6]. However, alternative approaches such as composting, aerobic treatment, and more commonly, anaerobic digestion, have been investigated. Filtration techniques such as ultrafiltration, nanofiltration and reverse osmosis can significantly reduce COD. Nevertheless, these methods face serious problems such as fouling, low flow rate operation and high associated costs [6, 14]. Optimisation designs such as the Central Composite Design (CCD) are commonly applied in flocculation and coagulation processes [4, 3, 8]. They offer almost as much information as a three-level factorial design, while requiring significantly fewer trials. One statistical approach that has proven to work very well for optimizing different systems and processes is the Response Surface Methodology (RSM) [12]. It assesses the relationship between the variables provided and the replies output. One particular type of design that is commonly used is the central composite design. This method is used in various processes to achieve the greatest efficiency, and enables experiments to be designed to obtain the greatest number of data with the fewest possible experiments, by analyzing the effect of simultaneous variation in several variables [5].

2 Methods and Materials

2.1 *Olive Mill Wastewater Sample Collection*

Wastewater from a fresh olive mill was collected in a 30-L sample. This OMW was acquired by through batch processing. After being removed from the decantation cell, the samples were put in sealed plastic containers.

2.2 *Preparation of the Coagulant*

We dissolved the *Moringa oleifera* powder in distilled water to make the coagulant. The mixture was then stirred for thirty minutes. The resultant suspension was then passed through Whatman No. 1 filter paper. Every day, the solutions were carefully developed and placed in a 4 °C refrigerator.

2.3 *Flotation and Coagulation Process*

To minimize the impact of OMW effectively, we initiated the process by performing flotation to eliminate grease. This natural flotation method entailed placing the raw OMW sample in a 20 L container and allowing it to settle for a 24-h period. Subsequently, we removed the oil phase that had risen to the top, leaving us with the crude liquid phase, now devoid of oil, for further physicochemical analysis. Following this pretreatment, we conducted coagulation experiments. 500 mL of the OMW that had been pretreated were put in 1-L beakers. To align with our experimental design points, we adjusted the pH by introducing lime. Afterward, we introduced the specified coagulant with varying concentrations. Stirring for 5 min at 150 rpm. Subsequently, we systematically adjusted the duration and speed of gentle agitation as per our experimental design protocol. We performed sampling after 2 h of sedimentation.

2.4 *Analysis of Physico Chemical*

Using "Accumet basic ab15," we continuously monitored the pH during the procedure, following the February 2001 NF T 90-008 standard (T 90-008). We used a HACH model 2100N turbidimeter in accordance with NF EN ISO 7027 standard (T 90-033) to measure turbidity. It's important to note that the turbidimeter we used has a 4000 NTU maximum. The content of phenolic compounds was assessed using a colorimetric method based on the Folin-Ciocalteu reagent [16].

Table 1 Experimental scale levels as well as independent test variables

Natural variables (x_j)	Symbol	Coded variables X_1, X_2, X_3, X_4				
		− 2	− 1	0	1	2
x_1 = *Moringa oleifera* concentration (g/L)	[Mo]	20	35	45	55	65
X_2 = pH	pH	6	7	8	9	10
x_3 = Agitation speed (rpm)	Ss	20	25	30	35	40
x_4 = Agitation time (min)	St	15	20	25	30	35

2.5 Experimental Design

This investigation's main goal was to determine the ideal coagulation conditions by assessing important factors like polyphenol removal, turbidity, and COD efficiency. The pH was between 6 and 10, and the range of coagulant concentrations was between 20 and 65 g/L. We stirred for a continuous 25 min at a constant speed of 35 rpm. For every variable, we identified five different levels based on initial evaluations. Table 1 lists the factors, related symbols, and designated levels for each variable.

The fundamental strategy of this investigation revolves around a thorough examination of regression variance using the ANOVA method. The primary objective is to evaluate the suitability of quadratic models. We formulated the regression equations by identifying the most pertinent terms, guided by predefined confidence levels of 95%, 99%, or 99.9%. These realizations were then applied to examine the interactions and linear effects between the variables being studied, which led to the determination of the perfect conditions. The experimental matrix for this study includes thirty-one experiments in total.

3 Results and Discussion

3.1 OMW Characterization

A summary of the physical and chemical properties of the floating and untreated OMW samples is provided in Table 2. These samples first appear to be cloudy and have a turbidity level that is approximately five thousand eight hundred forty-six NTU. Turbidity is efficiently decreased by 36.7% when natural flotation is used. In general, sediments, lipids, and oils are separated from the liquid phase using the basic flotation process. According to Khennoussi et al. [13], this mechanism causes materials with a relatively low density—roughly equal to or slightly more than that of the liquid medium—to rise to the surface. The unprocessed OMW contains a considerable organic matter concentration, as seen by its COD test of 177.6 g/L. Natural flotation, in this context, leads to a significant reduction in COD, amounting to approximately 27%.

Table 2 Raw and floated OMW characterization

	Raw (OMW)	Floated (OMW)	Removal efficiency (%)
Turbidity (NTU)	5846 ± 3.07	3702 ± 2.07	36.67
COD (g/L)	177.6 ± 8.06	129.6 ± 6.08	27.03
Polyphenols (g/L)	5.9 ± 0.19	4.5 ± 0.15	23.72

Using a central composite design (CCD), the relationship between turbidity, COD, and polyphenols and the pH, coagulant concentration, agitation time, and agitation speed was ascertained. As Table 1 illustrates, every feature in this design was positioned at five distinct levels. In particular, a centred composite plane was selected, where each component at each of these five levels varied between + α and − α. These 31 experiments were conducted in duplicate in order to validate the experiment and evaluate the experimental variance (Table 3).

3.2 *Analysis of Variance (ANOVA)*

An analysis of variance (ANOVA) was carried out to evaluate the adequacy of the model, with a particular focus on the interactions between the variable factors and their corresponding responses, following the approach described by Adelodun et al. in 2019. Table 4 presents the results derived from this ANOVA. The results shown in Table 4 demonstrate that the model is significant at a 99.9% confidence level, meaning that the expected responses agree closely with the experimental findings. The model's significance was assessed using the F-test, in which the experimental value that is greater than the F-critical value (which is fixed at 5.27) signifies the importance of the model. It's important to note that the models (FTurbidity > Fcritical, FCOD > Fcritical, FPolyphenols > Fcritical) have significance in this context. The model's relevance is further supported by the reduced *p*-values (< 0.05). The models provide considerable statistical relevance for Turbidity, COD, and polyphenol elimination, with *p*-values less than 0.0001. In conclusion, the model has demonstrated its ability to produce predictions within the chosen range of variables.

3.3 *Estimating Coefficients and Statistically Analyzing the Effects of Factors*

The variance analysis results for COD, turbidity, and polyphenol removal are summarized in Table 5. Smaller *P*-values (< 0.05) and larger F ratios imply stronger importance of the relevant coefficient terms; a 95% confidence level was used [10]. For each of the three selected answers, the table presents the estimated regression coefficients, F-statistic ratios, and P-values for the linear, quadratic, and interaction effects of the

Table 3 Results of treatment for polyphenols (mg/L), COD (mg/L), and turbidity (NTU)

Run		Factors				Responses		
Randomised	Actual	[Mo] (g/L)	pH	Ss (rpm)	St (min)	Turbidity (NTU)	COD (mg/L)	Polyphenols (mg/L)
20	1	35	7	25	20	346 ± 1.62	90,800 ± 8.07	2011.02 ± 0.05
18	2	35	7	25	30	246 ± 1.2	115,200 ± 10.03	2304.41 ± 0.03
25	3	35	7	35	20	318 ± 1.17	111,200 ± 8.60	3131.31 ± 0.08
12	4	35	7	35	30	290 ± 1.8	76,400 ± 7.23	3549.13 ± 0.06
9	5	35	9	25	20	540 ± 2.43	52,800 ± 4.32	2180.89 ± 0.02
24	6	35	9	25	30	338 ± 1.63	76,800 ± 6.78	2630.85 ± 0.03
2	7	35	9	35	20	282 ± 1.22	120,000 ± 10.02	2634.99 ± 0.04
13	8	35	9	35	30	262 ± 1.52	57,600 ± 6.12	2456.38 ± 0.03
10	9	55	7	25	20	510 ± 2.92	33,600 ± 4.62	2089.07 ± 0.07
17	10	55	7	25	30	258 ± 1.32	100,800 ± 8.07	2497.70 ± 0.05
3	11	55	7	35	20	726 ± 2.73	76,800 ± 6.30	2653.35 ± 0.06
15	12	55	7	35	30	446 ± 2.35	38,400 ± 4.20	2584.94 ± 0.02
21	13	55	9	25	20	564 ± 2.8	43,200 ± 6.04	2272.723 ± 0.03
5	14	55	9	25	30	230 ± 1.63	100,800 ± 8.2	2543.62 ± 0.04
1	15	55	9	35	20	384 ± 1.7	102,800 ± 6.05	2731.31 ± 0.05
22	16	55	9	35	30	168.8 ± 1.21	81,600 ± 6.80	2006.43 ± 0.03
7	17	25	8	30	25	97.8 ± 1.09	77,600 ± 8.53	3351.69 ± 0.07
4	18	65	8	30	25	328 ± 1.8	72,000 ± 4.52	2929.29 ± 0.04
6	19	45	6	30	25	332 ± 2.04	62,400 ± 3.42	2685.95 ± 0.06
23	20	45	10	30	25	139.2 ± 1.01	52,800 ± 6.62	2736.46 ± 0.03
19	21	45	8	20	25	438 ± 1.21	87,200 ± 5.93	2093.66 ± 0.02

(continued)

Table 3 (continued)

Run		Factors				Responses		
Randomised	Actual	[Mo] (g/L)	pH	Ss (rpm)	St (min)	Turbidity (NTU)	COD (mg/L)	Polyphenols (mg/L)
11	22	45	8	40	25	418 ± 2.07	120,000 ± 8.09	2933.88 ± 0.05
16	23	45	8	30	15	534 ± 2.08	86,000 ± 7.23	2465.56 ± 0.06
14	24	45	8	30	35	197.8 ± 1.20	81,600 ± 6.52	2438.01 ± 0.04
8	25	45	8	30	25	396 ± 3.52	81,600 ± 7.02	4017.45 ± 0.07
28	26	45	8	30	25	365 ± 2.21	82,301 ± 6.82	4010.2 ± 0.08
30	27	45	8	30	25	370 ± 1.22	80,443 ± 5.94	4050.3 ± 0.06
31	28	45	8	30	25	390 ± 2.33	81,196 ± 8.95	4099.4 ± 0.09
26	29	45	8	30	25	345 ± 1.52	82,263 ± 7.60	4063.1 ± 0.08
27	30	45	8	30	25	387 ± 2.03	81,364 ± 6.63	4078.5 ± 0.07
29	31	45	8	30	25	355 ± 2.01	80,000 ± 10.2	4073.2 ± 0.09

[Mo]: *Moringa oleifera* concentration; *Ss* Agitation speed; *St* Agitation time

Table 4 ANOVA study of polyphenols, COD, and turbidity

Source	Degrees of liberty	Sum of squares	Medium square	Report F	Prob. > F
Turbidity					
Model	14	488,937.66	34,924.1	15.0095	< 0.0001***
Error	16	37,228.73	2326.8		
Total corrected	30	526,166.39			
R^2	0.9292	R^2 adjusted	0.8673	DW	1.61
COD					
Model	14	$1.39 \times 10^{+10}$	993,038,888	21.1445	< 0.0001***
Error	16	751,431,423	46,964,464		
Total corrected	30	$1.47 \times 10^{+10}$			
R^2	0.9487	R^2 adjusted	0.9039	DW	2.31
Polyphenols					
Model	14	14,939,529	1,067,109	31.0099	< 0.0001***
Error	16	550,590	34,412		
Total corrected	30	15,490,119			
R^2	0.964455	R^2 adjusted	0.933354	DW	1.71

***Significant to 0.1%

components. The study found that agitation speed and concentration of *Moringa oleifera* have a strong linear effect on the elimination of COD. This is evident from the high F-experimental values, which greatly exceed the F-critical value. Adelodun et al. [1] previously highlighted the influence of *Moringa oleifera* on COD, and Zheng et al. [19] demonstrated the direct impact of agitation speed on COD removal. The analysis further revealed that four interactions ([Mo]pH, pHSs, [Mo]St, and SsSt) significantly affect the response, with high F-experimental values. However, these interactions have a positive coefficient value, indicating an unfavorable impact on COD removal. The quadratic effects pH and SsSs are highly substantial, with *P*-values of 0.0003 and 0.0005, respectively. As for turbidity removal, With the exception of the linear effect of agitation speed, all linear term coefficients have a significant impact on the response, according to the ANOVA results shown in Table 5 at a 95% confidence level. This observation aligns with the findings of Espinoza-Quiñones et al. [7]. It's interesting to note that varying agitation speed did not result in a noticeable change in turbidity, which is consistent with the research by Jebun et al. [11]. Moreover, two-level interaction effects ([Mo]*pH, [Mo]Ss, pHSs, and [Mo]St) were identified as significant model terms. Additionally, the quadratic terms were considered necessary, except for the quadratic effect StSt, which was found to be non-significant. Concerning polyphenolic content, the analysis of ANOVA demonstrated that two linear factors ([Mo] and agitation speed) significantly influenced the response ($p < 0.05$). Regarding factor interactions, the response was positively impacted by effects ([Mo]Ss and pHSs) with *p*-values less than 0.05. Furthermore, every quadratic term showed extremely statistically significant effects.

3.4 Verification of the Constructed Models

Figure 1 illustrates the results of a thorough verification process that was conducted to confirm the correctness of the selected central composite design model. The link between the observed and expected values for polyphenols, COD, and turbidity is shown in this figure. For each of the three responses, these charts show a significant correlation between the experimental and anticipated values. Notably, the data points closely align with a straight line, confirming the model's suitability for these responses. Furthermore, a random distribution of residuals is shown in Fig. 2, which plots residual values against line numbers, demonstrating the independence of the residuals. The regression models exhibit strong R2 coefficient determination values, ranging from 0.93 to 0.96, for turbidity, COD, and polyphenol removal efficiency. A high R^2 value that is near to 1 indicates a great agreement between the experimental and projected values, according to Bakraouy et al. [2]. The results of the adjusted determination coefficients show that only a small portion of the total variation—13.27%, 9.61%, and 6.67% for the three responses, respectively—remains unexplained by the regression models. A score near 2 for the Durbin-Watson statistic indicates a well-fitting model. The DW statistics in this instance nearly approximate 2, which supports the model's good fit.

3.5 Using RSM for Optimization

The solution's pH level has a significant impact on the speed of contaminants reduction. Seghairi et al. [15] showed in their work that coagulation processes happen quickly at difined pH levels for different types of water. In a related context, the *Moringa oleifera* plant has demonstrated its effectiveness in removing turbidity and contaminants from wastewater, as reported by Villaseñor-Basulto et al. [18]. Using the response surface methodology (RSM) and a centred central composite design (CCD) model, the variables (pH and *Moringa oleifera* concentration) related to the removal of COD and polyphenols were optimized and optimal values were found. In the context of removing COD and polyphenols, the influence of pH and *Moringa oleifera* concentration demonstrates a positive relationship, indicating that higher concentrations of both pH and *Moringa oleifera* lead to more substantial reductions in COD and polyphenols. When considering the optimization process, it becomes clear that at a pH level of 6 and a *Moringa oleifera* concentration of 65 g/L, there is a significant reduction in COD, reaching an efficiency of 88%. Similarly, for polyphenols, a decrease of 86% is observed at a *Moringa oleifera* concentration of 65 g/L and within a pH range spanning from 6 to 10 (Fig. 3).

Table 5 Effects testing and model coefficient estimates for polyphenols, COD, and turbidity

Terme	COD			Turbidity			Polyphenols		
	Coefficient estimé	$F_{statistics}$	*P*-value	Coefficient estimé	$F_{statistics}$	*P*-value	Coefficient estimé	$F_{statistics}$	*P*-value
Constante	81,309.571	–	< 0.0001***	372.57143	–	< 0.0001***	4056.0214	–	< 0.0001***
[Mo]	– 5583.333	15.9305	0.0011**	46.883333	22.672	0.0002***	– 98.52694	6.7704	0.0193*
pH	– 1116.667	0.6372	0.4364 NS	– 31.53333	10.2563	0.0055**	– 52.61325	1.9306	0.1837 NS
Ss	4850	12.0206	0.0032**	– 8.133333	0.6823	0.4209 NS	204.08249	29.0479	< 0.0001***
St	316.66667	0.0512	0.8238 NS	– 87.65	79.2423	< 0.0001***	33.903428	0.8017	0.3839 NS
[Mo]*pH	10,325	36.3187	< 0.0001***	– 50.95	17.8505	0.0006***	51.360193	1.2265	0.2845 NS
[Mo]*Ss	– 525	0.0939	0.7632 NS	30.05	6.2094	0.0241*	– 129.4823	7.7953	0.0131*
pH*Ss	7875	21.1277	0.0003***	– 62.2	26.6037	< 0.0001***	– 175.9699	14.3975	0.0016**
[Mo]*St	7125	17.295	0.0007***	– 45.7	14.3613	0.0016**	– 68.52043	2.183	0.159 NS
pH*St	– 1275	0.5538	0.4675 NS	– 6.95	0.3321	0.5724 NS	– 77.12925	2.766	0.1157 NS
As*St	– 20,625	144.9234	< 0.0001***	21.55	3.1934	0.0929 NS	– 123.5595	7.0985	0.0170*
[Mo]*[Mo]	– 1631.56	1.6208	0.2212 NS	– 30.13452	11.1602	0.0041**	– 260.0367	56.1904	< 0.0001***
pH*pH	– 5931.56	21.4225	0.0003***	– 24.45952	7.3526	0.0154*	– 367.3599	112.1441	< 0.0001***
Ss*Ss	5568.4405	18.8799	0.0005***	23.640476	6.8684	0.0185*	– 416.7171	144.3031	< 0.0001***
St*St	618.44048	0.2329	0.6359 NF	8.1154762	0.8094	0.3816 NS	– 432.213	155.2347	< 0.0001***

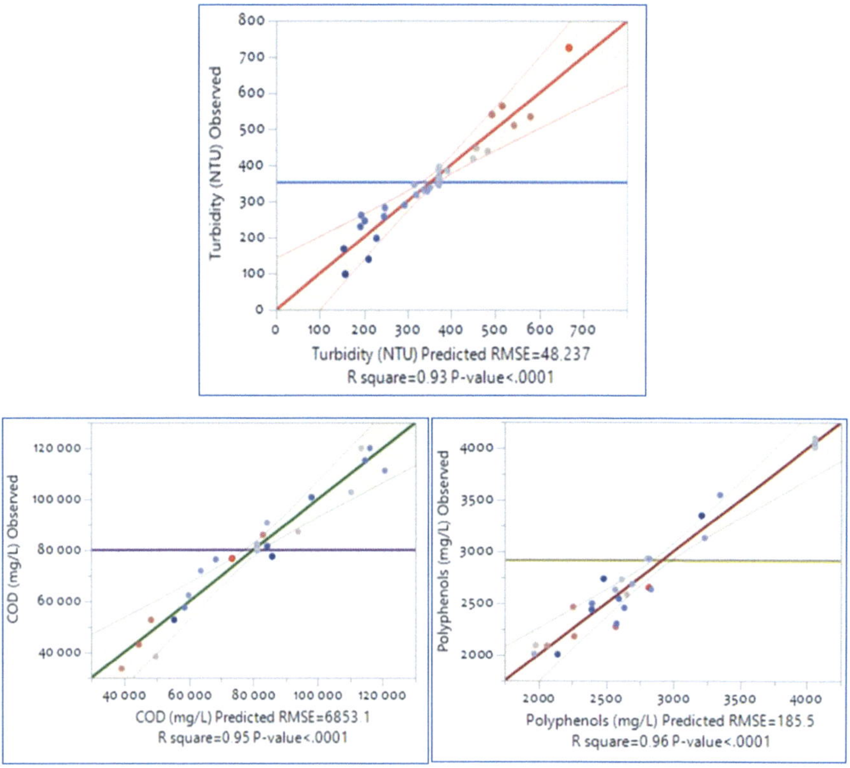

Fig. 1 Correlation between the measured and anticipated amounts of polyphenols, COD, and turbidity

The results of the optimization process for the treatment of olive mill wastewater (OMW) are presented in Table 6. The treatment procedure demonstrates highly effective purification performance, an 88% COD removal rate, and a remarkable removal rate exceeding 86% for phenolic compounds. These findings highlight the efficiency of utilizing response surface methodology (RSM) in combination with the design of experiments approach for optimizing the operational parameters required for treating OMW discharges using a natural coagulant. The optimum conditions for the use of *Moringa oleifera* to treat municipal wastewater include a pH of 6.07, a concentration of 189.69 mg/L, and a settling time of 215.42 min, according to the research done by Adelodun et al. [1]. Under these ideal conditions, the treatment process achieves a remarkable turbidity removal rate of 98.28% and a COD removal rate of 78.52% (Fig. 4).

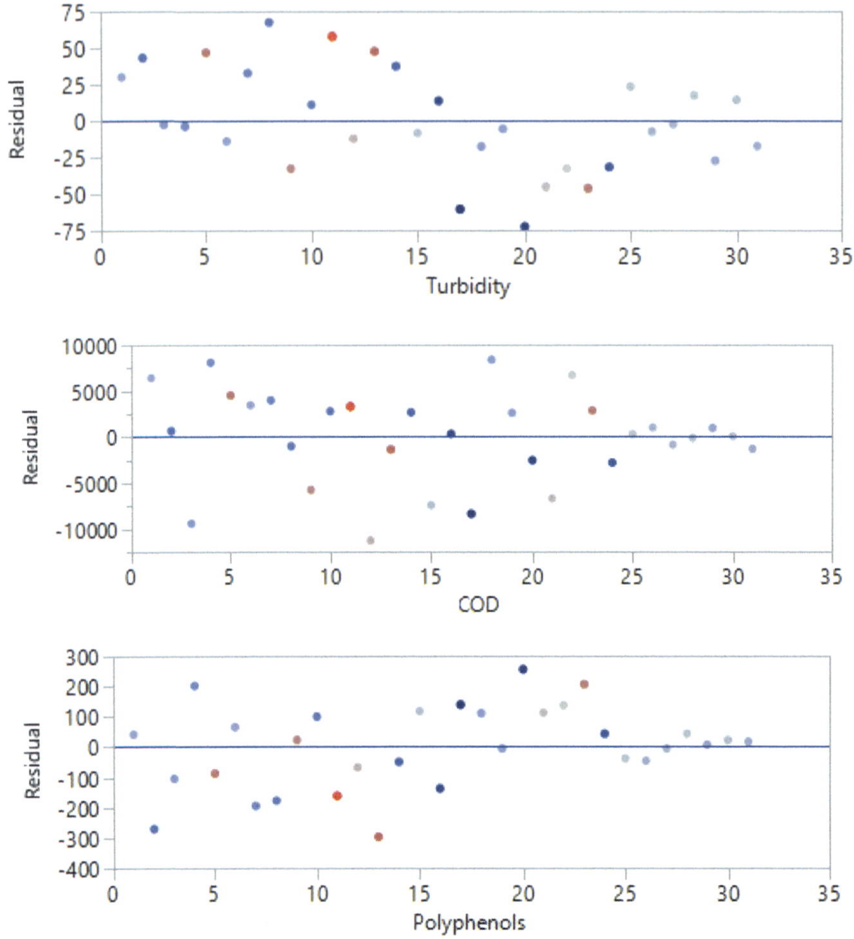

Fig. 2 Residual values based on the sequence of polyphenols, COD, and turbidity

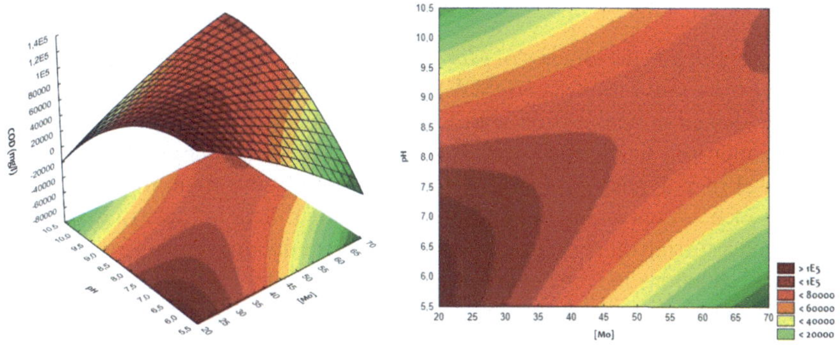

Fig. 3 Plots of central composite design responses for COD based on RSM

Table 6 Optimal values of turbidity, COD, and Polyphenols according to the optimal operating conditions

Responses		Parameters			
		[Mo] (g/L)	pH	Ss (rpm)	St (min)
Turbidity	96%	35	6–8	35	25
COD	88%	65	6	35	25
Polyphenols	86%	65	6/10	30	25

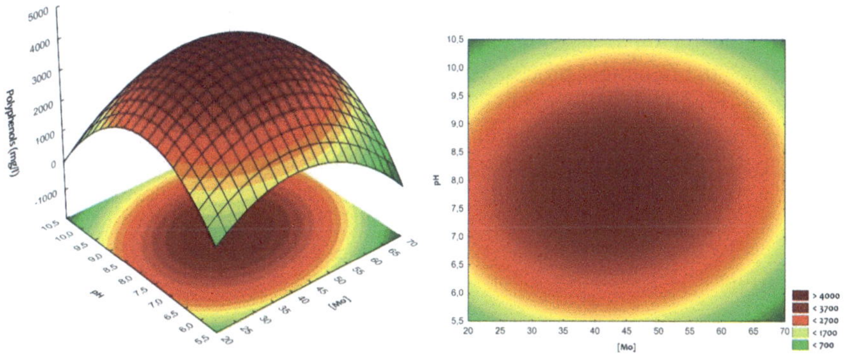

Fig. 4 Central composite design responses surface plots for Polyphenols according to RSM

4 Conclusion

This study highlights the efficacy of *Moringa oléifera* as a highly effective coagulant for treating olive mill wastewater (OMW). Additionally, it demonstrates that a central composite design in conjunction with response surface methodology (RSM) is an excellent approach to optimize treatment parameters, particularly with regard to the removal of polyphenols and COD. In terms of optimization, the study finds that a pH of 6, a concentration of 65 g/L of *Moringa oleifera*, a stirring speed of 35 rpm, and a stirring duration of 25 min are the most effective settings for obtaining significant reductions in COD by 88% and polyphenols by 86%. The study reports that the removal efficiencies for polyphenols and COD are respectively 88% and 86%. These findings show how the Response Surface methodology, in combination with the Central Composite Design technique, can be utilized to efficiently optimize operating parameters for the treatment of wastewater discharged from olive mills.

5 Recommendations

In order to maximize the treatment of olive mill wastewater (OMW) utilizing *Moringa oleifera*, several recommendations may be considered. Examining pre-treatment procedures like filtration or sedimentation can improve *Moringa oleifera's* coagulant properties. The overall efficiency may be increased by combining *Moringa oleifera* with additional types of treatment, such as advanced oxidation techniques or biological treatments. It is essential to assess seasonal variations in OMW characteristics to fine-tune treatment parameters year-round. Conducting economic feasibility studies and environmental impact assessments will ensure the financial viability and ecological safety of the treatment process. In addition, before full-scale implementation, pilot testing should be performed to identify practical challenges. Finally, to keep an ongoing compliance to legal requirements and to avoid negative impacts on water bodies, extended surveillance of treated wastewater is essential.

References

1. Adelodun B, Ajibade FO, Ogunshina MS, Choi K-S (2019) Dosage and settling time course optimization of *Moringa oleifera* in municipal wastewater treatment using response surface methodology. DWT 167:45–56. https://doi.org/10.5004/dwt.2019.24616
2. Bakraouy H, Souabi S, Digua K, Dkhissi O, Sabar M, Fadil M (2017) Optimization of the treatment of an anaerobic pretreated landfill leachate by a coagulation–flocculation process using experimental design methodology. Process Saf Environ Prot 109:621–630. https://doi.org/10.1016/j.psep.2017.04.017
3. Benalia A, Baatache O, Derbal K, Khalfaoui A, Amrouci Z, Pizzi A, Panico A (2024) The use of central composite design (CCD) to optimize and model the coagulation-flocculation process

using a natural coagulant: application in jar test and semi-industrial scale. J Water Process Eng 57:104704. https://doi.org/10.1016/j.jwpe.2023.104704
4. Bouyakhsass R, Souabi S, Rifi SK, Bouaouda S, Taleb A, Madinzi A, Kurniawan TA, Anouzla A (2023) Applicability of central composite design and response surface methodology for optimizing treatment of landfill leachate using coagulation-flocculation. Chem Eng Res Des 197:669–684. https://doi.org/10.1016/j.cherd.2023.08.001
5. Bouyakhsass R, Souabi S, Rifi SK, Taleb A, Pala A, Madinzi A (2023) Optimization of coagulation-flocculation for landfill leachate treatment: an experimental design approach using response surface methodology. Environ Nanotechnol Monit Manage 20:100841. https://doi.org/10.1016/j.enmm.2023.100841
6. Domingues E, Fernandes E, Gomes J, Castro-Silva S, Martins RC (2021) Olive oil extraction industry wastewater treatment by coagulation and Fenton's process. J Water Process Eng 39:101818. https://doi.org/10.1016/j.jwpe.2020.101818
7. Espinoza-Quiñones FR, Módenes AN, Theodoro PS, Palácio SM, Trigueros DEG, Borba CE, Abugderah MM, Kroumov AD (2012) Optimization of the iron electro-coagulation process of Cr, Ni, Cu, and Zn galvanization by-products by using response surface methodology. Sep Sci Technol 47(5):688–699. https://doi.org/10.1080/01496395.2011.629396
8. Getahun M, Befekadu A, Alemayehu E (2024) Coagulation process for the removal of color and turbidity from wet coffee processing industry wastewater using bio-coagulant: optimization through central composite design. Heliyon 10(7). https://doi.org/10.1016/j.heliyon.2024.e27584
9. Hodaifa G, Gallardo PAR, García CA, Kowalska M, Seyedsalehi M (2019) Chemical oxidation methods for treatment of real industrial olive oil mill wastewater. J Taiwan Inst Chem Eng 97:247–254. https://doi.org/10.1016/j.jtice.2019.02.001
10. Jaafar A, Darchen A, Hamzi SE, Lakbaibi Z, Driouich A, Boussaoud A, Yaacoubi A, El Makhfouk M, Hachkar M (2021) Optimization of cadmium ions biosorption by fish scale from aqueous solutions using factorial design analysis and Monte Carlo simulation studies. J Environ Chem Eng 9(1):104727. https://doi.org/10.1016/j.jece.2020.104727
11. Jebun N, Al-Mamun A, Alam MdZ, Ahmad Raus R (2018) Optimization of Flocculation Process for a New Myco-Coagulant to Reduce Water Turbidity. In: Yacob NA, Mohd Noor NA, Mohd Yunus NY, Lob Yussof R, Zakaria SAKY (eds) Regional conference on science, technology and social sciences (RCSTSS 2016). Springer, Singapore, pp 271–281
12. Karyab H, Ghasemi M, Ghotbinia F, Nazeri N (2023) Efficiency of chitosan nanoparticle with polyaluminum chloride in dye removal from aqueous solutions: optimization through response surface methodology (RSM) and central composite design (CCD). Int J Biol Macromol 249:125977. https://doi.org/10.1016/j.ijbiomac.2023.125977
13. Khennoussi A, Chaouch M, Chahlaoui A (2013) Traitement des effluents d'abattoir de viande rouge par électrocoagulation-flottation avec des électrodes en fer. rseau 26(2):135–150. https://doi.org/10.7202/1016064ar
14. Manikandan S, Subbaiya R, Saravanan M, Ponraj M, Selvam M, Pugazhendhi A (2022) A critical review of advanced nanotechnology and hybrid membrane based water recycling, reuse, and wastewater treatment processes. Chemosphere 289:132867. https://doi.org/10.1016/j.chemosphere.2021.132867
15. Seghairi N, Mimeche L, Bouzid A, Ayachi Y (2017) Traitement des eaux usees par coagulation-floculation en utilisant le sulfate d'aluminium comme coagulant. J Water Environ Sci 1:230–234
16. Singleton VL, Rossi JA (1965) Colorimetry of total Phenolics with phosphomolybdic-phosphotungstic acid reagents. Am J Enol Vitic 16(3):144–158
17. Tiwari AK, Pal DB (2022) Chapter 11—Nutrients contamination and eutrophication in the river ecosystem. In: Madhav S, Kanhaiya S, Srivastav A, Singh V, Singh P (eds) Ecological significance of river ecosystems. Elsevier, pp 203–216

18. Villaseñor-Basulto DL, Astudillo-Sánchez PD, del Real-Olvera J, Bandala ER (2018) Wastewater treatment using *Moringa oleifera Lam* seeds: a review. J Water Process Eng 23:151–164. https://doi.org/10.1016/j.jwpe.2018.03.017
19. Zheng H, Zhu G, Jiang S, Tshukudu T, Xiang X, Zhang P, He Q (2011) Investigations of coagulation–flocculation process by performance optimization, model prediction and fractal structure of flocs. Desalination 269(1):148–156. https://doi.org/10.1016/j.desal.2010.10.054

Sustainable Operation of Textile Wastewater Using Calcium Hypochlorite Oxidation Followed by Waste Iron Slag-Aided Micro-filtration

Md. Sahil Rafiq, Mohammad Shakhawat Hosen Apurba, Mohammad Moshiur Rahman, and Nadim Reza Khandaker

Abstract Bangladesh's primary source of foreign exchange earnings is the textile sector, employing over three million women. The industry's unexpected growth has led to adverse repercussions on the environment, shown by the release of untreated textile wastewater into aquatic bodies. Compliance concerns emerge due to the exorbitant expenses associated with treatment, which result from the utilization of expensive imported chemicals and reliance on energy-intensive equipment. A series of technology development projects were conducted at the bench and field scale to find an environmentally friendly approach for treating textile effluent utilizing materials and chemicals that are readily available in the local supply chain. The process involved chemical oxidation using calcium hypochlorite followed by waste blast furnace iron slag- aided micro-filtration. Complete removal of color, reduction in Chemical Oxygen Demand (COD), Total Dissolved Solids (TDS), and Total Suspended Solids (TSS) were achieved, producing waster in compliance with the discharge limits for both as a pretreatment option and as final polishing post-tertiary treatment. The developed process, reliant on locally available resources, has the potential to serve as a sustainable textile wastewater treatment option for resource-challenged Bangladesh and other resource-limited economies in South Asia. Implementing the developed process may help achieve Sustainable Development Goals.

Keywords Textile · Local supply chain · Sustainable development goals · Wastewater

Md. Sahil Rafiq · M. S. H. Apurba · M. M. Rahman · N. R. Khandaker (✉)
North South University, Dhaka, Bangladesh
e-mail: nadim.khandaker@northsouth.edu

Md. Sahil Rafiq
e-mail: sahil.rafiq@northsouth.edu

M. S. H. Apurba
e-mail: m.151102@s.pust.ac.bd

S. Yadav et al. (eds.), *Emerging Technologies in Biological and Hybrid Wastewater Treatment: Lessons from Developed to Enhancing Practices in Developing Countries*,
Earth and Environmental Sciences Library,
https://doi.org/10.1007/978-3-031-74515-7_12

1 Introduction

The Bangladeshi economy's swiftly expanding sector presents a distinctive opportunity for profitable expansion into emerging specialized markets. Bangladesh's textile and apparel industry is instrumental in the country's export earnings, as it encompasses a diverse array of activities, including the production of knitwear, casual clothing, and fashionable high-street fashion, in addition to spinning and weaving [1]. The Ready-made Garments (RMG) sector is the foundation of the Bangladeshi economy, functioning as the largest export-oriented industry and exhibiting substantial growth over the past 25 years. This sector plays a critical role in the generation of employment opportunities and the enhancement of the incomes of the underprivileged. This sector directly employs approximately four million individuals, while over twelve million individuals are indirectly involved. Over the past 25 years, there has been a significant increase in the number of textile and apparel units, rising from 180 to over 4000 [2]. Agriculture has historically served as the basis of the Bangladeshi economy and the primary source of income for the rural populace.

The nation focuses a significant emphasis on the growth of its export industry, alongside agriculture. The export portfolio of Bangladesh has undergone significant changes and improvements over time [2]. When Bangladesh was established, the primary export industries were tea and jute. Nevertheless, the economic importance of the jute sector has diminished as a result of recurring challenges, including flooding, declining jute prices, and reduced global demand. Bangladesh's sole multi-billion-dollar manufacturing and export sector is the RMG industry. The sector has consistently generated employment at double-digit rates and contributed to foreign exchange earnings. Figure 1 illustrates a bar graph that illustrates the export revenues generated by Bangladesh's RMG industry over the past decade. The revenue from apparel exports in Bangladesh was $19.08 billion in the fiscal year 2011–12 and increased to $31.45 billion by the end of the decade, as demonstrated by the visual representation.

According to data from the Export Promotion Bureau (EPB), Bangladesh's garment exports reached $46.99 billion in the fiscal year 2022–23, representing a

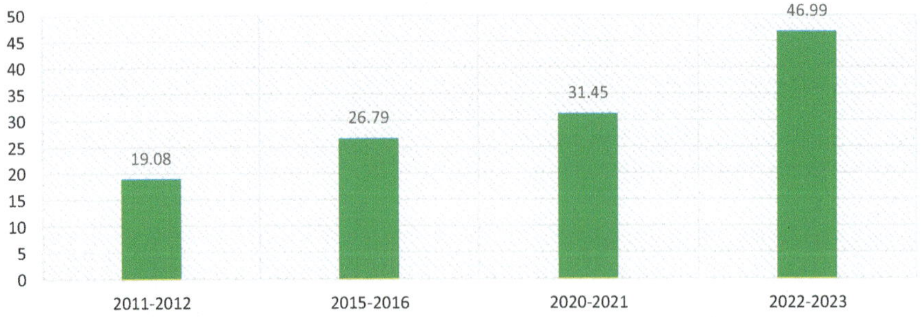

Fig. 1 Export earnings from the RMG sector are measured in billions of billion dollars by the Export Promotion Bureau (EPB) and BGMEA [3]

significant growth of 10.27% [4]. Remarkably, this performance surpassed the fiscal year target by 0.41%. The RMG industry recorded a revenue of $42.61 billion in the previous fiscal year, 2021–22. The knitwear subcategory within the RMG industry contributed $25.73 billion to the FY23 revenues, with a growth rate of 10.87%. Woven garments generated $21.25 billion, showing a growth of 9.56%. In contrast, the home textiles segment had a decrease, with exports amounting to $1.09 billion, indicating a negative growth rate of 32.47%. Moreover, these factors emphasis the dynamic and flourishing nature of Bangladesh's RMG industry, showcasing its impressive trajectory of expansion and progress. Textile wastewater is a significant environmental concern due to its high concentration of diverse pollutants and chemicals [5]. The issue is worsened by the clustering of water-intensive and highly polluted industrial facilities in Greater Dhaka, the densely populated capital of Bangladesh [6]. In order to mitigate the pollution caused by industrial effluents, it is imperative to restrict the release of contaminants at their source [7].

In addition to Bangladesh, other South Asian nations also make significant contributions to the export of ready-made clothes. India is presently the fifth-largest manufacturer of technical textiles globally, with a market value of around $24.3 billion [8]. The ambitious objective is to increase the size of this industry to an impressive $300 billion by 2047, coinciding with India's centenary celebration. The textile and apparel sector in India has strengths across the whole value chain, including fibers, yarns, fabrics, and finished products. India recorded its highest-ever textiles exports in the fiscal year 2021–22, exceeding the milestone of $44 billion. India is the largest global producer of cotton, accounting for 23% of the total global output. Moreover, it can boast of possessing the most extensive area of land dedicated to cotton farming, covering 39% of all global cotton-growing regions. India produced 9.5 million bales of raw jute in the fiscal year 2022–23. The Indian domestic textiles and apparel industry had a total value of $154 billion in 2022. It is expected to experience steady growth, aiming for a Compound Annual Growth Rate between 10 and 12%, with the goal of reaching $350 billion by 2030 (Textile Industry in India—Garment & Apparels Market in India).

Pakistan ranks 8th among the leading textile product exporters in Asia and holds the 4th position worldwide in cotton output, making it the third largest consumer of cotton. The textile industry in Pakistan accounts for a significant 46% of the country's total manufacturing sector and is a major source of employment for almost 40% of the population. It is worth mentioning that just a small 5% of Pakistan's textile companies are listed on the stock exchange. In all, there are 423 textile enterprises now operating in the country. Pakistan benefits from a wide-ranging assortment of raw materials, which includes a comprehensive selection of both artificial and natural yarns and fabrics, such as cotton, rayon, and various others. Pakistan benefits greatly from its diverse range of raw materials, which has a substantial impact on reducing costs and improving operational efficiency in the textile industry [9]. The garment export business in Sri Lanka is an essential and vital component of the country's economy, showing substantial expansion over the past forty years. The sector's value in 2021 amounted to $5.42 billion, exhibiting a notable year-on-year export growth of 22.93%. To address economic difficulties related to obtaining resources, the clothing industry is striving to reach $8 billion in exports by 2025, with a particular focus on increasing investment in domestic supply chains.

At present, the garment sector is the dominant export business in Sri Lanka, accounting for almost 44% of total exports and creating roughly 33% of the manufacturing jobs in the country [10]. Nepal's textile market sector predominantly exports its products to India, with an approximate export value of USD 116.11 million, accounting for a market share of 35.5%. Furthermore, the United States serves as a substantial export market, giving a noteworthy USD 116.11 million to the industry, which accounts for 35.5%. Turkey has exports worth USD 46.98 million, accounting for 14.36% of the total. Germany and the United Kingdom provide contributions of USD 23.63 million (7.22%) and USD 22.26 million (6.81%), respectively, to the textile industry in Nepal's apparel and garment sector [11]. Afghanistan's textile market sector predominantly exports its products to the worldwide market, with an approximate export value of USD 115.83 million, representing a market share of 33.6%. South Asia is a notable participant in the worldwide market, as it contributes significantly with exports totaling USD 115.83 million, which also represents 33.6% of the market. Pakistan closely trails behind with exports of USD 70.05 million, accounting for a 20.32% share.

The East Asia & Pacific region contributes USD 31.4 million (9.11%) and China contributes USD 30.47 million (8.84%) to the industry [12]. The majority of Bhutan's textile exports, valued at over USD 0.88 million, are directed towards the Indian market, accounting for a significant market share of 99.5%.

Nepal is a significant beneficiary of Bhutanese textile exports, receiving USD 0.88 million, which accounts for a considerable 99.5% stake in the sector. On the other hand, the United States, Germany, and Switzerland have minimal involvement in the export market, with contributions of only 0.11%, 0.07%, and 0.06% respectively [13]. The textile market in the Maldives predominantly exports its products to Germany and France, with both countries having an equal market share of 50%. Thailand's proportion of exports is rather tiny, amounting to 25%. In contrast, exports to Malaysia and Oman are negligible, making up only 0% of the market share [14].

2 Existing Treatment Processes and Their Limitations in Terms of Sustainable Operations

2.1 Brief Overview of the Textile Processes

The textile manufacturing industry in Bangladesh is vital for generating foreign currency and providing a significant contribution to the country's GDP [15]. The rapid growth of the textile dyeing industry has propelled it to become the dominant sector in Bangladesh, despite its substantial production of wastewater. Textile production is a complex and detailed process that involves three main stages: the transformation of fibers into yarn, the conversion of yarn into fabric, and the final

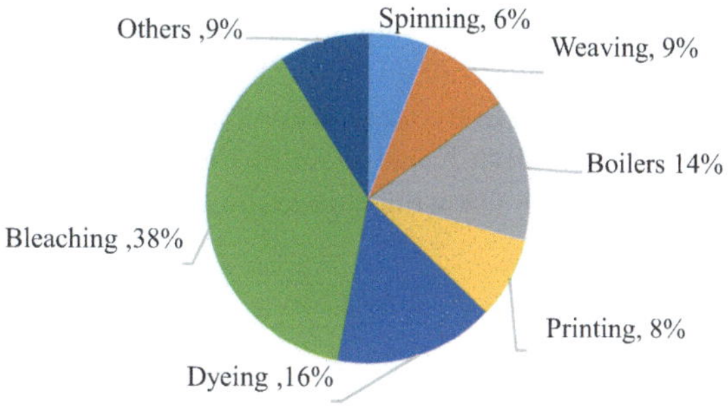

Fig. 2 Water consumption in wet textile processing [16]

step of creating textiles. These textiles are later used to make garments and various other things. The primary water-intensive activities and procedures that generate wastewater throughout these stages comprise sizing, de-sizing, bleaching, mercerizing, and dyeing [16]. Figure 2 displays the proportion of water usage in several textile wet processes. During each stage of the process, a wide variety of chemicals is used, including acids, alkalis, chlorinated inorganic compounds, sodium hypochlorite, organic substances such as dyes, bleaching agents, finishing chemicals, starch, thickening agents, surfactants, wetting and dispersing agents, as well as different metal salts [17].

The volume and makeup of wastewater releases depend on several factors, including the type of material being treated and the methods used. Changes in equipment, choice of chemicals, or other process factors can potentially affect the properties of the wastewater that is released. As previously stated, it is important to emphasis that chemical and wet processing in the textile sector alone contribute to 70% of pollution [18].

2.2 *Environmental Impact Arising from Untreated Textile Effluent*

Textile factories in Bangladesh are mainly situated near aquatic ecosystems. These factories often release a wide range of harmful substances, including both organic and inorganic compounds, directly into these aquatic environments without any treatment. Consequently, industrial establishments frequently release their trash directly into rivers without taking into account the inevitable ecological harm. The textile and dyeing industries generate wastewater, known as effluent, as a result of their manufacturing processes. This effluent consists of a range of pollutants [19]. The decomposition of organic components has a detrimental effect on water quality by leading to a reduction in quantities of dissolved oxygen. Non-biodegradable organic compounds remain in the aquatic ecosystem for long periods of time and can potentially infiltrate the food chain. Inorganic pollutants mostly comprise metallic salts, as well as alkaline and acidic chemicals. These non-living components participate in different chemical and biological reactions within the river ecosystem, resulting in a decline in the overall quality of the water. Table 1 presents a comparison of the wastewater quality in five textile companies before and after undergoing treatment [20]. The Bangladesh Department of Environment has recently implemented laws that define limitations on the colour and temperature thresholds for releasing treated wastewater [21].

2.3 Existing Methods for Treating Textile Wastewater

Effluent treatment techniques can be categorised into three primary classifications: physical, chemical, and biological methodologies. The earliest treatment steps largely consist of physical processes such as screening, sedimentation, flotation, and flocculation. These treatments are specifically intended to remove fibrous residues, unabsorbed chemicals, and solid particles. However, they are not as effective in removing colour. The following successive stages focus on reducing the amount of organic matter and involve a combination of physicochemical separation and biological oxidation methods. While depending entirely on biological therapy is inadequate for completely eradicating colour, chemical approaches might be costly. Physicochemical separation employs chemisorptions forces to separate colloidal organic molecules from the liquid phase. The importance of tertiary therapy stages has increased, however, they can significantly raise treatment costs. These processes are crucial for removing colour, but it is important to acknowledge that no individual treatment procedure can address all colour variances. It is crucial to protect biological processes from sudden increases in industrial pollutants, especially those containing dangerous substances. The initial stages involve activities such as equalisation, neutralisation, and disinfection. The underwhelming performance of the settling tank can be attributed to significant variations in flow, temperature, and waste liquor composition. These changes lead to the formation of convection currents and layering, which disrupt the normal sedimentation process. It is crucial to have a suitable balance in the flow and composition, as well as to include cooling or heat recovery. Chlorine disinfection safeguards microorganisms in the biological phase by preventing potential harm from toxic contamination [22]. Figures 3 and 4 illustrate the standard configurations of physiochemical and biological treatment plant systems.

Preliminary or Pretreatment: The preliminary or pretreatment stage of wastewater treatment involves the elimination of significant solid substances, such as cloth pieces, twigs, pebbles, and fats, that could potentially cause equipment damage or operational difficulties. The initial treatment primarily consists of two essential procedures: (i) Screening, used in effluent treatment facilities to remove relatively large solid waste, and (ii) Equalisation and Skimming, which aim to remove grease and oil, establish consistency, and reduce a portion of the Biochemical Oxygen Demand (BOD) and Chemical Oxygen Demand (COD). The main treatment is on removing items that can either float or settle, such as suspended solids and chemical organic

Table 1 Wastewater quality before and after treatment from five textile industries

Water quality parameters		Textile A	Textile B	Textile C	Textile D	Textile E	ECR, 2023 Standard [21]
p^H	BT	8.66 ± 0.02	7.95 ± 0.03	8.65 ± 0.03	8.05 ± 0.01	7.48 ± 0.03	6–9
	AT	7.88 ± 0.03	7.56 ± 0.02	8.48 ± 0.03	7.85 ± 0.01	7.36 ± 0.02	
BOD_5 (mg/l)	BT	340 ± 14.42	336 ± 6.65	425 ± 6.02	316 ± 9.07	298 ± 14.57	30
	AT	188 ± 5.60	81 ± 4.16	211 ± 7.45	185 ± 7.67	167 ± 9.70	
COD (mg/l)	BT	1131 ± 25.65	1044 ± 16.78	1316 ± 17.14	1422 ± 20.22	973 ± 7.22	200
	AT	197 ± 11.93	136 ± 6.60	492 ± 7.53	277 ± 6.55	113 ± 3.22	
TDS (mg/l)	BT	3855 ± 31.41	4226 ± 26.68	3227 ± 35.85	3923 ± 32.14	2775 ± 24.33	2100
	AT	2481 ± 22.71	2608 ± 31.29	2982 ± 36.57	2296 ± 32.22	1176 ± 34.72	
TSS (mg/l)	BT	245 ± 17.68	145 ± 6.50	159 ± 6.03	263 ± 11.72	116 ± 7.03	100
	AT	164 ± 10.21	76 ± 7.23	103 ± 4.72	158 ± 4.04	85 ± 3.60	

BT before treatment, *AT* after treatment

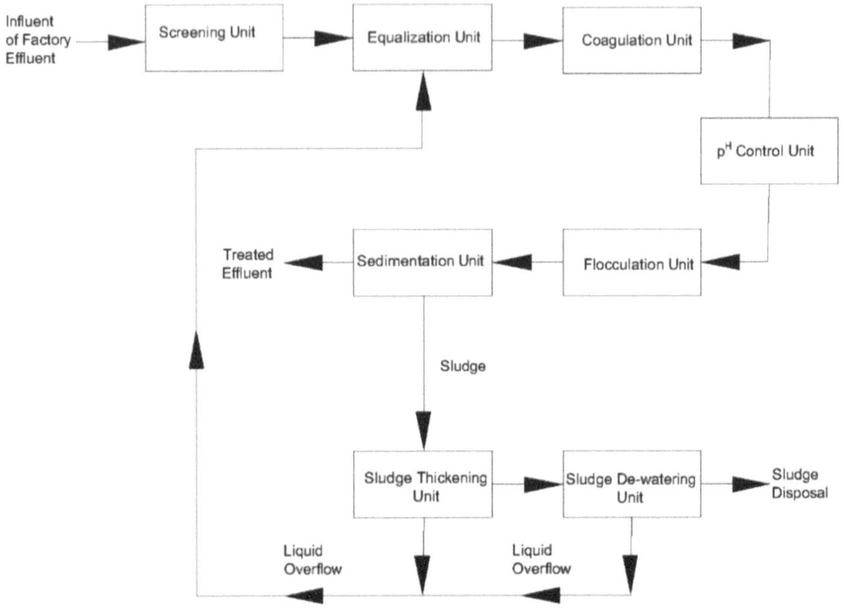

Fig. 3 Typical schematic representation of a physiochemical wastewater treatment facility [27]

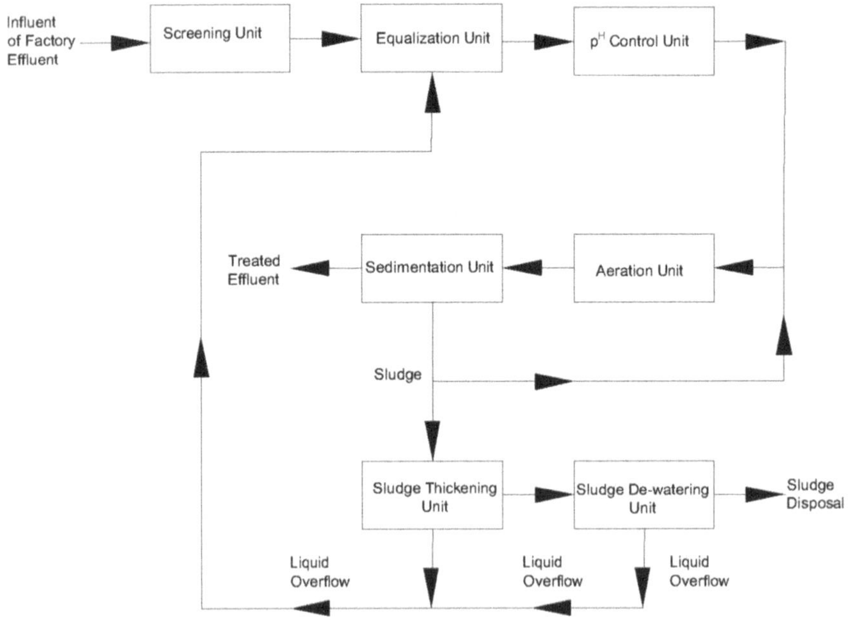

Fig. 4 Typical schematic representation of a biological wastewater treatment facility

waste. The primary treatment includes the following procedures: Coagulation is the mechanism by which suspended particles are induced to combine and form bigger entities. Flocculation is the process of promoting the aggregation of particles into larger clusters, known as flocs, through the use of coagulants. The objective of neutralisation is to adjust the pH level to fall within the range of 6.5–8.5. Sedimentation is utilised to separate and remove fine suspended particles from a solution [23]. Secondary treatment is conducted to eliminate biodegradable organic substances, encompassing the decrease of both BOD (biochemical oxygen demand) and COD (chemical oxygen demand), as well as the decomposition of organic waste. Biological treatment can occur in different ways, including aerobic and anaerobic processes [24].

Tertiary treatment: Due to the intricate and ever-changing nature of wastewater produced by the textile industry, it is improbable that a solitary treatment method can suffice for comprehensive effluent treatment. Tertiary treatment focuses on eliminating any remaining suspended or dissolved solids. Many wastewater treatment plants utilise membrane technologies such as microfiltration, ultrafiltration, nanofiltration, and reverse osmosis. In Bangladesh, cost-effective options such as granular media filters and activated carbon filters are commonly used for this particular purpose [25, 26].

2.4 *Limitations of Existing Treatment Methods*

Given the complex and variable characteristics of wastewater from the textile industry, it is unlikely that a single treatment approach will be suitable for managing all aspects of wastewater effectively. Physicochemical treatments are the preferred choice in textile industry due to their cost-effectiveness and space requirements, making them more favourable than biological or tertiary treatments. Table 2 highlights some limitations connected with physical, chemical, and biological treatment techniques.

3 Defining the Novel Process of Sustainable Operation of Textile Wastewater Using Calcium Hypochlorite Oxidation Followed by Waste Iron Slag-Aided Micro-filtration

3.1 *Brief Overview of the Proposed Process*

The aim of this study was to establish an environmentally acceptable method for treating textile effluent using materials that are easily accessible in the local area. Due to the presence of diverse chemicals in textile wastewater, a singular treatment

Table 2 Limitations of existing treatment processes in textile wastewater treatment

Process	Limitations
Biological	Dyes are widely regarded as hazardous and exhibit a notable degree of resistance to natural degradation processes. Acid and reactive dyes have high solubility in water and show very little adsorption onto sludge [28]
Coagulation	The efficacy of elimination is contingent upon the pH level and results in the generation of a significant amount of sludge. It lacks effectiveness in removing highly soluble colours and produces unsatisfactory outcomes when dealing with azo, reactive, acid, and cationic dyes [29–31]
Activated Carbon adsorption	The efficacy of the elimination technique is influenced by pH levels and is insufficient when addressing dispersion sulphur and vat dyes. System restoration involves significant costs and leads to the depletion of adsorbent material, requiring expensive disposal methods [32]

approach is inadequate for the complete elimination of colour, chemical oxygen demand (COD), and total dissolved solids (TDS). In order to address the potential difficulties associated with costs and compliance with regulations regarding imported decoloring agents. Our research efforts are directed towards establishing a decoloring technique that utilizes materials from the local area [27]. This policy seeks to reduce our dependence on foreign-imported chemicals. In order to assess the efficacy of the novel concept, we conducted a bench scale pre-treatment of wastewater from the equalization tank and also a tertiary treatment of treated wastewater from the outlet of the effluent treatment plant in textile facilities producing woven and knit fabrics.

3.2 Experimental Protocol

Bench scale Pre-treatment

The study assessed the effectiveness of a proposed method for treating wastewater from textile production, specifically in a denim manufacturing facility, utilizing a small bench-scale experimental system. The suggested procedure comprises multiple stages: initially, the wastewater is subjected to chemical oxidation by the use of calcium hypochlorite, followed by a sand filtration including a bed of waste iron nails. These filings act as a reservoir of iron in various oxidation states, such as Fe^0, Fe^{2+}, and Fe^{3+} [33]. Subsequently, the water undergoes rapid sand filtering. The discarded iron filings in this procedure serve two functions: they function as a coagulant during the quick sand filtering and also counteract any surplus hypochlorite that may persist following the oxidation stage. The sand acts as the filtration medium to eliminate any remaining particulate debris during the coagulation process. Gravel serves as a strong foundation for supporting both the layer of iron filings underneath and the layer of granular sand on top. This configuration avoids the iron particle matter from being transported across the bed by the water flow and guarantees that the sand remains intact and does not erode with the flowing water. The experiment involved mixing textile wastewater with calcium hypochlorite in a 1000 ml container and allowing them to react for a period of 5–7 min in a reactor. Afterwards, the treated wastewater was left undisturbed for duration of 1.5 h, mimicking the processes conducted in a clarifier. A total of fifty millilitres of the resultant transparent solution were collected for analysis, while the remaining transparent solution was filtered using a multimedia filter, taking advantage of the force of gravity. The investigation persisted until a decline in filtration rate occurred, signifying bed breakthrough. To view a graphical illustration, please consult Fig. 5 supplied.

Pilot scale tertiary-treatment

At first, the wastewater was subjected to calcium hypochlorite oxidation in a reaction vessel with a volume of 30,000 ml. The mixture was rapidly stirred for roughly 3–5 min. Afterwards, the effluent underwent a 30-min settling phase in a pilot-scale clarifier. The supernatant was sent through a multimedia filter system that used leftover blast furnace iron slag, gravel, and sand. A 100 ml sample was collected for further analysis. The waste blast furnace iron slag had a dual function: firstly, it operated as a coagulant in the filtration process, and secondly, it efficiently neutralized hypochlorite produced during the oxidation of the wastewater. The sand in this filtering system acted as the main medium for trapping any leftover particle

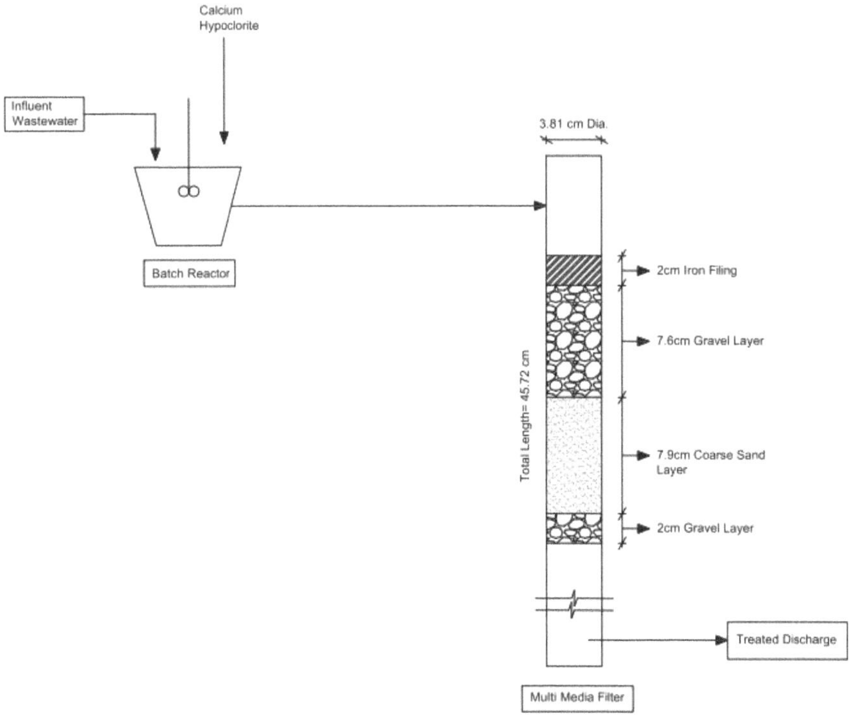

Fig. 5 Schematic representation of the bench-scale configuration utilized in the experimental project

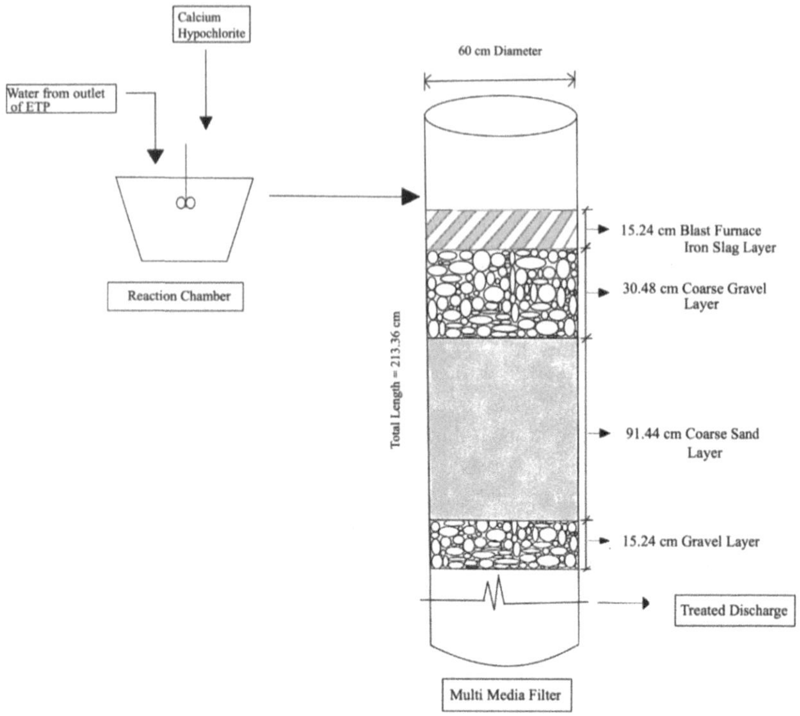

Fig. 6 Schematic representation of the pilot-scale configuration utilized in the experimental project

debris. The presence of multiple layers of gravel was essential for providing support to the filter bed and avoiding the displacement of sand due to the flow of water. The multimedia filter system, in this particular configuration, effectively captured iron particle matter and provided protection against sand erosion. A comprehensive representation of the complete process was illustrated in a flow diagram, as seen in Fig. 6. The investigation was done to evaluate the performance of the filtration system until a decline in the filtration rate was detected.

3.2.1 Examine Wastewater from Textile Dyeing and Processing

Bench scale Pre-treatment

In this experimental endeavor, we employed wastewater generated during the fabrication of blue denim textiles. The wastewater sample was obtained from the equalization tank of a specific wastewater treatment facility designed exclusively for the treatment of wastewater generated during the production of denim textiles. We collected a composite sample that represented a complete 24-h period in order to evaluate the efficiency of the treatment process. This sample accurately reflected the proportionate composition of the wastewater produced throughout the plant's continuous operations.

Pilot Scale Tertiary treatment

The wastewater samples were collected at the effluent treatment facility's output immediately after the secondary clarity procedure. These samples represent the beginning water source for the tertiary treatment phase. To evaluate the effectiveness of the treatment technique, we collected composite samples at regular intervals throughout a 24-h period, corresponding to the operating hours of the facility. This study aimed to evaluate the effectiveness of using calcium hypochlorite-assisted rapid sand filtration in removing color and meeting the discharge standards set by the Department of Environment (DOE) in Bangladesh in 2023.

3.2.2 The Oxidation Process Using Calcium Hypochlorite

Bench scale Pre-treatment

An optimisation technique was conducted on the wastewater generated during the dyeing of blue denim textiles by administering calcium hypochlorite. The augmentation process consisted of systematically injecting powdered calcium hypochlorite into a well-mixed batch reactor. The 1000 ml batch reactor was stirred manually using a wooden spatula until the effluent turned from blue to transparent. The entire mixing operation was thoroughly monitored and found to necessitate around five to seven minutes in total. To guarantee the accuracy and uniformity of the outcomes, a grand total of 24 batch tests were carried out.

Pilot Scale Tertiary treatment

To optimize the efficacy of calcium hypochlorite dosage, we performed an oxidation procedure on the treated wastewater obtained from textile dyeing. We gradually added calcium hypochlorite powder to the wastewater until it transitioned from an orange colour to a completely colourless state. The chemical conversion occurred in a completely mixed batch reactor with a liquid capacity of 30,000 ml. The contents of the batch reactor were thoroughly blended using a propeller mechanism, which systematically incorporated the calcium hypochlorite. The conspicuous change in the wastewater's hue, shifting from orange to colourless, signified the effective culmination of the oxidation process. In order to guarantee the dependability and uniformity of the outcomes, a comprehensive set of 50 batch tests were carried out.

4 Performance of Bench Scale Treatment Process

The initial phase of the treatment process was the use of calcium hypochlorite to chemically convert the wastewater generated during the dyeing of blue denim garments. The oxidation process was conducted using a continuous batch system equipped with a high-speed mixing reactor. The mean quantity of locally available calcium hypochlorite needed to completely eliminate color from each liter of effluent

Table 3 Analysis of the composite textile wastewater sample collected over a specific time period, taken from the inlet of the wastewater treatment facility and the multilayer filter

Parameter	Inlet average ± std	Outlet average ± std
pH	12 ± 0.2	8.2 ± 0.2
Color, Pt-co	243 ± 11	0
Turbidity, NTU	120 ± 10	10 ± 2
Total suspended solids, mg/l	159 ± 4	8 ± 2
Total dissolved solid, mg/l	422 ± 2	157 ± 4
Chemical oxygen demand, mg/l	590 ± 79	9 ± 3
Residual Cl^-, mg/l	Not detected	Not detected

was approximately 17.6 ± 0.5 g. Initially, the chemical oxygen demand level was 590 ± 79 mg/L (as indicated in Table 3), which decreased to 13.5 ± 1.5 mg/L (as seen in Fig. 7). In addition, the starting concentrations of total dissolved solids (422 ± 2 mg/L) and total suspended solids (159 ± 4 mg/L), as stated in Table 3, were decreased to 6.9 ± 0.7 mg/L and 8 ± 0.2 mg/L, respectively (as shown in Fig. 7c and Fig. 7e). The results are displayed in Table 3. The breakthrough profile of several water quality metrics is shown in Fig. 7, and the visual depiction of the colour change can be viewed in Fig. 8.

5 Performance of Pilot-Scale Treatment Process

The first phase in the treatment process involves the utilisation of calcium hypochlorite to oxidise the wastewater generated during the dyeing of blue denim fabrics. The oxidation process was conducted in a continuous batch configuration utilising a high-speed mixing reactor. The average amount of locally accessible calcium hypochlorite needed to eliminate colour completely from each litre of effluent was around 6.75 ± 0.5 g. The oxidation process led to a substantial decrease in the chemical oxygen demand of the wastewater, reducing it from an initial level of 154 ± 12 mg/L (as specified in Table 4) to 9 ± 3 mg/L (as indicated in Fig. 9a). In addition, the initial concentrations of total dissolved solids (1971 ± 8 mg/L) and total suspended solids (60 ± 8 mg/L), as stated in Table 4, fell to 157 ± 4 mg/L and 8 ± 2 mg/L, respectively, as shown in Fig. 9c, e. The findings are displayed in Table 4, while the significant characteristics of several water quality metrics are depicted in Fig. 9. Figure 10 displays the visual depiction of the colour alteration.

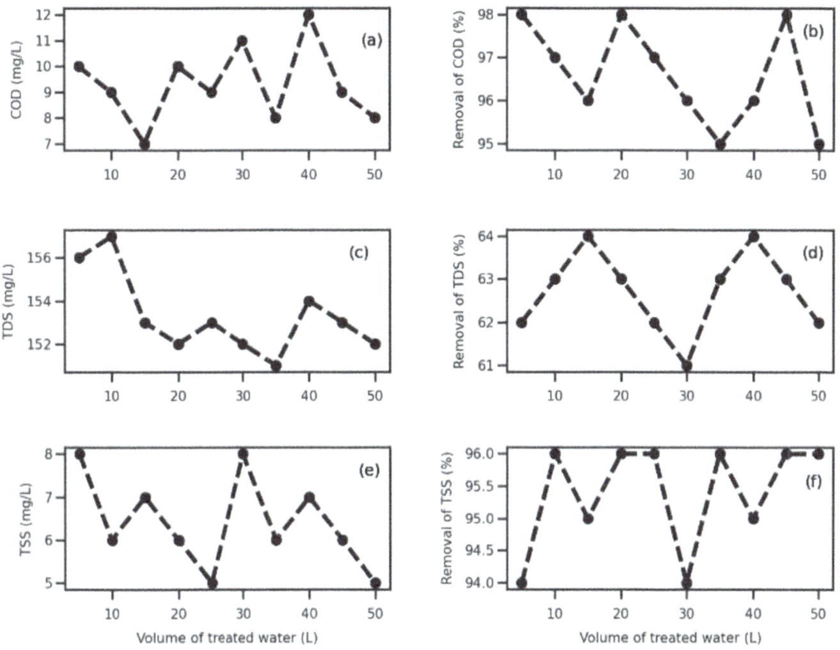

Fig. 7 Breakthrough profile of effluent water quality parameters at bench scale **a** COD, **b** Removal of COD%, **c** TDS, **d** Removal of TDS %, **e** TSS, **f** Removal of TSS %

Fig. 8 Side-by-side image of the composite sample of treated water from a complete cycle compared to the untreated raw wastewater, illustrating the noticeable difference in color

Table 4 Analysis of the composite textile wastewater sample collected over a specific time period, taken from the outlet of the wastewater treatment facility and the multilayer filter

Parameter	Inlet average ± std	Outlet average ± std
pH	8.1 ± 0.1	8.2 ± 0.2
Color, Pt-Co	504 ± 18	0
Total suspended solids, mg/l	60 ± 8	8 ± 2
Total dissolved solids, mg/l	1971 ± 8	157 ± 4
Chemical oxygen demand, mg/l	154 ± 12	9 ± 3
Residual OCl^-, mg/l	–	Not detected
Turbidity, NTU	43 ± 3	8 ± 2

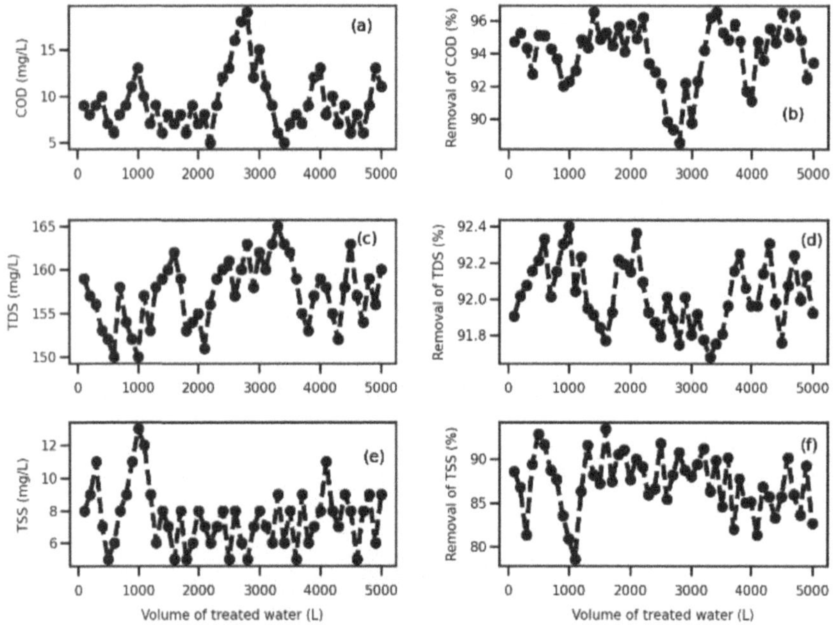

Fig. 9 Breakthrough profile of effluent water quality parameters at pilot scale **a** COD, **b** Removal of COD %, **c** TDS, **d** Removal of TDS %, **e** TSS, **f** Removal of TSS %

6 Discussions

6.1 Process Implication in the Circular Operation

Calcium hypochlorite was used as an oxidising agent during the treatment process to remove colour and enhance water quality. This was achieved by using two coagulants simultaneously, as described by Khandaker et al. [34] and Teter et al. [35]. This was

Fig. 10 Side-by-side image of the composite sample of treated water from a complete cycle compared to the untreated raw wastewater, illustrating the noticeable difference in color

accomplished by passing the water through sand filters, while also introducing iron into the process. Iron nails were used for pretreatment, and waste blast furnace iron slag was used for the pilot-scale treatment. Chloride ions (Cl^-) are potent oxidising agents that react with both organic and inorganic compounds, hence influencing the colour of water. These compounds frequently consist of double bonds or other groups that are quickly oxidised. Chromophores, which are groupings of atoms, are responsible for absorbing specific wavelengths of light in wastewater. This absorption of light is what gives the water its colour. Bleach's chlorine component decomposes the chromophores, modifying their structure and reducing their ability to absorb light. In addition, the ferric ions (Fe^0, Fe^+, Fe^{2+}) found in iron nails and slag contribute to the water treatment process by generating salts, causing metal anions to precipitate, and enhancing coagulation and lime softening [34, 35]. The process of dual cation polishing generates wastewater that falls within the acceptable discharge limits, while also being sustainable and requiring less energy [27]. This is achieved by using locally available materials and chemicals [36].

6.2 Benefits of Energy, Economics and Environmental Impacts of the Proposed Process

The financial assessment of the experimental small-scale technique, as opposed to depending on imported chemicals for pretreatment, unquestionably illustrates that the proposed strategy offers superior cost advantages for operators of wastewater treatment plants. A study conducted at a textile processing plant's wastewater treatment facility reveals that the cost of chemically removing colour from textile wastewater is 0.22 US dollars per cubic metre [37]. The expense associated with pretreating this specific wastewater, considering the prevailing market rate of calcium hypochlorite in the nearby region and the quantity needed for removing color, totals to 0.37 US dollars per cubic meter. Although the initial cost of pretreatment may appear higher than the usual treatment process, the eventual cost reductions resulting from reduced energy use make the pretreatment technique economically viable. The proposed approach involves the oxidation process, which results in a reduction in COD, leading to energy conservation of 1.8 kilowatt-hours per cubic meter (KWh/m^3) when using the specified pretreatment methodology. As a result, this causes a decrease in cost by 0.24 USD per cubic metre, which is comparable to 0.13 USD per kilowatt-hour of electricity used for industrial purposes. Therefore, the total decrease in treatment expenses is equal to 0.053 US dollars. As previously stated, implementing this strategy will result in significant reductions in carbon emissions due to the decreased electricity use during the aeration process. According to the Environment and Carbon Year End Report 2019–2020 [38], the cost of emitting one kilogramme of CO_2 is 0.27 UD dollars. By following the suggested technique, the reductionss of 80 kilogrammes of CO_2 emissions per cubic meter of treated wastewater will result in a savings of 1.84 US dollars per cubic meter. When factoring in the costs of reducing CO_2 emissions in the financial assessment and prioritising the goal of maintaining sustainable operations, the overall reduction in operating expenses would be around $1.89 per cubic meter of treated wastewater. Another beneficial aspect of this pre-treatment method is that all the materials used can be manufactured locally [39, 40], reducing concerns about potential disruptions in the supply chain caused by difficulties in cross-border shipping and changes in tariffs that could impact chemical prices and treatment costs.

7 Conclusions

The objective of the project outlined in this chapter was to offer an environmentally viable solution for the treatment of textile wastewater in Bangladesh. The primary goal was to develop a technique that relies on locally available materials, therefore promoting long-term sustainability. The experiments produced good outcomes in attaining complete removal of color and significant reductions in COD, TDS, and TSS. This treatment method exemplifies a bottom-up approach to supply chain management, emphasizing the utilization of easily accessible resources instead of depending on imports of costly chemicals. The government fosters the development of domestically-produced and environmentally-friendly technologies by imposing tariffs on imported equipment. The water treated using this method complies with the discharge standards established by the Bangladesh Department of Environment. The project's performance showcases the viability of utilizing locally sourced solutions for wastewater treatment, hence fostering an environmentally conscious approach to industrial operations in resource-constrained nations. This contributes to the achievement of Sustainable Development Goals in countries limited with resources.

8 Recommendations

Given the proven efficacy of the suggested technique in the pretreatment of textile wastewater at the bench and post treatment of textile wastewater at the pilot scale, it is recommended that this novel approach be expanded for use in actual facilities. In particular in economies with limited resources, policymakers should take this case study into consideration as a workable substitute for treating wastewater from the textile industry, utilising locally accessible waste materials, and using less energy.

Acknowledgements The authors sincerely appreciate the help provided by textile industries. We express our profound gratitude for the substantial efforts made by Md. Fayyaz Ahmad and Farzana Ahmed Mim, whose responsibilities as field assistants and technicians were indispensable in the successful culmination of this project.

References

1. (PDF) Contribution of the RMG Sector to the Bangladesh Economy (n.d.) Retrieved 23 Sept 2023, from https://www.researchgate.net/publication/5224225_Contribution_of_the_RMG_Sector_to_the_Bangladesh_Economy
2. Haider MZ (2007) Competitiveness of the Bangladesh ready-made garment industry in major international markets. Asia-Pac Trade Investment Rev 3, 3–27. References—Scientific Research Publishing (n.d.) Retrieved 23 Sept 2023, from https://www.scirp.org/(S(lz5mqp453edsnp55rrgjct55.))/reference/referencespapers.aspx?referenceid=2799721
3. Export Promotion Bureau-Government of the People's Republic of Bangladesh (n.d.) Retrieved 23 Sept 2023, from http://www.epb.gov.bd/
4. Apparel exporters expect turnaround by June next | The Business Standard. (n.d.) Retrieved 23 Sept 2023, from https://www.tbsnews.net/economy/rmg/apparel-exporters-expect-turnaround-june-next-508206
5. Kallawar GA, Bhanvase BA (2023) A review on existing and emerging approaches for textile wastewater treatments: challenges and future perspectives. Environ Sci Pollut Res Int 31(2):1748–1789. https://doi.org/10.1007/s11356-023-31175-3
6. Haque IMMS, Rahman MM, Rafiq MS, Shakhawat, M., & Nadim Reza Khandaker (2024) Energy recovery potential in Bangladesh from elevated temperature textile processing wastewater: an analysis of energy recovery, energy economics and reduction in carbon dioxide emission. Model Earth Syst Environ 10(2):2661–2673. https://doi.org/10.1007/s40808-023-01933-w
7. Rafiq M, Apurba MSH, Khandaker NR (2024a) Implementing sustainable practices in wastewater treatment and material waste management in the textile and apparel sector of Bangladesh. In: 7th International conference on civil engineering for sustainable development 2024 (ICCESD 2024), Khulna
8. Khan SR, Dhar D, Navaid M, Pradhananga M, Siddique F, Singh A, Yanthrawaduge S (2009) Export success and industrial linkages. Palgrave Macmillan US. https://doi.org/10.1057/9780230622128
9. Textile | Board Of Investment (n.d.) Retrieved 23 Sept 2023, from https://invest.gov.pk/textile
10. 75 (n.d.) Sri Lanka—Textiles. Retrieved 23 Sept 2023, from https://www.trade.gov/country-commercial-guides/sri-lanka-textiles
11. Textile Industry in Nepal—Apparel and Garment Sector of Nepal (n.d.) Retrieved 23 Sept 2023, from https://www.textileinfomedia.com/textile-industry-in-nepal

12. Afghanistan textile industries: Know textile business sector of Afghanistan (n.d.) Retrieved 23 Sept 2023 from https://www.textileinfomedia.com/textile-industry-in-afghanistan
13. Infomedia T (2023) Textile industry in Bhutan: Bhutan textile business sector and overview of Bhutan textile industries. Textileinfomedia.Com. Retrieved 23 Sept 2023, from https://www.textileinfomedia.com/bhutan
14. Infomedia T (2023) Textile industry in Maldives: textile business sector and business size of Maldives. Textileinfomedia.Com. Retrieved 23 Sept 2023, from https://www.textileinfomedia.com/maldives
15. Rabbi M (2016) Heavy metals content in inlet water, treated and untreated waste water of garments industries at Gazipur, Bangladesh
16. Munnaf A, Tusher T, Kabir M, Molla M (2014) Investigation of water quality parameters discharged from textile dyeing industries. 257–263
17. Dey S, Islam A (2015) A review on textile wastewater characterization in Bangladesh. Resources and Environment, 5, 15–44. References—Scientific Research Publishing (n.d.) Retrieved 23 Sept 2023, from https://www.scirp.org/(S(lz5mqp453edsnp55rrgjct55.))/reference/referencespapers.aspx?referenceid=2728892
18. Process Analysis Impacts of Textile Manufacturing | Environmental Impacts of Textile Manufacturing | Solid Waste Pollution in Textile Manufacturing | Fibre2fashion.com—Fibre2Fashion (n.d.) Retrieved 23 Sept 2023, from https://www.fibre2fashion.com/industry-article/3771/process-analysis-environmental-impacts-of-textile-manufacturing
19. Irin A, Islam M, Kabir M, Hoq E (2017) Heavy metal contamination in water and fishes from the Shitalakhya river at Narayanganj, Bangladesh. Bangladesh J Zool 44:267. https://doi.org/10.3329/bjz.v44i2.32766
20. Ahasanur Rabbi M, Hossen J, Mirja Sarwar Md, Kanti Roy P, Binte Shaheed S, Mehedi Hasan M (2018) Investigation of waste water quality parameters discharged from textile manufacturing industries of Bangladesh. Current World Environ 13(2):206–214. https://doi.org/10.12944/CWE.13.2.05
21. Bangladesh RMG (2023) Bangladesh sets new standards for textile wastewater discharge. https://rmgbd.net/2023/03/bangladesh-sets-new-standards-for-textile-wastewater-discharge/
22. Periyasamy AP, Ramamoorthy SK, Rwawiire S, Zhao Y (2018) Sustainable wastewater treatment methods for textile industry. In Muthu SS (ed) Sustainable innovations in apparel production. Springer Nature Singapore, pp 21–87. https://doi.org/10.1007/978-981-10-8591-8_2
23. Koltuniewicz AB (2017) Process intensification: definition and application to membrane processes. In: Figoli A, Criscuoli A (Eds) Sustainable membrane technology for water and wastewater treatment. Springer Singapore, pp 67–96. https://doi.org/10.1007/978-981-10-5623-9_3
24. Cho J, Ramínez JAL, Shon HK, Park K, Park M (2023) Sustainable water treatment using nanofiltration and tight ultrafiltration membranes. In Zhang HX (ed) Water sustainability. Springer US, pp 11–23. https://doi.org/10.1007/978-1-0716-2466-1_909
25. Kharat DS, Kumar L, Kumar P, Mahwar R, Ansari PM, Sengupta B (2007) Advance methods for treatment of textile industry effluents
26. Sonune A, Ghate R (2004) Developments in wastewater treatment methods. Desalination 167:55–63. https://doi.org/10.1016/j.desal.2004.06.113
27. Rafiq MS, Apurba MSH, Khandaker NR (2024c) Enhancing textile wastewater sustainability through calcium hypochlorite oxidation and subsequent filtration with assistance from waste blast furnace iron slag. Water Environ J 1–11. https://doi.org/10.1111/wej.12948
28. Chandrasekhar SS, Sahu N, Sridhar S (2018) Processing of dairy industrial effluent and kitchen wastewater by integration of microbial action with membrane processes. In Sridhar S (ed) Membrane technology (1st ed, pp 495–516). CRC Press. https://doi.org/10.1201/9781315105666-24
29. Beydilli MI, Pavlostathis SG, Tincher WC (1998) Decolorization and toxicity screening of selected reactive azo dyes under methanogenic conditions. Water Sci Technol 38(4):225–232. https://doi.org/10.1016/S0273-1223(98)00531-9

30. Kulkarni SJ (2017) Coagulation for wastewater treatment: a review on investigations and studies. Int J Sci Res Sci Technol 3(3):501–505. https://doi.org/10.32628/IJSRST1733182
31. Riga A, Soutsas K, Ntampegliotis K, Karayannis V, Papapolymerou G (2007) Effect of system parameters and of inorganic salts on the decolorization and degradation of Procion H-exl dyes. Comparison of H2O2/UV, Fenton, UV/Fenton, TiO_2/UV and TiO_2/UV/H_2O_2 processes. Desalination 211(1):72–86. https://doi.org/10.1016/j.desal.2006.04.082
32. Brown JP (1981) Reduction of polymeric azo and nitro dyes by intestinal bacteria. Appl Environ Microbiol 41(5):1283–1286. https://doi.org/10.1128/aem.41.5.1283-1286.1981
33. Roberts LC, Hug SJ, Ruettimann T, Billah MM, Khan AW, Rahman MT (2004) Arsenic removal with Iron(II) and Iron(III) in waters with high silicate and phosphate concentrations. Environ Sci Technol 38(1):307–315. https://doi.org/10.1021/es0343205
34. Khandaker NR, Brady PV, Teter DM, Krumhansl JL (2004) Arsenic removal in conjunction with lime softening (United States Patent US6802980B1). https://patents.google.com/patent/US6802980B1/en
35. Teter DM, Brady PV, Krumhansl JL, Khandaker NR (2006) Method of removing arsenic and other anionic contaminants from contaminated water using enhanced coagulation (United States Patent US7138063B1). https://patents.google.com/patent/US7138063B1/en
36. Rafiq MdS, Apurba MSH, Khandaker NR (2024) Harnessing untapped resources for sustainable energy production from municipal solid waste in recourse challenged economies: a case study of Rajshahi City corporation, Bangladesh. J Sustain Dev Peace J 2(1):35–57. https://doi.org/10.3126/jsdpj.v2i1.63240
37. Ebadul Haque M, Sumsuddha M, Ashrafee T (2014) Project report on study on chemicals consumption and operational cost of textile ETP plants [Technical Report]. Daffodil International University. http://dspace.daffodilvarsity.edu.bd:8080/handle/20.500.11948/dspace.daffodilvarsity.edu.bd:8080/handle/20.500.11948/1371
38. Environment and Carbon Year End Report 2019–2020 (2020)
39. Alam SNS (2012) world24info: a study on: HRM practices in Garments Industries and its Impact. http://world24info.blogspot.com/2012/02/research-study-on-hrm-practices-in.html
40. Rafiq MS, Apurba MSH, Khandaker NR (2024b) Voir dire of waste-to-energy conversion from municipal solid waste in Dhaka North City corporation through incineration: a practical venture or an inefficient endeavor? In: 2024 7th International conference on development in renewable energy technology (ICDRET), pp 1–6. https://doi.org/10.1109/ICDRET60388.2024.10503690

Application of Anaerobic Process and Energy, Materials Recovery for Textile Processing Wastewater

Md. Sahil Rafiq, Mohammad Shakhawat Hosen Apurba, Mohammad Moshiur Rahman, Zaman Mirdha, and Nadim Reza Khandaker

Abstract Scientific innovators have provided the inhabitants of Earth with a better quality of life. However, sometimes the road to self-destruction is paved with good intentions. Humankind is under existential threat. The burning issue on planet Earth is now the mitigation of climate change and sustaining our lifestyle with less energy, materials, and other factors of production. This chapter aims to guide process designers in treating textile processing wastewater in a less energy-intensive manner with a lower carbon footprint and reusing the wastes that have been rejected in various biological and chemical processes. The second law of thermodynamics explains how much materials or energy are wasted in the transformation processes. The objective is to recover the potential material or energy and reuse it to achieve Sustainable Development Goals. Circular economy of wastewater treatment and how waste by-products from textile processing wastewater treatment plants can be turned into value-added components to build sustainability. In this regard, wastewater recycling will be treated substantially by discussing technologies and processes available and the economy—examples of smart operating facilities where wastewater is turned into assets.

Keywords Textile · Anaerobic · Recovery · Energy · Materials

Md. S. Rafiq · M. S. H. Apurba · M. M. Rahman · Z. Mirdha · N. R. Khandaker (✉)
North South University, Dhaka, Bangladesh
e-mail: nadim.khandaker@northsouth.edu

Md. S. Rafiq
e-mail: sahil.rafiq@northsouth.edu

M. S. H. Apurba
e-mail: m.151102@s.pust.ac.bd

S. Yadav et al. (eds.), *Emerging Technologies in Biological and Hybrid Wastewater Treatment: Lessons from Developed to Enhancing Practices in Developing Countries*,
Earth and Environmental Sciences Library,
https://doi.org/10.1007/978-3-031-74515-7_13

1 Introduction to Circular Economy and Sustainable Wastewater Treatment

With scientific and technological advances, humanbeings have made remarkable changes in their quality of life. Historically, major developments such as using wood and coal for steam transformed industries, railroads, energy conversion, agriculture, metallurgy, textiles, and pharmaceuticals. These discoveries fundamentally altered human society by bringing it a period of prosperity and advancement. But as was vividly noted, this era of technological advancement simultaneously signaled the start of a major problem: the deterioration of the planet's ideal environment. As a result of our increasing use of fossil fuels, especially coal, we have begun to release carbon dioxide and other greenhouse gases (GHG) into the atmosphere, resulting in air pollution. Despite our progress in soil and water pollution, the climate remained a "tragedy of the commons," as [31] explainedpeople are exposed to carbon dioxide (CO_2) out of the air slowly, but itcomes out a lotwas from the impact on our global climate. As civilization advanced, so did our reliance on fossil fuels. We have discovered and extracted many fossil fuels, from coal to petroleum hydrocarbons, and initiated industrial processes, all of which have led to further increases in wind turbines. As per recent research, this accumulation of unwanted carbon dioxide upsets the environment's delicate balance, resulting in rising temperatures, rising water levels, atmospheric pressure delivery, and other side effects [36]. The Earth began to warm, resulting in sea level rise, species loss, drought, and extreme temperatures. This ongoing crisis has become a matter of global concern, as the Brundtland Report explains, and poses a serious threat to the well-being of present and future generations [14]. Many researchers are trying to find out the solutions of those global concerns and threats through their contributions to insustainability and circular economic models. A study in Poland shows that biogenic materials can be recycled and demonstrates the importance of economic systems in wastewater treatment plants. This study shows that a WWTP serving 70,000 people can recover 1126 Mg of organic carbon annually and generate 12.6 GWh of energy through biogas production. By co-digesting kitchen waste with sludge, it is projected that national, 470 mm^3 of biogas and 1270 GWh of energy will be produced yearly from the 13.1 million Mg of municipal waste generated in Poland in 2020. These findings underscore the large function of natural waste management in attaining sustainable and green useful resource usage in wastewater remedy techniques [21].

Another study critically evaluates wastewater treatment plants (WWTPs) in Mexico. It found that, of the 4231 municipal WWTPs, 32% were out of service by 2021 for various reasons, including high operating costs, lack of maintenance, and water leakage. The study noted an interesting trend: over a decade, the number of inactive plants doubled, with 929 plants closing their doors. The study has proposed the operationalization of WWTPs primarily based only on circular economy thoughts and commercial models to overcome the technical governance, environmental, and economic challenges that WWTPs in Mexico confront. Long-term resource recovery, price maintenance, and environmental advantages need to observe the combination of ecologically creative and financially possible eras into WWTP operations, fortifying wastewater management practices in Mexico [29]. Addressing the high treatment costs associated with imported chemicals, a pilot project tested a sustainable method using chemical oxidation with calcium hypochlorite and sand filtration with blast furnace iron slag in a treatment plat. The treated water demonstrated total dissolved solids (TDS) levels of 157 ± 4 mg/L, total suspended solids (TSS) levels of 8 ± 2 mg/L, and chemical oxygen demand (COD) levels of 9 ± 3 mg/L, with reductions of 92%, 87%, and 94%, respectively. This method effectively removed color from the water. These all meet the requirements of the Environmental Conservation Regulations (ECR) 2023 and provide an effective solution for those economies with limited resources [70].

So, the most urgent venture facing our world now is mitigating the results of weather trade even as retaining our popularity of existence with decreased power use, greater green use of sources, and much less terrible environmental effects. One of the critical techniques for this enterprise is to switch from fossil fuels to zero-carbon renewable strength assets, as recommended by the UN [19]. However, there are sizeable nearby variations in the availability and acceptability of renewable assets. These assets include a range of opportunities, waste recycling, photovoltaic panels, hydropower, wind power, geothermal strength, ocean strength, bioenergy, and more. All of those rely on elements including wind speed, land topography, sun radiation, and geographic vicinity. Figure 1 illustrates the vital role sustainable wastewater treatment technologies play in the context of a circular economy, emphasizing their significant contribution to the conservation of natural resources. As defined by the Commission (1987), sustainability is the key to addressing this crisis.

Sustainability is the ability to reconcile social welfare, environmental protection, and economic advancement to meet current needs without endangering future generations. The delicate balance can be reached thanks to the three pillars of sustainability:

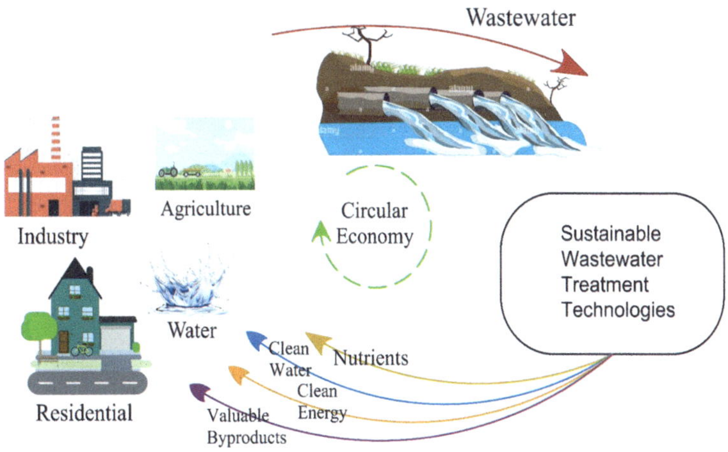

Fig. 1 Circular economy framework for wastewater

environmental justice, social protection, and economic success [58]. In the commercial enterprise, financial sustainability relates to an employer's potential to control its assets responsibly and generate profits over a long time while minimizing its environmental effect and promoting social welfare. Prominent examples of organizations embracing monetary sustainability consist of Unilever, which launched into an approach in 2010 to align sustainability with its financial performance. Unilever's initiatives covered increasing bundle recycling, advocating for the use of recycled substances, and raising the focus on accountable intake. Similarly, Suez, as detailed in its Sustainable Development Report 2020, notably reduced its emissions related to power consumption by way of ninety-five percent via shopping and producing renewable power [92].

In Fig. 2, we can observe a visual representation showing various pathways for water, energy, and materials, all of which are part of the overall circular economy. This diagram shows the various ways of recovering and reusing these resources, in line with the principles of sustainability and conservation.

A study emphasizes the importance of understanding local waste generation patterns and environmental elements for designing powerful waste management structures to mitigate health risks and environmental risks. Additionally, the study highlights the potential for producing electrical energy from MSW in Rajshahi City, Bangladesh, primarily based on records collected from 150 households and chemical analyses conducted by professionals [71]. In some cases, waste-to-energyconversion is not always the best solution, like Dhaka North City Corporation (DNCC) is dealing with increased energy demands and trash production. Every day, 4220 piles of rubbish are produced, of which 2774 tons are disposed of in the Amin Bazar Landfill. Despite 80% waste collection, the waste-to-energy incineration is tough due to a low heating value (LHV) of 2.753 MJ/kg for municipal strong waste (MSW), in comparison to 5.64–6.94 MJ/kg for recycled fractions. This discrepancy arises from recyclers

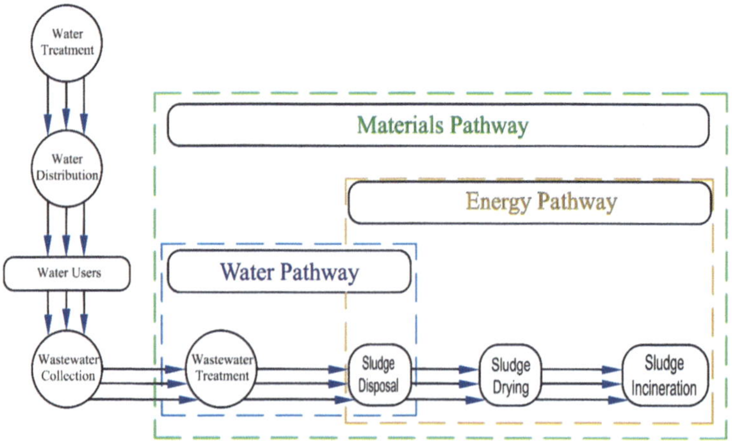

Fig. 2 Circular economy framework for water, energy and material recovery

extracting excessive-warmness value materials, that's often overlooked by way of government, complicating waste conversion plans which include the proposed 3000 tons/day incinerator to supply 42.5 MW of electricity [72]. The pursuit of sustainability also emphasizes how essential anaerobic treatment is for coping with wastewater. The anaerobic remedy is an environmentally pleasant method of treating wastewater that lets in for resource restoration and the manufacturing of renewable power. As we search for methods to gradually down weather change and create a stable planet so one can ultimate for future generations, we can study the efficacy of anaerobic solutions and their position in a sustainable future inside the sections that follow [5, 39, 74].

2 Introduction of Circular Economy in Treating Textile Processing Wastewater

2.1 The Critical Role of the Textile Industry in the Economic the Setting of Bangladesh

The economic system of Bangladesh is incomplete without the textile sector, which has contributed immensely to the growth and development of the country. Bangladesh's garment and textile industry has not only helped the country become a leading exporter to this region, but has also contributed significantly to job creation, foreign investment, and poverty reduction. This part of this chapter aims to determine the importance of the garment industry to the textile industry in the financial sector of Bangladesh. The garment industry of Bangladesh is very important to the national economy, contributes significantly to employment, and ranks high in terms of employment numbers and economic output. Four million individuals, largely women, are reportedly employed by it. About 83% of the country's entire export earnings come from it [1]. Bangladesh is the sector's 2nd-largest exporter of textiles, with total sales of $42.613 billion within the financial 12 months 2021–2022, in step with facts from the Bureau of Commerce and Industry Development [23]. It is important to acknowledge Bangladesh's infrastructure's financial significance.

The aforementioned phenomenon is a catalyst for stimulating economic growth, generating employment opportunities, acquiring foreign exchange earnings, and alleviating poverty. The sector's notable trajectory from its modest origins to attaining the status of a prominent global textile industry exemplifies the tenacity and resolve of the Bangladeshi populace. Bangladesh's persistent efforts to expand and modernize its textile industry position it favorably to sustain its substantial presence in the international textile market, thereby fostering the nation's economic growth and progress. As wealready know, our textile industries rely heavily on water and energy, using groundwater at 27.33 ± 1.46 °C, which is heated for washing, dyeing, and other processes. The effluent wastewater releases at 40.4 ± 8.7 °C without regaining its' thermal energy. A research program investigated the potential for energy production from wastewater, utilizing the capacity of groundwater and aquifer heating and wastewater treatment plants. The study presented a sustainable energy model, linking the results to fuel consumption, economiccosts, and CO_2 emission reductions. It estimated that 102 kilowatts per year could be produced, 8.6 billion cubic feet of natural gas, and 687,794 tons of coal, resulting in an annual energy cost of $1.35 billion. This will reduce CO_2 emissions by 24.8 million tons per year, more than 20% of Bangladesh's annual CO_2 emissions, and 2% of global textile and textile industries' emissions fostering a sustainable transition towards an integrated circular economy in the textile sector [30].

2.2 *Environmental Challenges and Sustainability of Textile Industries*

The textile sector has been widely criticized for its prominent role in global pollution. The textile supply chain has far-reaching ecological consequences. To give an example, this chain produces 1.7 billion tons of CO_2 emissions annually, which accounts for around 10% of all greenhouse gas (GHG) emissions worldwide [59]. Simultaneously, there are profound concerns regarding the excessive consumption of freshwater in the wet textile processing stages, which results in the discharge of wastewater and the pollution of ecosystems [33]. These industries contribute to over 2% of the world's Gross Domestic Product (GDP) and sustain the livelihoods of millions. However, they also exhibit a significant environmental footprint, with emissions exceeding 1.2 billion tons of carbon dioxide, 22 million tons of microfibers, and the generation of 21 billion tons of waste. Alarmingly, 20% of this waste constitutes highly hazardous wastewater effluent originating from dye houses [85]. Furthermore, significant problems arise from inefficient material usage in the textile industry, where 400 kg of finished goods is produced from 512 kg of materials, or 78% of the total materials used. The research highlights the significance of recycling wastewater into valuable resources, implementing commercially feasible concepts for pollution control, and using treated wastewater in manufacturing processes in order to lessen dependency on groundwater extraction and limit wastewater discharges to the environment. These findings also present the importance of the textile industry in implementing sustainable practices to increase productivity and protect the environment [73].

As it stands, the fashion industry's sustainability rating is a mere 32 out of 100, according to available assessments [82]. If the industry continues with its as-usual approach, it is projected that by 2050, it will deplete a staggering 26% of the global carbon budget allocated for staying within the 2 °C global temperature increase limit, a target set by the United Nations [2].

In conclusion, "ecoefficiency" encapsulates the capability of an organization to not only provide goods and services competitively but also to do so while consistently and comprehensively reducing their environmental footprint and efficiently managing resource consumption throughout their product lifecycles. In simpler terms, it revolves around "generating greater value with reduced environmental consequences" [61]. Achieving environmental sustainability is crucial for the textile sector, which uses a lot of resources during the protracted manufacturing process, including water, power, and different chemicals [44]. Cleaner production embodies a comprehensive and proactive environmental approach within the business realm. Its primary goal is to enhance overall efficiency and reduce potential hazards to both individuals and the environment, all while ensuring the production of high-quality goods [97].

3 Anaerobic Acetogenic Processes for Treating Textile Processing Wastewater

3.1 Novel Process of Anaerobic Acetogenic Processes for Treating Textile Processing Wastewater

The textile industry in Bangladesh is indeed a thriving region, but it consists of widespread environmental and aid-related results. Textile processing, being quite useful resource-based, significantly utilizes each water and power source. The unregulated launch of untreated wastewater poses a grave situation. The most severely impacted major rivers are the Karnafuli in Chittagong and the Buriganga, Shitalakhya, and Turag in Dhaka. Poor water quality in these rivers has been well reported by researchers [3, 10, 69, 79], compelling the Bangladeshi government to address this pressing problem. An environmentally appropriate way to treat textile water infection is by the use of anaerobic acetogenic therapy as a cure or pretreatment prior to aerobic polishing. This procedure uses less energy since it requires less continual aeration than aerobic wastewater treatment systems. It also prevents the production of methane gas. The inability to meet the energy costs associated with extended aeration processes results in the release of untreated textile wastewater effluent, jeopardizing the lives of millions of people. To ensure sustainability in the textile processing industry, comprehensive experiments have been conducted as part of an extensive research protocol [83, 84].

3.2 Common Wastewater Treatment Process in Bangladesh

This chapter primarily discusses the importance of wastewater treatment to Bangladesh's apparel sector. Because of the remaining chemicals, the effluent that

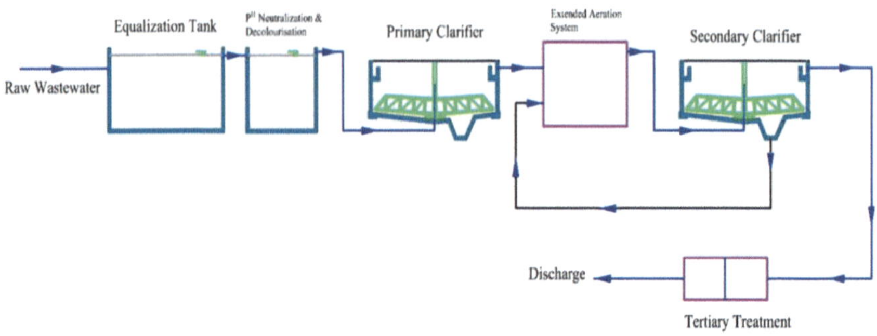

Fig. 3 Typical wastewater treatment process in textiles

is treated at the Effluent Treatment Plant (ETP) has a significant load of both COD and BOD. To ensure proper treatment, the treatment plant should incorporate an inspection system to remove fabrics and threads, as well as an equalization tank to regulate wastewater flow, temperature, pH levels, and other water-related parameters. The sequence for treating textile wastewater (Fig. 3) involves pH adjustment, the removal of colloidal solids, the elimination of color (dye), and the degradation of soluble biochemical oxygen. In the textile industry, there are two commonly employed process trains: one involving chemically assisted clarification followed by biological aerobic treatment, and the other employing extended aeration. The chemically assisted treating system needs minimal chemical utilization and is considered an extra sustainable alternative in Bangladesh. However, it's really worth noting that the strong residues generated from the clarifiers may additionally contain risky compounds originating from dyes and other chemicals used within the fabric dyeing system, necessitating secure disposal measures. Regrettably, inside Bangladesh, some short-sighted operators tend to bypass the treatment device, leading to the discharge of untreated wastewater.

Technologically in a textile industry, a typical wastewater treatment system (shown in Fig. 3) consists of several basic steps. First, wastewater requires primary treatment, where physical processes such as screening and sedimentation remove all solid particles and contaminants. Secondary treatment then uses biological processes to break down organic contaminations. In contrast, tertiary treatment can further purify the waterway by removing residual organic matter and particles before discharge or reuse.

This simple flowchart in Fig. 4 illustrates the general treatment process in the textile industry, which typically begins with initial screening and preliminary settling to remove the most complex materials. The wastewater then undergoes a secondary treatment using a biological method i.e., Anaerobic Acetogenic Pretreatment, to destroy the contaminated organic matter. Finally, tertiary treatment can further polish the effluent, removing residual solids and chemicals before it can be discharged or reused.

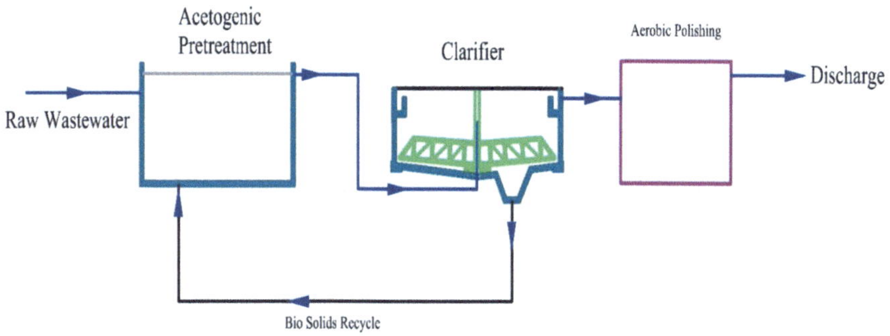

Fig. 4 Simplified flow chart of the altered treatment process for the textile processing wastewater treatment plant

3.3 Approach to Treating Textile Wastewater Using Anaerobic Acetogenic Pretreatment

Partially treating textile wastewater using anaerobic reactors, in preference to depending entirely on aerobic reactors, has established benefits. This method effectively decolorizes many dyes and decreases the high natural load [20]. Anaerobic–aerobic sequential treatment systems have shown the capability to remove 85% of color and 90% of COD [77]. The anaerobic conversion process is a complex microbial-mediated multistep process in the absence of oxygen. In this process, organic solids are converted into methane and carbon dioxide by different metabolic processes involving a variety of microbial communities [86]. This procedure comprises multiple stages, including hydrolysis, fermentation, and acetogenesis, culminating in the generation of methane through the conversion of acetate and the creation of hydrogen via the reduction of carbon dioxide using hydrogen.

In the anaerobic acetogenic method, the focal point is on improving the acetogenic step even as limiting the methanogenic step. This entails optimizing reactor conditions for the acetogenic breakdown of complex organics to provide hydrogen, carbon dioxide, and fatty acids, thus reducing BOD_5 and biodegradable COD. Acetogenic conditions are maintained by operating the reactor with short hydraulic and solids retention times, typically 2–4 days, under non-aeration conditions, occasionally using shock aeration to control the growth of methanogenic microorganisms (Fig. 5).

The subsequent stage in the advancement of anaerobic processing is dedicated to ensuring waste stability, reducing the production of carbon dioxide, and enhancing the yield of hydrogen. Studies by Logan et al. [48], Liu et al. [47], Fascetti [25], Fang and Liu [24], and others suggest that hydrogen is a fuel that is good for the environment and may be used to power fuel cells [24, 25, 47, 48]. According to Ferchichi et al. [26], ongoing research efforts have focused on refining the acetogenic process and examining possible uses for waste that is high in carbohydrates to maximize hydrogen production [26]. Applying acetogenic reactors to complex

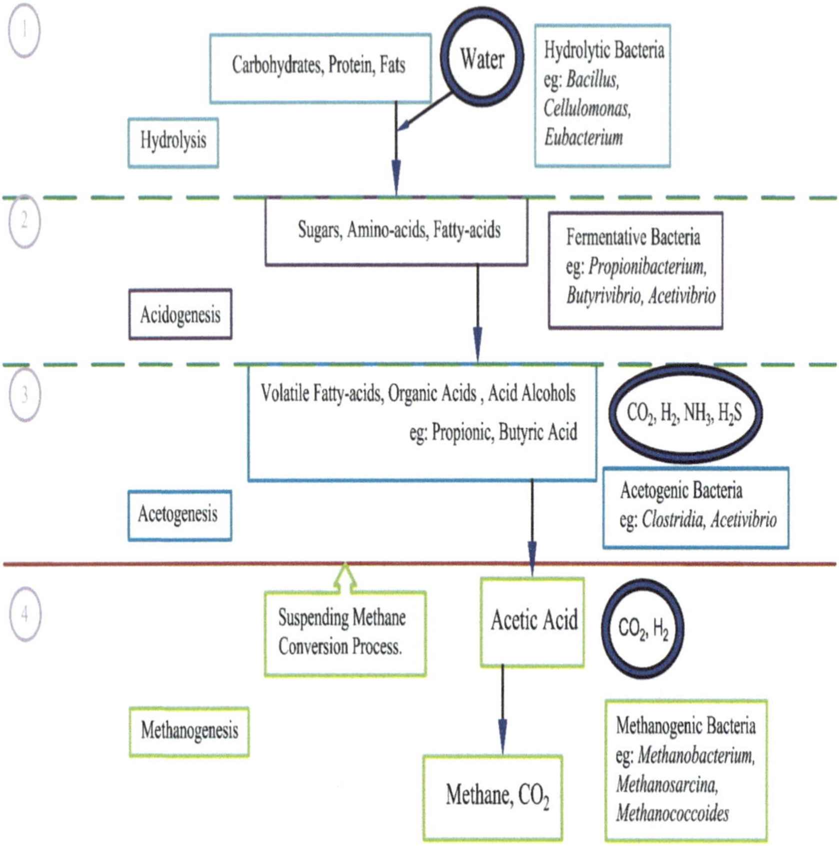

Fig. 5 Diagram showing how the anaerobic process can be modified to enhance anaerobic acetogenesis [56]

chemical wastewater, such as those from textile processing, is a novel approach. The purpose is to harness the hydrolysis step of acetogenesis to break down shade and other complex natural compounds observed in fabric effluent (Kamble). This innovative method, if a success, should offer a unique answer for coloration and COD elimination from wastewater in textile production.

However, optimizing an acetogenic reactor for complex wastewater like textile dyeing wastewater challenges the conventional wisdom that such reactors should operate with a large biomass content to counterbalance sluggish kinetics in transforming complex organics. To deal with complicated, low-strength wastewater, like textile wastewater, the acetogenic manner needs to be cautiously researched and developed, in assessment to traditional aerobic sewage treatment systems with lengthy retention intervals. To determine whether the method is appropriate in this new setting, preliminary acid testing must be conducted. Subsequently, the focus shifts to organizing kinetics and gaining a profound knowledge of important method variables for the eventual application at a commercial scale. Historically, anaerobic acetogenic reactors have mainly been employed for dealing with waste streams plentiful in effortlessly degradable components, as exemplified within the case of food waste [48]. Utilizing the anaerobic acetogenic method to treat complex textile wastewater is an innovative strategy.

Figure 5 demonstrates how the anaerobic process is modified to enhance the process. This shows the modifications made in anaerobic digestion to improve the generation of acetate, a crucial component in wastewater treatment. Parameters such as p^H, temperature, and loading variables are incorporated to optimize the operational parameters. The aim of these modifications is to establish the best possible conditions for the development and functioning of acetogenic bacteria. Furthermore, this involves incorporating particular substrates or co-substrates to enhance acetate generation and promote acetogenesis process.

3.4 A Case Study on Denim Processing Wastewater Treatment Through Anaerobic Acetogenic Pretreatment

Analysis of wastewater from the denim industry revealed interesting characteristics. This wastewater showed an observed total COD of 372 ± 36 mg/L. In addition, it exhibited a high pH value, registering 8.5 ± 0.7, and exhibited a high color intensity, averaging 661 ± 65 Pt-Co. In addition, the BOD_5/COD ratio was estimated to be 0.62. Collectively, these observations indicated the presence of high levels of organic matter in wastewater, indicating biodegradability. Untreated wastewater was added to the acetogenic reactor in a volume of 500 mL without any prior changes. The reactor functioned with the pre-defined waste-feeding approach and adhered to a semi-batch manner for 9 days.

The outcomes of the experimental study indicated that, once the initial acclimation phase had concluded, the acetogenic culture displayed the capacity to eliminate the color within the wastewater. Furthermore, as the reactor continues to operate, a decrease in the effluent's critical parameter values (see Figs. 6, 7, 8 and 9) shows a discernible drop in the chemical oxygen demand. The meticulous removal of 90% of the color and the more than 80% drop in COD were indications of the method's efficacy. These results strongly suggest that acetogenic treatment can be a feasible and environmentally friendly alternative to the pretreatment of laundry wastewaters, which can be applied to the currently used methods, substitutes for chemically treated, color removal, and solids removal.

Figure 6 depicts a side-by-side contrast between untreated wastewater and wastewater that has undergone anaerobic acetogenic treatment, which serves as a clear and visible demonstration of the procedure's effectiveness. Notably, the pictureshows discernible variations in shade, transitioning from a darker hue inside the untreated water to a visibly clearer look within the handled water.

Fig. 6 Image comparing untreated and treated wastewater to demonstrate the procedure's effectiveness

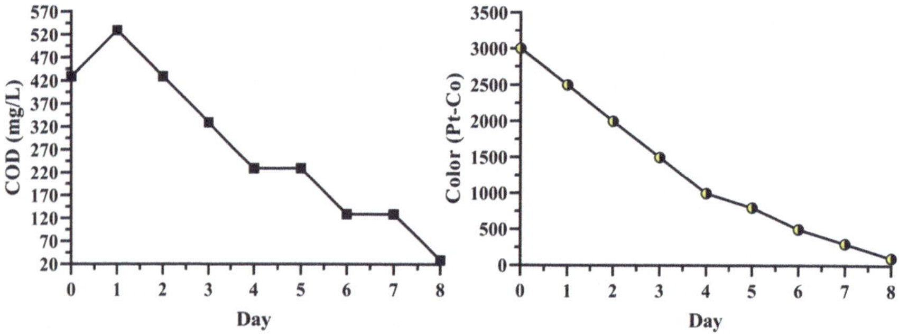

Fig. 7 COD (mg/L) and color analysis of the treated effluent generated during the acetogenic pretreatment of denim processing wastewater

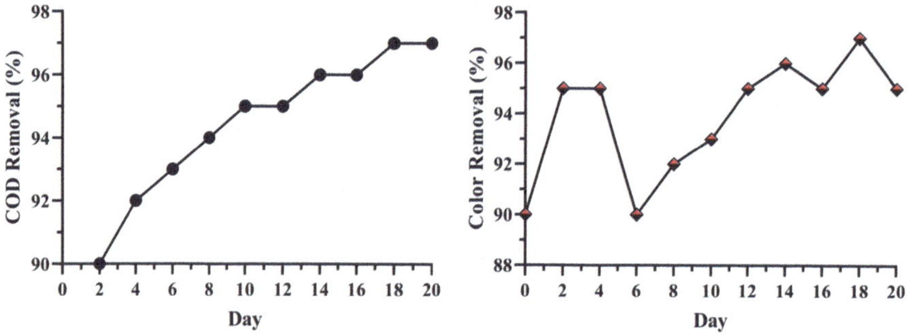

Fig. 8 Efficiency in removing cod and color using the acetogenic pretreatment process

In Fig. 7, the graph on the left shows the ongoing progress of acetogenic pretreatment applied to denim processing wastewater. Within this graph, the x-axis represents the passage of time in days, while the y-axis is chemical oxygen demand (COD), measured in milligrams per liter (mg/L). As the day develops, there's a mentioned and sustained downward trend in COD levels, indicating the opportunity for biological reduction of contaminants within the treated effluent.

Fig. 9 Picture of composite textile wastewater before and after treatment

On the right-hand facet of the diagram, we observe the sensitivity analysis of the color trade inside the purified effluent at some stage in the pretreatment of denim bath wastewater with acetogen. In this case, the x-axis shows the time trend over the day, while the y-axis records the outcome of the color (Pt-Co.) test. As the timeline advances, a substantial and continuous reduction in color intensity becomes evident. This empirical and visual decline in color intensity serves as a compelling indication of the treatment process's success in the elimination or reduction of colored compounds found within the wastewater. This observation strongly underscores the efficiency of the treatment process in its role in enhancing overall water quality. Additionally, reactor operating parameters indicated that system performance began to decline beyond a feed-to-microbe operating ratio of 0.05–0.1 [78].

3.5 *A Case Study on Composite Textile Wastewater Treatment Through Anaerobic Acetogenic Pretreatment*

The complex wastewater used in this study originated from a range of sources that were used in the dyeing and printing processes required to create different fabrics. The wastewater exhibits notable diversity and complexity. The composite textile wastewater exhibited distinct characteristics, including a color level averaging 3542 $\pm$ 352 Pt-Co, a p^H of 9.4 $\pm$ 0.5, a COD measuring 5176 $\pm$ 132 mg/l, and a BOD_5/COD ratio of 0.5.

In Fig. 8, the left part of the chart illustrates the effectiveness of the acetogenic pretreatment method in getting rid of COD, expressed as a percent. The x-axis information the passage of time in days, at the same time as the y-axis measures the proportion of COD removal. As time progresses, a consistent and brilliant increase in the proportion of COD removal will become evident. This constant sample underscores the growing performance of the acetogenic pretreatment technique in progressively eradicating organic pollution from the wastewater. Simultaneously, in the right facet of the graph, we study a comparable portrayal of performance, this time regarding the removal of color as a percentage via the acetogenic pretreatment procedure. The x-axis displays the progression of days, and the y-axis quantifies the proportion of shade elimination. Over time, there has been an outstanding and non-stop development in the percentage of shade elimination, presenting giant evidence of the effectiveness of the acetogenic pretreatment manner in decreasing the color content inside the wastewater as the treatment approaches unfold. Moreover, Fig. 9 visually displays the assessment of untreated and pretreated wastewater.

Figure 10 demonstrates how well the acetogenic pretreatment process removes Total Dissolved Solids (TDS). The graph has time on the x-axis, measured in days, and the percentage of TDS removal on the y-axis. As time goes by, we can see a steady and significant increase in the percentage of TDS removal. This shows that the acetogenic pretreatment method is getting higher at lowering the quantities of dissolved solids inside the wastewater. Furthermore, the acetogenic process achieved over 90% elimination of color, COD, and decreased TDS. This reduction in total dissolved solids is specifically tremendous, as many fabrics processing wastewater treatment effluents conflict to meet regulatory standards for TDS without resorting to pricey membrane structures [76]. The process's effectiveness is demonstrated by the notable elimination of color, COD, and TDS in Figs. 8, 9 and 10. The effectiveness of the treatments shown in earlier research is further supported by Fourier-transform infrared spectroscopy [56]. The presence of initial particles in the effluent of the normalized acetogenic culture entirely collapsed, as revealed by a comparison of treated wastewater from day 1 and day 20 of the acetogenic reactor operation. This result indicates that modified acetogenic processes can degrade pollutants commonly observed in wastewater from the textile industry.

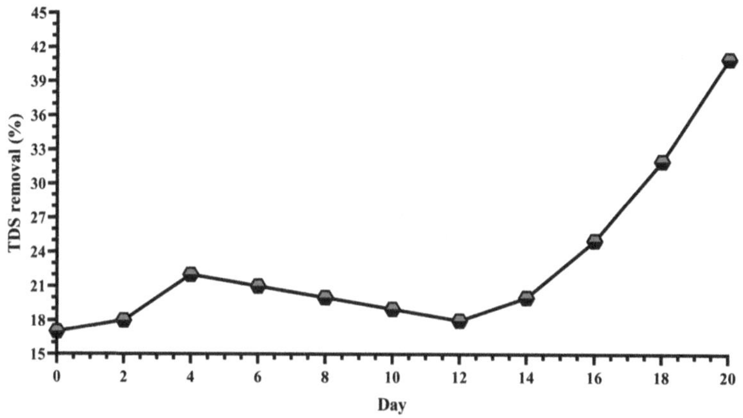

Fig. 10 Efficiency in removing TDS using the acetogenic pretreatment process

3.6 Energy Efficiency and Feasibility for Industrial Use

Reducing the amount of BOD and degradable COD in the aerobic treatment will be possible if an acetogenic pretreatment procedure is used. As a consequence, this decrease lowers the demand for aeration, which in turn lowers the amount of electricity needed to run the aeration blowers. The power consumption of aerators can vary, dependingon the particular device in use [81]. By analyzing the data collected from the production sites, and the laboratory evaluation of the initial denim wastewater scenario, we can now estimate the energy needs. With a daily flow of 723 m^3 and a BOD_5 of 227 mg/l, the daily BOD_5 input to the aerobic basin amounts to 163 kg. The treatment plant uses high-efficiency bubble aerators, each consuming energy at a rate of 4.0-kWh/day/kWh, which is BOD_5 per kilogram brought in daily the total daily energy demand is 655 kWh. Nonetheless, given an 86% BOD_5/COD removal effectiveness resulting from the acetogenic procedure, the BOD_5 load would be reduced to a mere 42 kg/day. This would necessitate 163 kWh/day for blower energy. This equates to a noteworthy conservation of 496 kWh/day in energy.

These energy savings carry substantial importance, particularly in an emerging economy such as Bangladesh, where the per capita yearly energy consumption is 320 kWh [81]. The acetogenic method has proven highly efficient in the initial treatment of wastewater from textile processing. It effectively addresses color removal and leads to big discounts in COD, BOD_5, and TDS levels. Unlike the continuous aeration process used in extended aerobic systems, it only requires periodic air purging, resulting in significant energy savings by eliminating the need for constant blower operation. With very little adjustments, this strategy easily fits within the textile facilities' current wastewater treatment infrastructure. Modern aerators in aeration basins can be modified for sporadic, unexpected aeration, which drastically shortens the aeration period from 24 h to a few minutes and saves a significant amount of money on electricity.

Moreover, those centers already have sludge management systems and secondary clarifiers installed, which reduces the need for additional capital expenditures. When acetogenic pretreatment is used, operational changes are typically necessary, which leads to significant cost savings in all strength and chemical types. In general, the acetogenic system proves to be an economical and environmentally friendly solution for handling wastewater from fabrics.

4 Resources and Potential Energy Recovery from Textile Processing Wastewater

4.1 Potential Energy Recovery

Water is used as the main carrier to facilitate the water process in the textile industry, where energy is very important for heating the process. This heat may be used by saturated steam or hot air, designed to accelerate the movement of the above-mentioned processes. During these procedures, a substantial quantity of surplus thermal energy and lukewarm wastewater is produced and discharged into disposal channels. In addition, cleaner production methods can effectively recycle and recover waste heat and hot effluent [94]. These techniques could involve incorporating heat recovery systems at various process outlets to capture thermal energy using heat exchangers. Afterwards, this recovered heat can be used to heat or pre-heat new water for future use. Utilizing waste heat requires additional efforts to alter traditional industrial processes or machinery. This is carried out in order to lower operating expenses of using natural gas or coal and manage pollution levels in the environment. These methods consist of bleaching, mercerizing, dyeing, printing, washing, and finishing [8, 46]. It is crucial to mention that these procedures require a large amount of energy and water, leading to significant energy and water usage [54, 68, 74]. In the textile wet processing industry, the main source of energy and fuel primarily comes from burning natural gas. This combustion serves various purposes, such as generating electricity and creating saturated steam, which is essential for activities like drying and curing [80]. The machines operate using the generated electricity.

The steam that is completely saturated is used to heat drying spaces, defrost clothes, and heat bath water. Despite the industry focusing heavily on production, the waste generated from these activities is often ignored. Around half of the energy used in manufacturing is believed to result in the release of waste heat. Nevertheless, it is entirely feasible to recover and utilize this wasted heat through straightforward alterations to disposal streams without compromising processes or product quality [88]. Moreover, these adjustments are in line with the increasing prevalence of cleaner production methods in the textile processing industry as mentioned by Ozturk et al. [63]. Furthermore, the importance of taking these practices into account has increased significantly because of the crucial requirement to preserve energy and water resources. Because of their growing rarity, these resources might not be sufficient to satisfy demand in the coming years [13, 22]. Furthermore, the increasing cost of energy and strong competition in the industry highlight the significance of integrating eco-friendly electricity and water technologies in industrial processes [91]. In textile wet processing, adopting Waste Heat Recovery (WHR) technologies can lead to significant reductions in environmental pollution and resource conservation, as supported by Hiben et al. [32]. By lowering emission rates, these advancements can greatly decrease the fuel consumption for heating process water and producing steam, ultimately leading to a decrease in environmental pollutants.

Utilizing heat exchangers in textile processing and the exhaust systems of power turbines and boilers can effectively reduce the temperature of waste streams, including wastewater and exhaust emissions, as highlighted by Gadhi et al. [28]. To recover waste heat using a particular heat exchanger, an experiment was carried out in a textile finishing factory. This specific heat exchanger was made up of 60 finned tubes and 97 stainless steel tubes that were arranged in a staggered layout and single-pass design without baffles. The thermal overall performance of this heat exchanger was assessed the use of nicely-mounted strategies, specifically the Dittus-Boelter and Gnielinski correlations. Performance evaluation was carried out using the ε-NTU method. Interestingly, the technology showed an incredibly quick payback period—just 3.5 months. To enhance the heat exchanger's performance, a thermal stack was created when warm air passed via the tubes.

In comparison, when a Double Pipe Heat Exchanger (DPHX) was considered as a substitute arrangement, it was found to be less effective for waste heat recovery. The selected heat exchanger, known as the Shell and tube heat exchanger (STHX), achieved an efficiency of 0.85 and recovered 74.2% of energy. Moreover, a validation procedure that compared target and real temperatures verified that the planned STHX was a reasonable choice for recovering waste heat. [41].

4.2 Resources Recovery

Various chemicals and dyes are regularly applied in the everyday operations of textile processes [93]. Reactive dyes are predominantly favored as anionic colorants utilized within the textile sector, primarily in the context of coloring procedures, particularly in the dyeing processes involving cotton materials [6]. Discharging dyeing wastewater not only leads to environmental harm but also disrupts natural cycles [51, 87]. Additionally, the existence of substances like chlorides, sulfates, heavy metals, and unattached dyes that include azo compounds in the wastewater are regarded as having cancer-causing properties [16, 17, 27].

Membrane separation techniques are commonly used in treating wastewater because of their capacity for adsorption, sieving, and electrostatic interactions [50, 64]. Different mechanisms, such as microfiltration (MF) with pore sizes from 1000 to 100 nm, are utilized in membrane filtration. Nqombolo et al. [60] discussed the 0.1 nm pore size of Reverse osmosis (RO), Zhao et al. [96] addressed nanofiltration (NF) with pore sizes under 2 nm, and Makertihartha et al. [50] explained ultrafiltration (UF) with pore sizes ranging from 100 to 2 nm. Microfiltration (MF) is utilized for removing suspended particles, whereas ultrafiltration (UF) is capable of removing both particles and large molecules [18]. However, it is not successful in removing water-soluble and low molecular-weight [51]. Conversely, nanofiltration (NF) can effectively separate divalent salts, various mineral salts, and compounds from low molecular weight substances [52]. Nanofiltration membranes, which have pores ranging from 0.5 to 2.0 nm, offer advantages like high permeability, lower osmotic pressure, and effective elimination of water-soluble dyes [9] (Fig. 11).

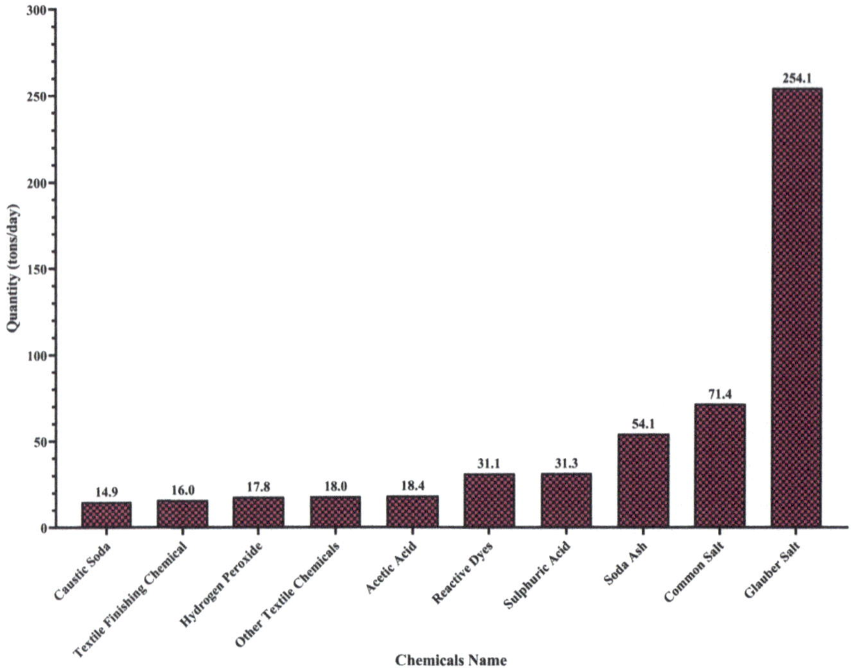

Fig. 11 Primary chemicals employed in the textile processing industry, including the precise quantity of each chemical required to process 1 kg of textile materials [89]

5 Utilizing Waste Derived Materials

5.1 Biogas

A biogas reactor with biomass recirculation and pH control may easily create biogas with a high methane concentration, up to 60% by volume, without the need for any additional treatment of the sugar-rich liquid. This implies that these unwanted wastewaters from the textile sector, which present both environmental and financial issues, can be used as a source of energy in a district heating plant, for example. Moreover, the COD of the wastewater is greatly reduced by more than 85% using this innovative method. As a result, this proposed technology combines both environmental and economic benefits by producing energy from a sustainable source. Considering the global scale of cotton pretreatment (with a global annual output of 25 million tons), this research could make a substantial contribution to reducing greenhouse gas emissions and, consequently, mitigating global warming in the distant future [43].

As previously mentioned, the anaerobic biodegradation process is a complex sequence of stages. In the initial step, which is hydrolysis, microorganisms release extracellular enzymes while operating in anaerobic conditions. These enzymes serve

Table 1 Typical composition of biogas [38, 95]

Constituents	% Composition
Methane, CH_4	55–65
Carbon dioxide, CO_2	35–45
Nitrogen, N_2	0–1
Hydrogen, H_2	0–1
Hydrogen Sulfide, H_2S	0–1
Oxygen, O_2	0–3

to break down the biodegradable solids into soluble forms. Subsequently, in a series of sequential steps, these soluble organic compounds are transformed into CH_4 and CO_2 gases, generally referred to as biogas [5].

Table 1 shows the typical composition of biogas.

5.1.1 Generation of Electricity from Biogas

Biogas finds application in the generation of heat or combined heat and power (CHP) systems. Within the European Union (EU), the utilization of biogas for energy recovery through CHP has been steadily on the rise [62]. From 2010 to 2011, there was an 18.2% increase in electricity and CHP production from biogas in the EU, reaching a total of 35.9 terawatt-hours (TWh). In this regard, Germany, the UK, Italy, France, the Netherlands, and Spain were the main leaders in 2011. Germany was the highest-producing country, producing 10.9 GWh of electricity in power plants and 8.5 GWh in CHP plants. By 2020, electricity and CHP production may exceed 65 TWh, according to data from EurObserv'ER. It is crucial to understand that variations in electrical efficiency may exist. Engines with a power below 5 kW may achieve around 20–25% efficiency, while engines exceeding 2 MW can reach 35–40% efficiency. CHP systems capture heat from the cooling system, leading to a higher efficiency rate, typically ranging from 85 to 95% [75]. On the other hand, there are considerable drawbacks to this technology, including environmental issues associated with noise and pollution. Typically, CHP manufacturers set the limit for hydrogen sulfide (H_2S) at 100–500 mg per normal cubic meter (mg/Nm^3) to prevent operational issues.

5.1.2 Introduction of Biogas to the Natural Gas Network

Biogas can be upgraded into biomethane and subsequently integrated into the natural gas grid [4]. A study found that generating biomethane produced 44.6 g of CO_2 equivalent per kilowatt-hour (CO_2 eq kWh) in specific greenhouse gas emissions. When compared to natural gas, emissions are decreased by 82%. Therefore, the production of biomethane acts as an important energy source that can reduce the consumption of natural gas and consequently decrease greenhouse gas emissions. Multiple nations, such as Switzerland, Sweden, Germany, France, and the United States, have already integrated biomethane into their natural gas distribution systems. Biogas must undergo purification and upgrading processes to ensure compatibility with the pipeline gas to inject biogas into the natural gas grid [45].

5.1.3 Usage of Biomethane from Biogas as Vehicle Fuel

Biomethane has the potential to be compressed and employed as a vehicle fuel. The transport sector's consumption of oil accounts for 51% of the overall oil consumption, whereas industry and residential sectors represent 34% and 6%, respectively. Reducing oil usage in transportation is highly significant for curbing greenhouse gas (GHG) emissions, as transportation contributes to 21% of the total GHG emissions [11].

Gas-powered vehicles offer several benefits over traditional vehicles, including a significant reduction in emissions like CO_2 (by over 95%), particles, NOx, and non-CH_4 hydrocarbons [66]. For biomethane to be a feasible substitute for vehicle fuel, its price must be 20–30% less than that of bioethanol and biodiesel. Presently, the commercial cost of biomethane, utilizing existing technology, falls in the range of USD 0.4–0.55 per liter, or its gasoline equivalent, which is produced from organic waste or liquid manure [12]. By the end of 2005, there were over 5 million natural gas vehicles in operation worldwide, and this number continues to increase annually [66]. Nevertheless, there is still a shortage of gas refueling stations. In the foreseeable future (by 2020), the mid-term (by 2030), and the long-term (by 2050, primarily for passenger/light-duty cars and heavy-duty city vehicles), biomethane is on track to emerge as an alternative fuel option [12].

5.2 *Possibility of Recycling and Reusing of Textile Sludge*

5.2.1 Utilizing Textile Sludge in Construction Materials

Sludge may be a beneficial resource in building, assisting in maintaining natural sources like clay and fixing waste management troubles. Sludge can be brought to bricks to provide several blessings, including pathogen elimination, oxidation of natural waste, a discount on frost damage, and immobilization of heavy metals in the course of burning.

Patel and Pandey [65] explored the feasibility of incorporating textile industry sludge as a partial replacement for construction materials [65]. They evaluated the sludge and used standard protocols to look at the wide variety of heavy metals it contained. After being dried, the sludge turned into ground right into a high-quality powder and used to make bricks. Brick fine changed into assessed in the observationregarding elements such as leaching houses, compressive electricity, and curing time. The outcomes established that brick compressive energy fell with increasing cement sludge percentage. Nonetheless, the electricity of those sludge-primarily based bricks turned into on par with that of other building substances, such clay- and concrete-based bricks. Moreover, the quantity of heavy metal that seeped through the sludge bricks remained within permissible bounds. Bricks come in a range of designs and colors, which makes them useful for creation. Reusing waste materials, such as sludge, to make bricks is a doable approach to reduce pollution.

In a different study, Priyadharshini and Manikannan [67] examined bricks made from dried textile waste [67]. It has been demonstrated that the sludge content and fire temperature affect brick quality. These sludge-based bricks met the requirements for compressive strength,thus, they could be used in construction.

The use of textile industry sludge in concrete—more precisely, M:20 grade concrete with OPC 43 grade cement—was investigated by Kilkarni [42]. Sludge was used as fine aggregate at several weight percentages (4–36% by weight), and workability was evaluated using compaction factors and slump values. Workability decreased as sludge content increased, according to the study, with up to 32% of sludge exhibiting successful usage in M:20 grade concrete [42].

Loganayagan et al. [49] evaluated sludge from wastewater treatment plants that process wastewater from the dyeing of knitwear and socks. Their findings showed that bricks with a composition of 90% clay soil and 10% sludge were suitable for load-bearing walls. The best options for partition walls were found to be bricks made of either 30% sludge and 70% clay soil or 20% sludge and 80% clay soil. Reducing color formation was also aided by the firewall used in brick production. Small impurities have shown promise as cost-effective substitutes, ranging from 1 to 15% for load-bearing bricks and up to 30% for split bricks [49].

Nagarajan and Sudhakar [57] investigated the potential of incorporating sludge into various concrete and mortar specimens. They found that 4% of the sludge material had good strength for nonstructural components after 28 days of curing. Higher sludge percentages (e.g., 8 and 12%) were less efficient than reference instances, although being able to accommodate large loads. According to the research, sludge from textile ETPs might be efficiently controlled by mixing it with ordinary concrete. Sludge-infused pavement blocks, hollow blocks, and mortar bricks had been tested to be sensible means to construct compound walls, partition partitions, and lighter-weight walkways, respectively. This tactic reduces the negative environmental effects of irrelevant sludge disposal in addition to imparting an alternative constructing useful resource [57].

5.2.2 Using Sludge in Farming and Making Fertilizer

Sludge has two main uses in farming: to provide important nutrients like phosphorus and nitrogen and to naturally enhance soil quality. Many countries have set rules for spreading sludge on farmland, taking into account factors such as heavy metals, pathogens, and pretreatment.

Kakati et al. [37] explored how textile effluent treatment plant sludge could be used as a fertilizer for growing green gram. They discovered that a mix of 10% sludge and 90% of the fertilizer's composition is farmyard manure, which significantly boosted plant metabolism, as shown by various biochemical measures. Sludge concentrations as high as 100%, however, had a detrimental effect on plant growth. They proposed that plant development may be efficiently supported by lower levels of sludge [37].

In a study [90], investigations were carried out involving terry towel textile sludge applied to fast-growing pulp wood tree species in diverse soil varieties. During these experiments, discrepancies in enzyme sports have been noted, with *Eucalyptus tereticornis*showing expanded catalase pastime, at the same time as *Casuarina junghuhniana* showcased improved peroxidase interest. The research emphasized the viable utility of sludge as an organic soil amendment however expressed reservations regarding its concentration of heavy metals and contaminants.

The idea of turning fabric sludge into a gradual-release natural fertilizer was investigated through Muthu [55]. It was found that considerable degrees of number one vitamins, along with potassium, phosphorus, nitrogen, and important micronutrients, have been present in fabric sludge. This advised that it would make an organic fertilizer that is excessive in vitamins [55], [7]. Kaushik and Garg [40] carried out a series of investigations centered on vermicomposting, employing a combination of textile mill sludge, cow dung, and agricultural residues. The study's outcomes discovered that a blend comprising 80% cow dung and 20% stable sludge from fabric generators created a good environment for the proliferation and reproduction of worms. However, higher proportions of sludge had unfavorable consequences. The resulting vermicompost displayed stepped-forward nutrient content material and a discounted C: N ratio [40]. To summarize, those research findings imply that sludge holds capability for agriculture and the manufacture of fertilizers. However, it is vital to meticulously assess sludge concentrations and the ability environmental outcomes.

5.2.3 Other Approaches to Managing Sludge

Biosolids were employed in the area of cement manufacturing, an industry recognized for its enormous energy usage. The enormous power prices associated with clinker production, regularly necessitating pretty high temperatures, may be alleviated via the usage of biosolids as an abundant and accessible supplementary power reservoir.

Moreover, the incorporation of biosolids into exhaust gas emissions presents a possibility to reduce NO_X emissions. This is accomplished by utilizing the ammonia found in the biosolids along with the thermal energy obtained from the hot gases to help convert NOx into harmless nitrogen gas [15].

The process used in the "sludge-to-fuel" (STF) method transforms the organic materials found in sludge into an oil that can be burned. This method employs a solvent and operates under standard atmospheric conditions, typically within a temperature range of 200–300 °C [53]. Alternatively, it can be executed at elevated temperatures along with high pressures, approximately 10 MPa [34]. In a specific system, a water-based reactor transforms mechanically dewatered sludge into various components, including oil, char, carbon dioxide, and wastewater. About 10% of the product is char, which is then disposed of in designated landfills, while the exhaust gas is treated and released into the atmosphere. The oil produced recovers about 90% of the heat value of diesel fuel and can be sold to external customers or refineries. The use of burned textile mill sludge as an adsorbent to extract dyes from wastewater was studied by Jahagirdar et al. [35]. Following the ash's calcination at 800 °C, the structure of the remaining ash was examined. He discovered that the sludge was porous and could be used to capture colors in wastewater treatment by using scanning electron microscopy.

6 Conclusion

After carefully analyzing how the Anaerobic Acetogenic Process is used to treat wastewater sustainably, we have concluded that this bankruptcy points the way toward a future that is both resource- and environmentally-efficient. We have explored the complex field of wastewater control, developing cutting-edge strategies that not only address environmental issues but also provide significant economic and sustainability benefits. We have visible unequivocally for the duration of our research the opportunity for resource and electricity healing from wastewater. Utilizing waste-derived products, especially sludge, and biogas, shows that we canmake big improvements. By the use of these techniques to deal with fabric wastewater, we may also move away from the detrimental effects of our industrial processes on the surroundings and in the direction of a resource control version that is cyclical, regenerative, and harmonious.

The insights in this chapter remind us that embracing the Circular Economy isn't just essential; it is also an opportunity. It's an option to redefine how we deal with waste to convert wastewater right into a precious aid and embark on a sustainable direction that benefits both industry and the planet.

7 Recommendations

7.1 Scope for Future Studies

- To treat wastewater from the textile sector, extensive research should be done to optimize the anaerobic treatment process, with a special focus on reactor design, substrate composition, and operating parameters.
- To optimize treatment efficiency and increase biogas production, anaerobic treatment along with several other technologies such as membrane filtration or electrochemical processes has been investigated.

- Experimentwith how cutting-edge techniques like photocatalysis and bioelectrochemical systems might be applied to clean textile effluent and remove solid contaminants.

7.2 Policy Planners and Decision-Makers

- Comprehensive policies are necessary to develop so that the industries are obliged to adopt sustainable solutions and energy/materials strategies in the textile sector, setting clear targets for reducing water consumption, energy use, and pollution.
- It is essential to set up a device for tracking and revealing sustainable and environmentally friendly wastewater treatment techniques that textile manufacturing facilities will employ.
- Collaborate with other industries and environmental commissions to expand industry-specific sustainability hints and provide technical help to facilitate the adoption of circular economy techniques.

7.3 Stakeholders

- Invest in pilot projects to demonstrate the feasibility and advantages of anaerobic treatment and resource restoration technology in textile processing plants, showcasing successful case studies to encourage wider adoption.
- Take part in capacity-building programs to teach personnel how to maintain and operate anaerobic treatment systems and energy recovery methods, encouraging a sustainable lifestyle within the organization.
- Form alliances with academic institutions and IT firms to research present-day solutions for wastewater treatment and waste stream monetization, pitching a closed-loop technique to practical resource management.

In conclusion, Stakeholders may also drive significant progress toward sustainable wastewater management inside the textile area by imposing specific recommendations which will enhance environmental safety, promote conservation, and facilitate the shift to a circular economic system model.

References

1. 10 Key Statistics on Bangladesh's Growing Textile and Clothing (2024) Retrieved 7 July 2024, from https://www.worldfashionexchange.com/blog/rise-of-bangladesh-textile-and-garment-industry/
2. A New Textiles Economy: Redesigning fashion's future (2024) Retrieved 7 July 2024, from https://www.ellenmacarthurfoundation.org/a-new-textiles-economy

3. Akhtar S, Naoshin N, Zaman M, Medha N (2021) Impact of river pollution on human health: a case study of Shitalakhya River, Dhaka, Bangladesh
4. Amato A, Tsigkou K, Becci A, Beolchini F, Ippolito NM, Ferella F (2023) Life cycle assessment of Biomethane vs. Fossil Methane production and supply. Energies 16(12):4555. https://doi.org/10.3390/en16124555
5. An Den Berg L (1982) Aaerobic digestion of wastes. Conserv Recycl 5(1):5–14. https://doi.org/10.1016/0361-3658(82)90028-5
6. Aspland JR (1997) Textile dyeing and coloration. AATCC
7. Bashar S, Rafiq M, Fairoze F, Das A, Khandaker N (2024) Waste to energy or bio-fertilizer production: municipal solid waste characterization in Khulna city for engineered applications. In: 7th International conference on civil engineering for sustainable development 2024 (ICCESD 2024), Khulna
8. Baydar G, Ciliz N, Mammadov A (2015) Life cycle assessment of cotton textile products in Turkey. Resour Conserv Recycl 104:213–223. https://doi.org/10.1016/j.resconrec.2015.08.007
9. Bes-Pía A, Iborra-Clar A, Mendoza-Roca JA, Iborra-Clar MI, Alcaina-Miranda MI (2004) Nanofiltration of biologically treated textile effluents using ozone as a pre-treatment. Desalination 167:387–392. https://doi.org/10.1016/j.desal.2004.06.151
10. Bhowmik A (2008) Buriganga pollution: reasons & prospects, pp 87–97
11. Bordelanne O, Montero M, Bravin F, Prieur-Vernat A, Oliveti-Selmi O, Pierre H, Papadopoulo M, Muller T (2011) Biomethane CNG hybrid: a reduction by more than 80% of the greenhouse gases emissions compared to gasoline. J Nat Gas Sci Eng 3(5):617–624. https://doi.org/10.1016/j.jngse.2011.07.007
12. Börjesson P, Mattiasson B (2008) Biogas as a resource-efficient vehicle fuel. Trends Biotechnol 26(1):7–13. https://doi.org/10.1016/j.tibtech.2007.09.007
13. Breach PA, Simonovic SP (2018) Wastewater treatment energy recovery potential for adaptation to global change: an integrated assessment. Environ Manage 61(4):624–636. https://doi.org/10.1007/s00267-018-0997-6
14. Brundtland Report | Sustainable Development & Global Environmental Issues | Britannica (2024) Retrieved 7 July 2024, from https://www.britannica.com/topic/Brundtland-Report
15. Calderbank A (1989) The occurrence and significance of bound pesticide residues in soil. In: GW Ware (ed) Reviews of environmental contamination and toxicology, vol 108. Springer New York, pp 71–103. https://doi.org/10.1007/978-1-4613-8850-0_2
16. Capar G, Yetis U, Yilmaz L (2006) Membrane based strategies for the pre-treatment of acid dye bath wastewaters. J Hazard Mater 135(1–3):423–430. https://doi.org/10.1016/j.jhazmat.2005.12.008
17. Carneiro PA, Umbuzeiro GA, Oliveira DP, Zanoni MVB (2010) Assessment of water contamination caused by a mutagenic textile effluent/dyehouse effluent bearing disperse dyes. J Hazard Mater 174(1–3):694–699. https://doi.org/10.1016/j.jhazmat.2009.09.106
18. Choi N-C, Cho K-H, Kim M-S, Park S-J, Lee C-G (2020) A hybrid ion-exchange fabric/ceramic membrane system to remove As(V), Zn(II), and turbidity from wastewater. Applied Sciences 10(7):2414. https://doi.org/10.3390/app10072414
19. ClimateChange | United Nations (2024) Retrieved 7 July 2024, from https://www.un.org/en/climatechange
20. Delée W, O'Neill C, Hawkes FR, Pinheiro HM (1998) Anaerobic treatment of textile effluents: a review. J Chem Technol Biotechnol 73(4):323–335. https://doi.org/10.1002/(SICI)1097-4660(199812)73:4%3c323::AID-JCTB976%3e3.0.CO;2-S
21. Dereszewska A, Cytawa S (2023) Circular economy in wastewater treatment plants—potential opportunities for biogenic elements recovery. Water 15(21):3857. https://doi.org/10.3390/w15213857
22. Ersayin E, Ozgener L (2015) Performance analysis of combined cycle power plants: a case study. Renew Sustain Energy Rev 43:832–842. https://doi.org/10.1016/j.rser.2014.11.082
23. Export Promotion Bureau-Government of the People\'s Republic of Bangladesh (2024) Retrieved 7 July 2024, from https://epb.gov.bd/

24. Fang HHP, Liu H (2002) Effect of pH on hydrogen production from glucose by a mixed culture. Bioresour Technol 82(1):87–93. https://doi.org/10.1016/S0960-8524(01)00110-9
25. Fascetti E (1998) Photosynthetic hydrogen evolution with volatile organic acids derived from the fermentation of source selected municipal solid wastes. Int J Hydrogen Energy 23(9):753–760. https://doi.org/10.1016/S0360-3199(97)00123-7
26. Ferchichi M, Crabbe E, Gil G-H, Hintz W, Almadidy A (2005) Influence of initial pH on hydrogen production from cheese whey. J Biotechnol 120(4):402–409. https://doi.org/10.1016/j.jbiotec.2005.05.017
27. Fersi C, Dhahbi M (2008) Treatment of textile plant effluent by ultrafiltration and/or nanofiltration for water reuse. Desalination 222(1–3):263–271. https://doi.org/10.1016/j.desal.2007.01.171
28. Gadhi TA, Ali I, Mahar RB, Maitlo HA, Channa N (2021) Waste heat and wastewater recovery in textile processing industry: a case study of adopted practices. Mehran Univ Res J Eng Technol 40(3):606–616. https://doi.org/10.22581/muet1982.2103.14
29. García-Castillo C, Maldonado-Villalpando E, Seguí-Amórtegui L, Guerrero-García-Rojas H (2024) Circular economy, eco-innovation and a business model for the operation of wastewater treatment plants in Mexico. Resources 13(7):87. https://doi.org/10.3390/resources13070087
30. Haque IS, Rahman MM, Rafiq MdS, Apurba MSH, Khandaker NR (2024) Energy recovery potential in Bangladesh from elevated temperature textile processing wastewater: an analysis of energy recovery, energy economics and reduction in carbon dioxide emission. Model Earth Syst Environ 10(2):2661–2673. https://doi.org/10.1007/s40808-023-01933-w
31. Hardin G (1968) The tragedy of the commons. Sci New Series 162(3859):1243–1248
32. Hiben YG, Kahsay MB, Lauwaert J (2023) An economic optimization model for a cost-effective solar-assisted industrial heating and cooling system: a case study in MOHA Soft Drinks and Sheba Leather Factories, Ethiopiadepartment (SSRN Scholarly Paper 4313642). https://doi.org/10.2139/ssrn.4313642
33. Hussain T, Wahab A (2018) A critical review of the current water conservation practices in textile wet processing. J Clean Prod 198:806–819. https://doi.org/10.1016/j.jclepro.2018.07.051
34. Itoh S, Suzuki A, Nakamura T, Yokoyama S (1994) Production of heavy oil from sewage sludge by direct thermochemical liquefaction. Desalination 98(1–3):127–133. https://doi.org/10.1016/0011-9164(94)00137-5
35. Jahagirdar SS, Shrihari S, Manu B (2015) Reuse of incinerated textile mill sludge as adsorbent for dye removal. KSCE J Civ Eng 19(7):1982–1986. https://doi.org/10.1007/s12205-015-0731-3
36. Kabir M, Habiba UE, Khan W, Shah A, Rahim S, Rios-Escalante PR, los D, Farooqi Z-U-R, Ali L, Shafiq M (2023) Climate change due to increasing concentration of carbon dioxide and its impacts on environment in 21st century; a mini review. J King Saud Univ Sci 35(5):102693. https://doi.org/10.1016/j.jksus.2023.102693
37. Kakati JP, Ponmurugan P, Rajasekaran N, Gnanamangai BM (2013) Effect of textile effluent treatment plant sludge on the growth metabolism of Green gram (Vigna radiata L). Int J Environ Pollut 51(1/2):79. https://doi.org/10.1504/IJEP.2013.053181
38. Kalia VC, Sonakya V, Raizada N (2000) Anaerobic digestion of banana stem waste. Bioresour Technol 73(2):191–193. https://doi.org/10.1016/S0960-8524(99)00172-8
39. Kamble S (1964) Anaerobic waste treatment fundamentals. Public Works. https://www.academia.edu/3244224/Anaerobic_waste_treatment_fundamentals
40. Kaushik P, Garg VK (2004) Dynamics of biological and chemical parameters during vermicomposting of solid textile mill sludge mixed with cow dung and agricultural residues. Bioresour Technol 94(2):203–209. https://doi.org/10.1016/j.biortech.2003.10.033
41. Kavak H, Timuralp C, Doner N (2020) A comparative study of waste energy recovery from a hot-oil central boiler in a textile finishing factory. Energy Sources Part A: Recovery, Utilization Environm Effects 1–12. https://doi.org/10.1080/15567036.2020.1738595
42. Kilkarni MGJ, Dwivedi AK, Jahgirdar SS (2012) Textile mill Sludege as fine aggregate in concrete

43. Opwis Klaus, Gutmann Jochen (2012) Generation of biogas from textile waste waters. Chem Eng Trans 27:103–108. https://doi.org/10.3303/CET1227018
44. Köhler AR, Som C (2014) Risk preventative innovation strategies for emerging technologies the cases of nano-textiles and smart textiles. Technovation 34(8):420–430. https://doi.org/10.1016/j.technovation.2013.07.002
45. Köppel W, Götz M, Graf F (2009) Biogas upgrading for injection into the gas grid quality aspects, technological and ecological consideration. Gwf-Gas Erdgas Int Issue 150:26–35
46. Lin S, Qi H, Hou P, Liu K (2023) Resource recovery from textile wastewater: dye, salt, and water regeneration using solar-driven interfacial evaporation. J Clean Prod 391:136148. https://doi.org/10.1016/j.jclepro.2023.136148
47. Liu H, Grot S, Logan BE (2005) Electrochemically assisted microbial production of hydrogen from acetate. Environ Sci Technol 39(11):4317–4320. https://doi.org/10.1021/es050244p
48. Logan BE, Oh S-E, Kim IS, Van Ginkel S (2002) Biological hydrogen production measured in batch anaerobic respirometers. Environ Sci Technol 36(11):2530–2535. https://doi.org/10.1021/es015783i
49. Loganayagan S, Rajkumar G, Pavthra A, Poonkundran M (2020) Experimental study on concrete by partial replacement of fine aggregate by textile effluent treatment plant sludge. IOP Conf Ser Mater Sci Eng 764(1):012043. https://doi.org/10.1088/1757-899X/764/1/012043
50. Makertihartha IGBN, Rizki Z, Zunita M, Dharmawijaya PT (2017) Dyes removal from textile wastewater using graphene based nanofiltration. 110006. https://doi.org/10.1063/1.4982336
51. Mamun Kabir SM, Mahmud H, Schöenberger H (2022) Recovery of dyes and salts from highly concentrated (dye and salt) mixed water using nano-filtration ceramic membranes. Heliyon 8(11):e11543. https://doi.org/10.1016/j.heliyon.2022.e11543
52. Marcucci M, Ciardelli G, Matteucci A, Ranieri L, Russo M (2002) Experimental campaigns on textile wastewater for reuse by means of different membrane processes. Desalination 149(1–3):137–143. https://doi.org/10.1016/S0011-9164(02)00745-2
53. Millot N, Huyard A, Faup GM, Michel JP (1989) Sludge liquefaction by conversion to fuels. Water Sci Technol 21(8–9):917–923. https://doi.org/10.2166/wst.1989.0294
54. Mokhtar NM, Lau WJ, Ismail AF, Kartohardjono S, Lai SO, Teoh HC (2016) The potential of direct contact membrane distillation for industrial textile wastewater treatment using PVDF-Cloisite 15A nanocomposite membrane. Chem Eng Res Des 111:284–293. https://doi.org/10.1016/j.cherd.2016.05.018
55. Muthu SS (ed) (2015) Environmental implications of recycling and recycled products. Springer Singapore. https://doi.org/10.1007/978-981-287-643-0
56. Nadim R, Khandaker N, Rio F, Sarkar L, Sharmin A (2020) Acetogenic aerobic sequential batch reactors in series operation for textile wastewater treatment. 8:766–769
57. Nagarajan R, Sudhakar H (2011) Reuse of textile ETP sludge as construction material. New Building Materials and Construction World
58. Nations U (2024) Sustainability. United Nations; United Nations. Retrieved 7 July 2024, from https://www.un.org/en/academic-impact/sustainability
59. Niinimäki K, Peters G, Dahlbo H, Perry P, Rissanen T, Gwilt A (2020) The environmental price of fast fashion. Nat Rev Earth Environ 1(4):189–200. https://doi.org/10.1038/s43017-020-0039-9
60. Nqombolo A, Mpupa A, Moutloali RM, Nomngongo PN (2018) Wastewater treatment using membrane technology. In: Yonar T (ed) Wastewater and water quality. InTech. https://doi.org/10.5772/intechopen.76624
61. Nußholz J (2017) Circular business models: defining a concept and framing an emerging research field. Sustainability 9(10):1810. https://doi.org/10.3390/su9101810
62. O'Keefe DM, Chynoweth DP (2000) Influence of phase separation, leachate recycle and aeration on treatment of municipal solid waste in simulated landfill cells. Bioresour Technol 72(1):55–66. https://doi.org/10.1016/S0960-8524(99)00089-9
63. Ozturk E, Koseoglu H, Karaboyaci M, Yigit NO, Yetis U, Kitis M (2016) Sustainable textile production: cleaner production assessment/eco-efficiency analysis study in a textile mill. J Clean Prod 138:248–263. https://doi.org/10.1016/j.jclepro.2016.02.071

64. Padaki M, Surya Murali R, Abdullah MS, Misdan N, Moslehyani A, Kassim MA, Hilal N, Ismail AF (2015) Membrane technology enhancement in oil–water separation. A review. Desalination 357:197–207. https://doi.org/10.1016/j.desal.2014.11.023
65. Pandey (2009) Exploring the reuse potential of chemical sludge from textile wastewater treatment plants in India-A Hazardous Wa. Am J Environ Sci 5(1), 106–110. https://doi.org/10.3844/ajes.2009.106.110
66. Persson M, Jonsson O, Wellinger A (2006) Task 37—Biogas upgrading to vehicle fuel standards and grid injection. IEA Bioenergy 37:20–21
67. Priyadharshini B, Manikannan K (2018) Utilization of textile sludge in manufacturing e-bricks. Int J Civ Eng Technol 9:2266–2273
68. Qureshi RF, Qureshi K, Bhatti I, Khatri Z (2018) Statistical modeling for stability of emulsion liquid membrane for the removal of anionic dyes from textile wastewater. Mehran Univ Res J Eng Technol 37(3):631–638. https://doi.org/10.22581/muet1982.1803.16
69. Rabbani MdL (2017) Pollution sources assesment of turag River, Bangladesh. IOSR J Mech Civ Eng 14(02):84–91. https://doi.org/10.9790/1684-1402038491
70. Rafiq MS, Apurba MSH, Khandaker NR (2024a) Enhancing textile wastewater sustainability through calcium hypochlorite oxidation and subsequent filtration with assistance from waste blast furnace iron slag. Water Environ J wej.12948. https://doi.org/10.1111/wej.12948
71. Rafiq MdS, Apurba MSH, Khandaker NR (2024) Harnessing untapped resources for sustainable energy production from municipal solid waste in recourse challenged economies: a case study of Rajshahi city corporation, Bangladesh. J Sustain Dev Peace J 2(1):35–57. https://doi.org/10.3126/jsdpj.v2i1.63240
72. Rafiq MS, Apurba MSH, Khandaker NR (2024b) Voir Dire of waste-to-energy conversion from municipal solid waste in Dhaka North City corporation through incineration: a practical venture or an inefficient endeavor? In: 2024 7th International conference on development in renewable energy technology (ICDRET), pp 1–6. https://doi.org/10.1109/ICDRET60388.2024.10503690
73. Rafiq S, Apurba MSH, Khandaker NR (2024) Implementing sustainable practices in wastewater treatment and material waste management in the textile and apparel sector of Bangladesh
74. Rakib MI, Saidur R, Mohamad EN, Afifi AM (2017) Waste-heat utilization—the sustainable technologies to minimize energy consumption in Bangladesh textile sector. J Clean Prod 142:1867–1876. https://doi.org/10.1016/j.jclepro.2016.11.098
75. Reith JH, Wijffels RH (2003) Bio-methane & bio-hydrogen: status and perspectives of biological methane and hydrogen production. Dutch Biological Hydrogen Foundation
76. Sakamoto M, Ahmed T, Begum S, Huq H (2019) Water pollution and the textile industry in Bangladesh: flawed corporate practices or restrictive opportunities? Sustainability 11(7):1951. https://doi.org/10.3390/su11071951
77. Sarayu K, Sandhya S (2012) Current technologies for biological treatment of textile wastewater—a review. Appl Biochem Biotechnol 167(3):645–661. https://doi.org/10.1007/s12010-012-9716-6
78. Sarker MdSR, Rahman MM, Khandaker N (2018) Defining the kinetics of the novel application of anaerobic acetogenics for treating textile dyeing wastewater. Model Earth Syst Environ 4(3):1259–1270. https://doi.org/10.1007/s40808-018-0470-9
79. Sarwar MI, Majumder AK, Islam MN (2010) Water quality parameters: a case study of Karnafully river Chittagong, Bangladesh. Bangladesh J Sci Ind Res 45(2):177–181. https://doi.org/10.3329/bjsir.v45i2.5722
80. Sharma S (2013) Energy management in textile industry. Int J Power Syst Oper Energy Manage 191–195. https://doi.org/10.47893/IJPSOEM.2013.1087
81. Sikora A (ed) (2021) Anaerobic digestion in built environments. IntechOpen. https://doi.org/10.5772/intechopen.92507
82. Studio QID (2024) Pulse of the fashion industry 2019. Global Fashion Agenda. Retrieved 7 July 2024, from https://globalfashionagenda.org/resource/pulse-of-the-fashion-industry-2019/
83. Sultana MS, Islam MS, Saha R, Al-Mansur M (1970) Impact of the effluents of textile dyeing industries on the surface water quality inside D.N.D Embankment, Narayanganj. Bangladesh J Sci Ind Res 44(1), 65–80. https://doi.org/10.3329/bjsir.v44i1.2715

84. Textile effluents, their treatment and discharge in Bangladesh (2015) Textile News, Apparel News, RMG News, Fashion Trends. https://textiletoday.com.bd/textile-effluents-treatment-discharge-bangladesh
85. Thakker AM, Sun D (2023) Sustainable development goals for textiles and fashion. Environ Sci Pollut Res 30(46):101989–102009. https://doi.org/10.1007/s11356-023-29453-1
86. Thiele JH, Zeikus JG (1988) Control of interspecies electron flow during anaerobic digestion: significance of formate transfer versus hydrogen transfer during syntrophic methanogenesis in flocs. Appl Environ Microbiol 54(1):20–29. https://doi.org/10.1128/aem.54.1.20-29.1988
87. Tufekci N, Sivri N, Toroz I (2007) Pollutants of textile industry wastewater and assessment of its discharge limits by water quality standards. Turk J Fish Aquat Sci 7:97–103
88. TuğrulOğulata R (2004) Utilization of waste-heat recovery in textile drying. Appl Energy 79(1):41–49. https://doi.org/10.1016/j.apenergy.2003.12.002
89. Uddin MA, Begum MS, Ashraf M, Azad AK, Adhikary AC, Hossain MS (2023) Water and chemical consumption in the textile processing industry of Bangladesh. PLOS Sustain Transformation 2(7):e0000072. https://doi.org/10.1371/journal.pstr.0000072
90. Umadevi M, Patil G, Subramanian A (2014) Catalase and peroxidase enzyme activities in pulp wood tree seedlings under textile sludge applied sandy loam and clay loam soils. J Biodivers Environ Sci 4:53–63
91. Vajnhandl S, Valh JV (2014) The status of water reuse in European textile sector. J Environ Manage 141:29–35. https://doi.org/10.1016/j.jenvman.2014.03.014
92. World Development Report 2020: Trading for Development in the Age of Global Value Chains (2020) [Text/HTML]. World Bank. Retrieved 7 July 2024, from https://www.worldbank.org/en/publication/wdr2020
93. Younas T, Tayyaba N, Ayub A, Ali S (2021) Textile fabric's and dyes. TekstilnaIndustrija 69(3):47–59. https://doi.org/10.5937/tekstind2103047Y
94. Zahid B, Faisal S, Siddiqui SH (2017) Analysis of caustic soda of different manufacturers in Pakistan for mercerization of cotton textiles. Mehran Univ Res J Eng Technol 36(4):857–866. https://doi.org/10.22581/muet1982.1704.10
95. Zhang R, Zhang Z (1999) Biogasification of rice straw with an anaerobic-phased solids digester system. Bioresour Technol 68(3):235–245. https://doi.org/10.1016/S0960-8524(98)00154-0
96. Zhao S, Tao Z, Chen L, Han M, Zhao B, Tian X, Wang L, Meng F (2021) An antifouling catechol/chitosan-modified polyvinylidene fluoride membrane for sustainable oil-in-water emulsions separation. Front Environ Sci Eng 15(4):63. https://doi.org/10.1007/s11783-020-1355-5
97. Ziegler A, Seijas Nogareda J (2009) Environmental management systems and technological environmental innovations: exploring the causal relationship. Res Policy 38(5):885–893. https://doi.org/10.1016/j.respol.2009.01.020

Higher Order Circularity Strategies in Wastewater Treatment: A Case Study from Vernacular Settlements of Sri Lanka

K. K. K. Sylva and R. Shanthini

Abstract Wastewater treatment methods aim to protect public health and the environment by removing contaminants and preventing water pollution of natural water bodies. Most of the existing treatment methods, inclusive of the costly high-tech methods, are mostly focused on single-objective, end-of-pipe solutions devised under linear production and consumption framework. Such approaches are categorized as lower-order strategies under the concept of circular economic principles which present also a set of higher-order circulatory strategies to effectively maximize resource efficiency whilst minimizing waste generation. This chapter focuses on opportunities for application of higher-order circulatory strategies in wastewater treatment, and demonstrates its viability via a case study from vernacular settlements of Sri Lanka, known as the *Ellangava* system. *An ellangava* system is an ensemble of multi-objective, ecologically-comprehensive, expansive subsystems weaved together towards effectively employing cost-effective ecological services to enhance water quality and to ensure its availability throughout the year. We extend the discussion to identify the advantages of subsystems contribution in achieving the ecosystem sustainability goals in wastewater treatment through a roadmap.

Keywords Wastewater · Higher-order circularity · Sri Lanka · *Ellangava* system

K. K. K. Sylva (✉)
Department of Engineering Management, Faculty of Engineering, University of Peradeniya, Peradeniya, Sri Lanka
e-mail: kamanisylva@eng.pdn.ac.lk

R. Shanthini
University of Peradeniya, Peradeniya, Sri Lanka
e-mail: rshanthini@eng.pdn.ac.lk

S. Yadav et al. (eds.), *Emerging Technologies in Biological and Hybrid Wastewater Treatment: Lessons from Developed to Enhancing Practices in Developing Countries*, Earth and Environmental Sciences Library,
https://doi.org/10.1007/978-3-031-74515-7_14

1 Introduction

Ecosystem resources are threatened every moment by anthropogenic activities. The climate change we are experiencing from the early twenty-first century impacts the ecosystem, increasing ocean acidification and atmospheric carbon dioxide concentrations [1]. Other pressures on the ecosystem, such as degradation, defaunation and fragmentation [1], are experienced by us every day. We are responsible for anthropogenic actions. Our increasing use of natural capital is outpacing the biosphere's regenerative capacity [2]. Our rapid discarding of non-biodegradable matter interrupts biological circular systems at an alarming speed beyond the Earth's absorptive capacity [3]. The anthroposphere, part of the environment influenced by human activities, interacts with natural cycles.

Out of the four fundamental cycles of water, carbon, nitrogen, and oxygen, the hydrological cycle, commonly known as the water cycle, plays an overarching role in the functioning of geological and biological Earth. It operates beyond water cycling and contributes to the circulation of solar energy, transporting sediments and chemical elements for the biosphere's sustainability [4, 5]. Solar radiation evaporates water from the sea and other water bodies from transpiration. The resulting steam rises in the atmosphere and cools and condenses to return to land as precipitation [4]. The water reaching the ground shapes the surface flow and penetrates through the soil layers to recharge aquifers and is deflected towards the land to be discharged into water bodies such as streams, lakes and wetlands [4, 5]. Atmospheric, surface and groundwater are interlinked in solid, liquid and gaseous phases to form the hydrological cycle [5]. Besides, "water is an active chemical agent, efficient transporter of mechanical energy and heat, and a carrier of dissolved and suspended substances" (P. 714) [5], helping sustain life in the biosphere with other vital cycles. However, anthropogenic activities adversely influence the efficient transporting property of water.

Sewage and other chemical waste are critical problems that have been addressed with many advanced solutions globally to prevent their adverse effects on the natural ecosystem. However, contamination due to human-generated waste is not preventable. Recent studies discuss the presence of heavy metals, endocrine disruptors, pathogens, pharmaceuticals and hydrocarbons in untreated sewerage, unfavourably affecting natural ecosystems and the biosphere [6]. According to Wear et al. [6], sewage pollution commonly impacts water-based ecosystems as water sources are considered the first choice for human waste disposal. As an active agent and efficient transporter, water will also become the medium of transport for these unsolicited matters. Nevertheless, it would become a fast carrier of these contaminants through soils, on the surface and up to the coastal regions, polluting the hydrological cycle. Humans invent and introduce advanced wastewater treatment facilities to prevent the transporting of contaminants. However, catastrophic impacts such as global warming, climate change, extreme weather conditions, and related flooding backfire to adversely affect humans' most advanced preventive actions.

According to Hughes et al. [7], climate-related impacts on wastewater systems are of three broader themes: "nuisance flooding spills and odour, water quality deterioration due to increased uncontrolled discharges and damage to infrastructure" (p. 1). They point out that increased rainfall, drought, and high temperatures could affect the wastewater treatment plants and processes. The listed adverse effects are enormous on the anthroposphere itself, and the adverse effects, such as increased contaminants in water, will carry over the biosphere through the hydrological cycle. Solutions to address the bouncing effects of anthropogenic activities are crucial.

Sustainability requires assessing top-down and bottom-up approaches [8]. However, in contributing to a sustainable ecosystem while anthropogenic systems function, reflective thinking is practical in understanding which direction to give priority and when. This chapter attempts to examine the trends in the development of wastewater treatment and reflect on the approaches for an amicable solution. The chapter highlights the strategies for wastewater treatment and discusses the recent tendency towards biological and hybrid systems as cost-effective and sustainable solutions. After that, we complement the circular economy concepts in water security and the need to use higher-order circularity strategies, especially for developing countries. We discuss a case of vernacular settlements of Sri Lanka for applying higher-order techniques. Next, we highlight the challenges to implementing higher-order circularity strategies in wastewater treatment for the present scenario and discuss an easily applicable road map to overcome the difficulties.

2 Approaches to Wastewater Treatment

Wastewater originates from both municipal sources and industrial sources. Wastewater from municipal sources considered in this section includes 'black water' consisting of excreta and related matter, 'grey water' resulting from cooking, washing and bathing-related activities, agricultural, horticultural and aquaculture effluents, and stormwater and surface runoffs that have risen to problematic levels owing to man-made structures. We briefly present popular treatment processes and systems relevant to wastewater from municipal sources that are dealt with in detail in several textbooks [9, 10]. In general, industrial wastewater treatment processes are specific to the type of industry generating the wastewater, and we will not discuss such in this section.

When the black water generation sources are spaced far apart, a single source's volume of black water is manageable. Surrounding soil structure is conducive, and onsite sanitation systems, such as sewage pit structures, work admirably well in treating black water [11]. A typical sewage pits system consists of an underground septic tank and an underground soak pit of varying designs. Black water is anaerobically treated in the septic tank, and the effluent from the septic tank is fed to the soak pit, which slowly distributes the said effluent to the surrounding soil. The septic tank and the soak pit need no regular maintenance or energy input, which is popular among Sri Lankan households. In case the effluent leaving the septic tank is not adequately treated in the septic tank, it is possible to treat the septic tank effluent in a second treatment unit, such as an up-flow anaerobic sludge bed reactor [12], before sending the effluent to the soak pit. It is also possible to treat high-strength grey water originating from a municipal facility, such as a restaurant, using a septic tank followed by a sequencing batch reactor [13].

Black water originating from households, hotels, restaurants, and public spaces in crowded urban environments has to be treated at centralized systems. The general approach to wastewater treatment in a centralized system could be categorized under physical, chemical and biological treatments. The components of physical, biological and chemical treatment processes in a wastewater treatment system can vary depending on the quantity and the characteristics of the wastewater to be treated and the final quality of the treated water to be attained.

The primary stages of wastewater treatment invariably are the physical treatment systems involving screening, comminution, and grit removal. Screening is a process by which coarse and settleable solids are removed by interception. Comminution is a process in which coarse solids are ground into almost uniform-sized solids. Grit removal is a process by which sand, stone and other heavy solid materials are removed. A popular method used for grit and particulate matter removal is sedimentation, which separates suspended particles that are heavier than water from water by allowing them to settle in a settling tank or basin under gravitational forces [14]. Sedimentation requires no energy input during operation, even though it is space and time-consuming. Descriptions and designs of the aforementioned treatment units can be found in Tchobanoglous and Burton [9] and others.

Secondary stages of wastewater treatment involve biological treatment methods which primarily imitate the natural microbial degradation of the dissolved and colloidal carbonaceous organic matter in terrestrial and aquatic environments [14]. Degrading agents are invariably microorganisms, primarily bacteria, readily available in nature. Microorganisms degrade the carbonaceous organic matter into gases and cell matter, and the cell matter is also known as digested/biological solids.

The most basic biological treatment process is achieved in aerobic stabilization ponds, which are shallow earthen basins providing environments conducive to the natural aerobic biological degradation of low-strength organic wastewater. In the case of high-strength organic wastewater, a deep earthen basin would provide the facultative and/or anaerobic environments required for the degradation of waste by the facultative and anaerobic microorganisms in addition to the aerobic degradation occurring at the shallow surface zone [15]. Though slow and space-consuming, the aforementioned degradation processes require minimal outside control and energy for their operations. Biological treatment processes for treating large quantities of high-strength wastewater in space and time-limited situations are designed to speed up the degradation process by controlling the process environments, such as temperature, acidity/alkalinity and amount of dissolved oxygen, to maximize the growth of microorganisms and, hence, the degradation of targeted waste in the wastewater. Some of the commonly used biological treatment processes are suspended-growth systems, such as activated-sludge processes and aerated lagoons, and attached growth systems, such as trickling filters and rotating biological contactors. Process descriptions, underlying microbiological processes and unit designs of the aforementioned treatment systems can be found in Sperling [16], Tchobanoglous and Burton [9] and Tchobanoglous et al. [10].

The anaerobically-operated septic tank [11], the up-flow anaerobic sludge bed reactor [12, 17] and the aerobically-operated sequencing batch reactor [13, 18], referred to in the second paragraph of this section, are also biological wastewater treatment units that have become popular as of late.

The digested/biological solids and the additional settleable solids resulting from the biological treatment processes of the secondary stage of wastewater treatment are known as secondary sludge, and removal of the secondary sludge from the biologically treated wastewater is an essential step towards the completion of the secondary stage of wastewater treatment. Sludge thickening is the primary sludge treatment process achieved by separating the supernatant liquid from the thickened sludge by physical treatment methods such as gravity, flotation, centrifugal, gravity-belt or rotary-drum thickening. The moisture content of the thickened sludge can be reduced by dewatering the sludge with the help of physical treatment methods such as filtration, centrifugation or low-cost sludge drying beds. Dewatered stabilized sludge can be disposed of by applying it to agricultural, forest or disturbed lands as a soil conditioner. Descriptions and designs of the aforementioned treatment systems can be found in Andreoli et al. [19], Drinan and Spellman [20] and Tchobanoglous and Burton [9].

Chemical treatments generally used in wastewater treatment systems are disinfection to destroy pathogens from wastewater, dewatered solids or treated water using chlorine, chlorine dioxide, ozone and other similar chemicals; coagulation and flocculation of tiny particles in the wastewater or treated water to facilitate the formation of large particles to promote separation by sedimentation using alum, ferric sulphate, ferric chloride, aluminium sulphate, among others; adjustment of pH of treated water to neutral level before disposal using sulphuric colour acid, hydrochloric acid, phosphoric acid, sodium hydroxide, calcium hydroxide (Lime) and others; controlling odour; oxidization of harmful constituents in wastewater, sludge or treated water. Descriptions and designs of the aforementioned treatment systems can be found in Tchobanoglous and Burton [9].

Nutrients, such as phosphates, nitrates, nitrites and ammonia, found in fertilizers, foods, soaps and detergents and also in human waste, are found in municipal wastewater and must be removed to control or eliminate potential overgrowth of non-targeted plants in the environments to which the treated water is released, and other related harm. A comprehensive review of conventional and advanced technologies suitable for nutrient removal from domestic wastewater can be found in Rout et al. [21]. Vymazal [22] recommended the use of low-cost constructed wetlands be used for the said purpose.

Micropollutants, such as pharmaceuticals, cosmetics and perfumes, detergents, steroid hormones, pesticides, heavy and trace metals (cadmium, lead, copper, iron, manganese, boron, silicon, germanium, and others), persistent organic pollutants (polychlorinated biphenyls, polybrominated diphenyl ethers, and dichlorodiphenyl-trichloroethane), micro- and nanoplastics, and artificial sweeteners, found in wastewater originate from households, public places, agricultural sources and surface runoffs. Most micropollutants in the environment in amounts that are detrimental to life forms are of anthropogenic origins [23]. Conventional wastewater treatments mentioned above and elsewhere, in general, are not designed for either the removal of or the treatment of micropollutants [24]. Margot et al. [25] recommended that the treated water from the wastewater treatment plant be subjected to powdered activated carbon absorption followed by either ultrafiltration or sand filtration for enabling good removal of most micropollutants and macropollutants without forming problematic byproducts, the strongest decrease in toxicity and total disinfection of the effluent. However, Margot et al. [25] identified the suggested treatment process as costly. A comprehensive account of the treatment of micropollutants in water and wastewater is given in Virkutyteet al. [26]. In addition to the advanced treatment processes, such as nanofiltration, adsorption, ion exchange, membrane bioreactors and abiotic degradation and volatilization, suggested by Virkutyteet al. [26], there was the low-cost option of wetland treatment with a long retention time of almost 30 days for both microbial degradation and photolysis of selected micropollutants. Kröpfelová et al. [27] reported that three horizontal-flow constructed wetlands in the Czech Republic, designed to treat municipal wastewater, removed 90 of aluminum, 78 of zinc (78) and 50–75% of uranium, antimony, copper, lead, molybdenum, chromium, barium, iron and gallium.

All treatment methods aim to protect public health and the environment by removing contaminants and preventing water pollution of natural water bodies. The cost-effectiveness of these methods could vary depending on the volume of wastewater to be treated, contextual conditions, and treatment goals. However, recent trends in wastewater treatment are inclined towards hybrid and biological treatment as sustainable solutions [21, 22, 26, 27].

3 Sustainability and the Concept of Circular Economy

The circular economy is a concept that emerged among many other sustainable development solutions for reducing ecosystem changes resulting in anthropogenic, unsustainable practices. Circular economic principles present a set of strategies to maximize resource efficiency and minimize waste generation. The circular economy concept promotes sustainability in contrast to the traditional linear economic model 'take-make-dispose' by the attempt to circulate already extracted virgin resources to add lifecycle value. Regeneration is promoted by circular economic attempts to restore ecosystem resources, increase biodiversity, and improve soil health. The tools of a circular economy examine the lifecycle of any process to rectify issues caused by a linear economy. Regenerative and sustainable societies can use the principles of a circular economy for wastewater treatment and to secure water needs.

The higher-order principles of a circular economy provide a more comprehensive strategic framework for transforming from linear thinking to a circular thinking model. They attempt to guide long-term policy development and systemic change to achieve sustainability. Whole system thinking, regenerative and resilient design, local and regional circularity, technological innovation, cross-sector collaboration, circular culture and mindset, transition roadmaps, education and awareness, sustainable consumption and lifestyle are some strategies that will lead to higher-order thinking of a circular economy. In comparison, a lower-order review of a circular economy orients towards the traditional approaches and linear thinking, where the take-make-dispose model is the priority. This thinking leads to resource depletion, waste generation and adverse environmental impacts. Linear production and consumption, profit orientation, and resistance to change can be in lower-order thinking patterns. Lower-order thinking is more focused on end-of-pipe solutions for any problem.

Among the R strategies introduced in a circular economy, the progression from 3R [28] to 10R [29, 30] tries to the higher levels of thinking of a circular economy. The Reduce in 3R is inclined towards higher-order thinking, while Reuse and Recycle are lower-order thinking, aiming to improve a product's life in the existing production systems. In the 10R strategies, Refuse, Rethink and Reduce are all aimed at alternative methods for reduction of the use of a product claiming for higher order circularity strategies.

4 Higher-Order Approaches of Water Quality Maintenance

Authors propose whole system thinking to achieve the goals of a circular economy [3, 31]. "A systems-based approach that can transcend disciplinary boundaries can be compelling in enabling well-informed and meaningful decision-making processes that challenge current norms, values and practices" (p. 24800) [31]. Systems thinking identifies subsystems that could act in favour of the whole system to achieve overall objectives [32]. The whole system thinking applied in wastewater treatment reduces wastewater production by anthropogenic activities. Anthropogenic factors affect water quality by elevating concentrations of heavy metals, mercury, coliforms and nutrient load in water [33]. Courses of water pollution from rural settings could relate to agricultural activities, runoff from croplands, barnyards and feedlots, wasteland application and storage facilities, construction sites, mining operations, sewage discharge, washing clothes and vehicles, bathing of animals in water bodies, timber harvesting and pastureland (poultry farms) [33]. However, the sources of pollutants differ in urban areas. They can be from industrial discharge, channelization, municipal discharges, landfill sites, domestic effluent, septic systems, livestock waste in residential areas and land use/land cover change [33]. Applying higher-order circular thinking to analyze the lifecycle processes can reduce pollutant sources.

Another higher-level circularity, regenerative and resilient designs, could be effectively used in city and town planning to have regenerative zones for wastewater treatment within cities. Resilient designs could promote water efficiency and reduction in water demand, and regenerative cities could promote natural wetlands, waterways, and estuaries in urban landscapes for biological water cleaning. Recycled water could recharge aquifers be used in third pipes and sewer mining. Developing green infrastructure, permeable pavements, green roofs, and supporting stormwater through urban wetlands will introduce circulation within regions while reducing the urban heat island effect. The use of black water sludge in agriculture has proven to reduce heavy metal content in the food cycle [34].

Refusing from the R strategies could be simulated under resilient designs. Treatment processes with high chemical treatment, which could cause further damage to the environment, can be refused. Chemicals used in the treatment process, such as aluminium-based coagulants, can be toxic to aquatic life. By reducing chemicals in treatment processes, we can avoid the tendency to have any residues of coagulants, flocculants, disinfection byproducts or residual chemicals used for pH adjustments. Instead, we can use organic products such as bio-coagulants and bio-flocculants [35]. By Rethinking, we can select the most appropriate methods for maximizing resource usage in treatment. Hybrid methods with biological and physical treatment may be advantageous in reducing further damage to the environment in wastewater treatment. Implementing durable and long-lasting equipment and infrastructure to minimize replacements and achieve maximum efficiency may result in Rethinking.

Local and regional circularity could prevent or minimize the polluted water entering the hydrological cycle. Increasing water scarcity compels reusing water in local, regional and global environments [36]. The level of circularity may depend on the level of water pollution, the income level of different region zones and available data on circularity [36, 37]. Sato et al. [37] state that wastewater treatment is limited in developing countries. Treatment facility investments have not considered the increasing populations and wastewater volumes. However, "purposefully crafted and skillfully implemented regulation and supportive policy measures" (p. 231) [36] may be advantageous for increasing the circularity of wastewater by treating or reducing the sources of wastewater. In addition, Sarov [38] discusses many economic benefits of using sludge, such as extraction of resources, residual biogas for heating and electricity, sand for construction and purified water for reuse or discharge into hydro-basins.

Technological innovation can significantly affect wastewater treatment processes, making them more efficient, cost-effective, and environmentally friendly. Advanced sensors and monitoring systems can provide data on water quality, flow rates, and the presence of contaminants [39]. Automation of systems can improve the efficiency of wastewater treatment and result in optimum dosing of purification agents, flow control and sludge handling [40]. Membrane technologies like nanofiltration and reverse osmosis can efficiently remove contaminants from wastewater [41].

Cross-sector collaboration, circular culture and mindset, transition roadmaps, education and awareness, sustainable consumption and lifestyle may help the policymakers, stakeholders, and the community to build a local, regional and global circularity culture. Higher-order approaches of circularity involve systemic change by redesigning the systems and supply chains to minimize waste and prioritize reuse to ensure long-term sustainability. In contrast, the lower-order approaches operate within existing linear economic models. However, the lower-order circularity approaches are essential for conserving resources and reducing waste in the existing systems.

5 Lower-Order Circularity Approaches for Wastewater Treatment

A circular economy's lower-order strategies focus on waste reduction and resource efficiency within existing systems. However, these essential strategies may not represent a complete transition to a circular economy. The lower-order strategies of a circular economy could help promote sustainability and resource conservation in wastewater management. Source separation and segregation allow separate wastewater streams. For example, separating grey and black water at the sources can enable targeted treatment for different types [42]. Blackwater contains organic load and pathogens, posing higher contamination and separation, allowing adequate treatment [42]. Energy recovery, such as biogas from anaerobic digestion, can offset energy consumption in treatment plants. Simultaneously, we can extract valuable nutrients such as phosphorous and nitrogen from wastewater to produce fertilizers to reduce chemical inputs in agriculture [43]. We can dewater or incinerate the sludge to recover agricultural and energy production resources. When chemical processes are unavoidable, we can recover chemicals from wastewater, such as nitrogen and phosphorus, for converting to fertilizer. Anaerobic digestion can be used for energy recovery to power treatment facilities or feed back into the grid [44]. However, the lower-order strategies could result in higher costs for implementation.

Reusing and recycling water for non-portable purposes such as irrigation, industrial processes, or toilet flushing reduces the demand for fresh water. Advanced treatment technologies such as membrane filtration and ultraviolet disinfection can improve treated water quality and eliminate the need for chemical additives [45]. Integrating natural and engineered systems such as constructed wetlands and vegetated swales can enhance water quality and promote biodiversity. Education and public awareness are essential at lower order, as well as responsible handling of wastewater and regulations and standards will ensure the reduction of discharge of harmful substances to the environment. We can Recycle wastewater with appropriate methods to help reduce the use of freshwater in industrial facilities. Employing closed-loop water systems within industrial processes to reduce the demand for freshwater can be highlighted at this level. Higher-order and lower-order strategies can effectively transmit linear thinking to a closed-loop water system. However, applying higher-order strategies secures water needs while reducing water pollution. The next section discusses a case study from the Sri Lankan vernacular settlements where higher-order circularity strategies are visible.

6 A Case Study for Water Security from Sri Lankan Vernacular Settlements

The tank-cascade system, a natural drainage basin-based irrigation technology [46], has been a significant part of Sri Lanka's cultural heritage for over two millenniums. This section presents the major components of the tank-cascade system and their functions, and their underlying circular economic concepts are elucidated in the following section.

The tank-cascade system is termed *ellanga* system or, briefly, *ellangava* in Sinhala, the language of the majority in Sri Lanka. *Ellangava* means hanging next to one another [47–49]. As Tennakoon [49] presents, "*ellan* means hanging and *gava* means next to one another" in Sinhala. We will use the term *ellangava* to refer to the tank-cascade system.

Central to ellangava design, the cascaded tank structure involves a series of interconnected tanks of varying sizes and functions via canals to capture, store, and distribute rainwater so that water levels in the natural streams and rivers do not dry up below critical levels. *Ellangavas* in Sri Lanka are millenniums old, integral to the country's rural landscape, and used for irrigation and water management across different elevations and landscapes.

The primary purpose of an *ellangava*is as follows:

An *ellangawa* is a very low-gradient sluggishly moving water path from a head-end summit[1] of a micro- or meso-valley to its lower-end at a large reservoir or a secondary stream (*Oya*),[2] filling a chain of tanks one after another on its way down, gently spilling over the tanks' excess water into long flat plains in between the tanks and in that whole process, the tanks accumulating, preserving, releasing and guiding water to downstream fields for irrigation [49].

Food and Agriculture Organization of the United Nations (FAO) has recognized the crucial role that *ellangava* long played and continues to play at the global, regional and national levels for food security and sustainable development in the context of climate change and designated *ellangava* as a Globally Important Agricultural Heritage System (GIAHS) on April 19, 2018 [50]. We present a schematic of a typical *ellangava* system in Fig. 1.

At a village level, the central feature of an *ellangava* subsystem is the large tank (or reservoir) known as the village-tank.[3] Water entering and leaving the village-tank undergoes, by ecologically appropriate design, several levels of natural control and

[1] Head-end summit is named 'mudunna' in Sinhala.

[2] Stream is named 'oya' in Sinhala.

[3] Village-tank is named 'pahalawewa' in Sinhala.

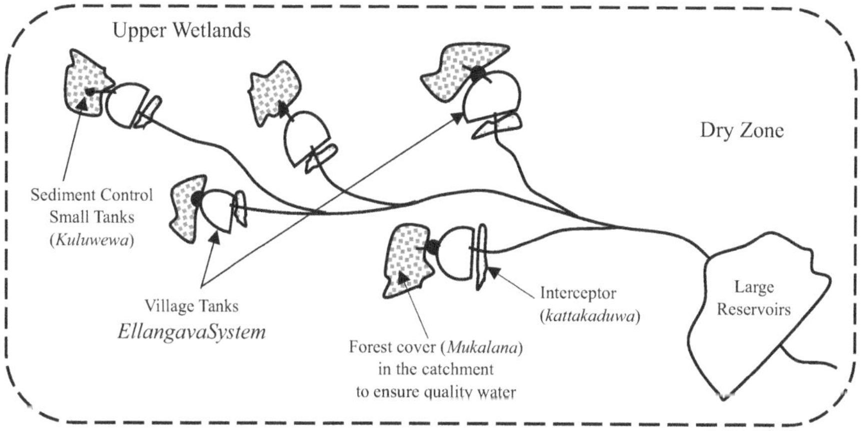

Fig. 1 Schematic of typical *ellangava* system in Sri Lanka

purification processes. We present a schematic of an *ellangava* subsystem at a village level in Fig. 2.

The upper part of a village-tank, the high land catchment area, necessarily features a community-protected dense forest cover,[4] which acts as a sponge to capture rainwater in the rainy season, recharges groundwater through infiltration, maintains the canal base flow, provides habitat for wild animals, and conserves bio-diversity [51]. The forest cover also provides non-timber forest products such as edibles (wild fruits, greens and nuts), barks and resins, and medicinal herbs [52].

Water moving through the forest cover gets collected in several small silt-trapping forest-tanks,[5] the primary purpose of which is to retain eroded soil particles entering the forest-tank and release sediment-reduced water to the village-tank [49]. In addition to silt reduction, forest-tanks also contribute to providing water for wild animals for their well-being and to keep the predators out of the village and its ecosystem, storing the rainwater and enriching the groundwater table, and providing water for the forest vegetation [52, 53].

It is also common to find man-made waterholes[6] serving similar purposes as the forest-tanks. In addition, waterholes provide water for cattle and act as a buffer to downstream water flow during heavy storms, thereby protecting the embankment [52].

Water entering the village-tank passes through a belt of large-size trees and a belt of meadow. The prime purpose of the large-sized trees in the tree belt[7] is perhaps to

[4] Forest cover is named ‘mukalana’ in Sinhala.

[5] Forest-tank is named ‘kuluwewa in Sinhala, and kulu or kullla refers to a traditional utensil used to separate impurities heavier than paddy or grains from paddy or grains.

[6] Manmade waterhole is named ‘godawala’ in Sinhala.

[7] Tree belt is named ‘gasgommana’ in Sinhala.

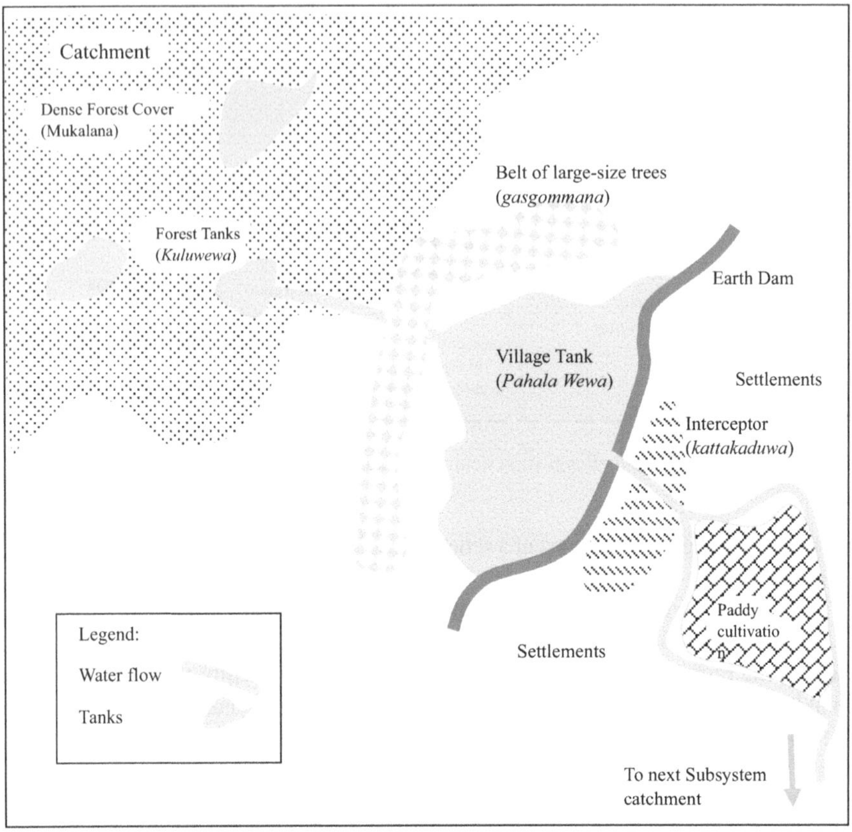

Fig. 2 A schematic of an *ellangava* subsystem at a village level

reduce evapotranspiration of the water in the village-tank by functioning as a wind-breaker. The meadow belt,[8] consisting of bushes and reeds, is a filter for screening sediments and grit removal. These belts also accommodate water when it spills from the tank due to excess rain. Sakalasooriya [53] observes that the roots of the large water-tolerant trees, such as white *marudah* trees[9] and white wood chaste trees,[10] naturally grown in the tree belt make water cages, creating breeding and living spaces for some fish species. Tennakoon [49] observes that the reeds grown in the meadow belt provide habitat for not only the wading birds but also for microorganisms, which are food for the shallow water-loving small varieties of indigenous fish and all-preying birds. The tree and meadow belts also provide wild fruits, food for cattle, medicinal herbs and wildlife habitat [54]. Barking deer, wild boar, sambur, wild buffalo and elephants are also found roaming in these belts [53] when relatively dry.

[8] Meadow belt is named 'perahana' in Sinhala.

[9] *Terminalia arjuna*, known as Kumbuk in Sinhala and Maruthu in Tamil.

[10] *Vitex leucoxylon*, known as Nabada in Sinhala and Nir nocchi in Tamil.

It is noteworthy that these upstream natural processes of intentional design not only render all the services provided by the current primary wastewater treatment methods mentioned in Sect. 2 but also render additional services such as maintaining water in the village-tank throughout the year by groundwater management, controlling flood, wildlife conservation, ecosystem services, and evaporation reduction of water in the village-tank.

Even though the prime objectives of the water stored in the village tank are irrigation and water management, it is also the primary water source for the villagers for their everyday activities, such as collecting water for consumption, bathing and cleaning animals. Interestingly, the village-tank has dedicated places for the aforementioned different uses [52]. White *marudah* trees, with its salt-absorbing capacity to purify water, are prevalent where water is used for human consumption. It is of interest to note that the results of a scientific study carried out by Rathnayaka et al. [55] showed that the roots of white *marudah* trees were effective in improving water quality in contaminated water and can be used to treat the water quality of total hardness, conductivity, ammonia, total suspended solids, alkalinity, and chloride concentrations.

Water leaving the village-tank flows through an area known as an interceptor,[11] which is a reserved land consisting of differing micro-climatic environments: the downstream side of the tank bund, waterhole, wetland, and dry upland [56, 57]. The interceptor ecosystem consists of diverse vegetation, from aquatic plants to tree species found in the dry forest. It can reduce tank seepage and prevent salts and ions in polluted water from seeping through the bund to the paddy field [56].

In a survey carried out using 31 interceptor sites in the central dry zone by Dharmasena [56], it was found that interceptor sites are the habitat for 13 plant species, and 88% of the plant species found are of natural origin. It was also found that 60% of the 77 plant species in interceptor sites provide various community needs such as timber, fuel wood, medicine and fencing materials [56].

FAO in Sri Lanka [54] reported that two villages, Udakadawala and Palugaswewa, located in the lower part of the Palugaswewa elangawa, were self-sufficient in rice. FAO in Sri Lanka [54] also reported the villagers earned significant income by cultivating onion, chilli, corn and pulses in the well-drained paddy fields during minor seasons and legumes, coarse grains, fruits and vegetables in uplands and home gardens. Moreover, as observed by FAO in Sri Lanka [54], the tank is a source of fish and other aquatic food sources, and the tree belt and the interceptor provide wild fruit and food, local medicine and habitat to many predators.

FAO in Sri Lanka [54] states that the villagers believe that many plant species are salt-absorbing. Hence, these plant species can be used for water purification in bio-remediation. The plant species referred to are: Vetakeya (*Pandanus thwaitesii*), Matgrass (*Cyperus pangore*), Ratton (*Calamus spp*), Palmyra (*Borassus flebelfifer*), Karanda (*Pongamia pinnata*), Thimbiri (*Diospyros malabarica*), Damba (*Syzygiumassimile*), Areconut (*Areca catechu*), Milla (*Vitex pinnata*), Tamarind (*Tamarindus*

[11] Interceptor is named 'kattakaduwa' in Sinhala.

indica), Beli (*Aegle marmelos*), Margosa (*Azadericta indica*), Kumbuk (*Terminalia arjuna*), Woodapple (*Feronia fimonia*) and Mee (*Madhauca longifolia*).

7 Appreciation of Higher-Order Circularity Strategies in *Ellangava*

In Sri Lankan vernacular settlements, the cascading system of tanks in the upper highlands links to large reservoirs built in dry zones to ensure water security throughout the year for agricultural purposes [3, 48]. The *ellangava* system depends on the interconnected system of small tanks, channels, and large reservoirs to form a rich water and soil conservation system [58]. The *ellangava* system is not typically designed to treat or clean wastewater. However, some salient features of the landscape provide evidence of maintaining water quality in the whole system. The whole system thinking is a concept promoted in higher-order circularity, and the *ellangava* system demonstrated this concept with its system connectivity.

The '*ellangagammana*' or the cascade village system consists of four major zones: tank and its elements, channels and paddy fields, protected forest and catchment and the households [47]. Out of these zones, some features of the zone of channels and paddy fields highlight the water quality maintenance features such as tree belt (*gasgommana*), filter (*perahana*), catchment forest (*mukalana*), silt trapping small tanks (*kuluwewa*) before the water enters a consecutive subsystem [47]. The water puddles (*godawala*) are depressions found in the protected forest and catchment zone for collecting rainwater to last during the year and feed village-tanks [49]. Filling large reservoirs from reserves in smaller tanks during the dry season exemplifies the regenerative and resilience feature of higher-order circularity in *ellangava* system. The water puddles in the catchment preserve water to recharge small tanks during the dry season. The regenerative and resilient design observed in the *ellangava* system is overridden by ad hoc political and transferred technology goals for creating large reservoirs for energy and industrial needs [58]. As stated by Mendis (p. 210) [58] "such practices are deemed to be less efficient and new techniques are to be introduced" has been the directive to change from a resilient system to modern large reservoirs.

The water purification features in the zone used by human households are white marudah trees[12] planted in *diyamankada* to purify water and *kattakaduwa* is used to reduce salinity and other impurities such as heavy metals. *Kattakaduwa* is the preserved land area between a tank bund and the downstream paddy field as the main purpose to safeguard the tank bund [56]. However, it has been proven that the area acts as a salinity interceptor belt [59]. These features demonstrate that secure water is reaching for domestic and agricultural purposes. Regenerative and resilient designs are incorporated throughout the water distribution channels for water quality maintenance.

[12] *Terminalia arjuna*, known as Kumbuk in Sinhala and Maruthu in Tamil.

The *ellangava* system does not use energy for cleaning water other than wetlands, demonstrating innovative design. The use of aquatic plants and macrophytes in these tanks is visible to absorb nutrients and contaminants from wastewater. The practice of phytoremediation [60] is visual in these subsystems. These wetlands help the pollutant load before the water enters another subsystem. A dense forest cover separates each subsystem to ensure water quality for the consequent subsystem. Any cultivation has been prohibited in the forest cover in ancient management systems. The forest cover has been infringed upon with changing patterns of settlements in the recent past [58].

The *ellangava* system is a classic cross-sector collaboration system that secures quality water throughout the year for the agricultural society while preserving soil and maintaining ecosystem stability and interconnectedness. Village planning of *ellangava* has been mainly based on the availability of water while paying attention to other considerations of ecosystem survival. Water puddles in the forest have been A particular area in paddy cultivation known as *kurulupaluwa* has been reserved for invading birds [48, 61, 62]. According to Sylva and Sylva [48], the water released from the village-tank is diverted to paddy fields through canals constructed with one bank and the other left to blend into the forest adjacent for the surface water to enter the canal and allow wild animals safe and free access to water demonstrating cross-sector collaboration with ecosystem [63].

Education and awareness are visible in society to maintain the village subsystem to adhere to the natural water purification process, such as water cleaning trees and ponds after human consumption. All materials used for construction have been eco-friendly, which does not pollute the water system [48]. Other growth trends have gradually overpowered nature-based educational needs [58]. However, as depicted in Table 1, priority has been given to higher-order circularity in water management in *ellangava* system.

The trends in changing livelihood and related needs have overridden the sustainable system and led to industrialization without securing the ancient system qualities. Consequently, many development activities and transferred technological influences have adversely affected the *ellangava* system of rich soil and water conservation. Small tanks were identified as inefficient and were submerged in large modern reservoirs [58]. The bottom-up constructions of small to large systems have been replaced with top-down large systems to fulfil growth needs.

Table 1 Circularity features of the vernacular settlement model for water security

Circularity concepts		Feature in the vernacular settlement
Higher order circularity	Whole system thinking	A cascade system of village-tanks in the upper wetlands to flow water to dry zone to fill large reservoirs
	Regenerative and resilient design	Use of rainwater and water puddles in the catchment to ensure water security of village-tanks during drought Filling of large reservoirs from village-tanks during the dry season Use of gravitation flow of water
	Local and regional circularity	Water diversion for agriculture, human consumption, sustainability of wildlife within each *ellangava* system and sending clean water to the next system
	Innovation in technology	Use of gravitational flow innovatively to circulate rainwater with necessary arrangements for water security at different levels
	Cross-sector collaboration	Soil protection and ecosystem sustainability while attending to water needs
	Circular culture and mindset	Top-down and bottom-up mindset for circulation
	Education and awareness	Knowledge of natural water cleaning systems throughout society
	Sustainable consumption and lifestyle	Forest covers the catchment to ensure quality water Green material for living, which will not add to pollution
Lower order circularity	Cleaning used water	Planting trees (*Kumbuk* and other trees) that naturally clean used water by the community
	Removal of contaminants	*Kattakaduwa* is used to reduce salinity and other impurities, such as heavy metals
	Repurposing	Sediment collection is used for agriculture

8 A Road Map for the Application of Vernacular Water Management Model to the Present Scenario and Challenges

The cascade system and its connected water quality maintenance is a cost-effective solution for the developing world as it does not need high investments in modern technology. The gravitational flow of water is the core concept used in water circulation, and the significant structures are limited to earth dams constructed for small tanks and large reservoirs. However, population growth and unplanned settlements have obstructed the water flow patterns and the village structure, gradually affecting the prosperous agricultural society.

Sustainable consumption patterns of the *ellangava* system, the "viable green society" (p. 152) [58], has been changing with present needs and transferred technological product usage. Without sufficient consideration, many small tanks have been submerged for the current hydropower needs in the wet zone of Sri Lanka [58]. High pollution levels in the last few decades have made natural cleaning systems less effective due to the large quantity of solid waste discharged into water bodies. At this level, the lower-order circularity strategies predominate to reduce water pollution. However, in transferring up-to-date technology, cost is a significant challenge for developing countries in the waste industry [64].

Regulatory directives on ensuring higher-order circularity to maintain water quality by naturally cleaning forest reserves, plants and wetlands should be prominent for sustainable water management. Other directives recommended for the present scenario are adequate solid waste disposal systems in human settlements and directing the waste for circulating without using water bodies for disposal, ensuring educational programs and community commitment. Ensuring bottom-up cleaning programs for already polluted water bodies might need high-tech solutions and biological cleaning methods depending on the context-specific pollution levels. However, developing countries could face funding problems for high-tech resource recovery techniques [64]. Figure 3 identifies a transition Road Map for applying higher-order and lower-order circularity concepts in water management for a circular water security system.

The cascade concept can be applied to the present scenario with some modifications to the elements of the system. The village is the smallest subsystem in the vernacular settlement, and the new system must be brought down to a smaller unit such as households or organizations, a factory or agricultural land, depending on whichever exists in a context. After identifying the smallest subsystem of operation, it is necessary to picture this unit in the whole system, such as a district, state, or country.

The next step will be to identify the linkage patterns of these subsystems to the whole system through water bodies. Finally, identifying the responsibility of each subsystem to the whole system to function smoothly is necessary. Considering overall issues related to a subsystem and bringing solutions will solve most of the problems of the whole system [32, 3]. Figure 4 depicts the cascade system used in newly identified subsystems as a simple roadmap for sustainable cities.

9 Conclusion

This chapter demonstrate, via a case study from the Sri Lankan vernacular settlements, known as e*llangava* system, that the application of circular economic higher-order strategies to wastewater treatment could efficiently secure water needs while cost-effectively reducing water pollution. *Ellangava* system is a gravity-operated tank-cascade system consisting of a series of interconnected tanks of varying sizes and functions via canals to capture, store, and distribute rainwater so that water levels in the natural streams and rivers do not dry up below critical levels. A typical *ellangava* system consists of a large tank (or reservoir), the upper part of which is a high land catchment area with a community-protected dense forest cover consisting of several small silt-trapping forest-tanks and/or waterholes, and the lower part of

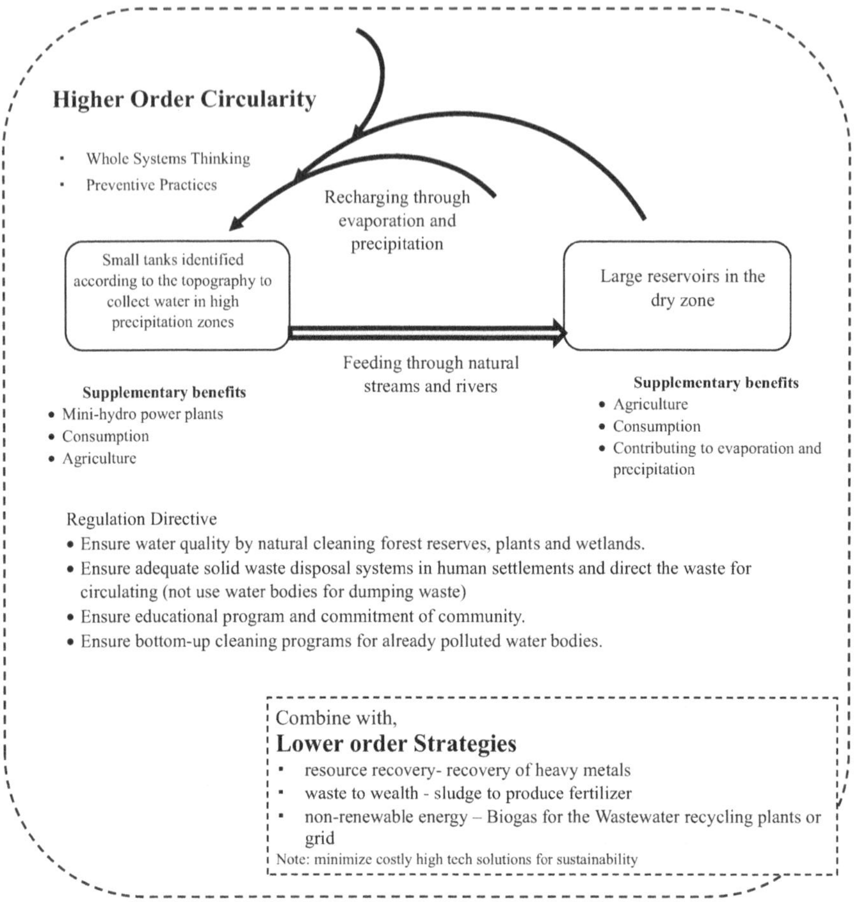

Fig. 3 A transition road map for the application of circularity concepts in water security

which is a reserved land consisting of diverse vegetation, from aquatic plants to tree species found in the dry forest. The aforementioned subsystems of the *ellangava* system offers multiple cost-effective ecosystem services resulting in groundwater management, flood control, water purification by pollutants removal, human livelihood support, wild animals habitat management, biodiversity conservation, wildlife conservation, and more. Some of the higher-order circular economic strategies identified with *ellangava* system are as follows: whole system thinking; regenerative and resilient design; local and regional circularly; innovation of cost-effective appropriate technology; cross-sector collaboration; circular culture and mindset; education and awareness of the stakeholders; sustainable consumption and lifestyle; employment of ecosystem services to water purification; repurposing. Application of the aforementioned vernacular waste management model to deal with the current challenges faced in black water, grey water and stormwater management is outlined in this chapter

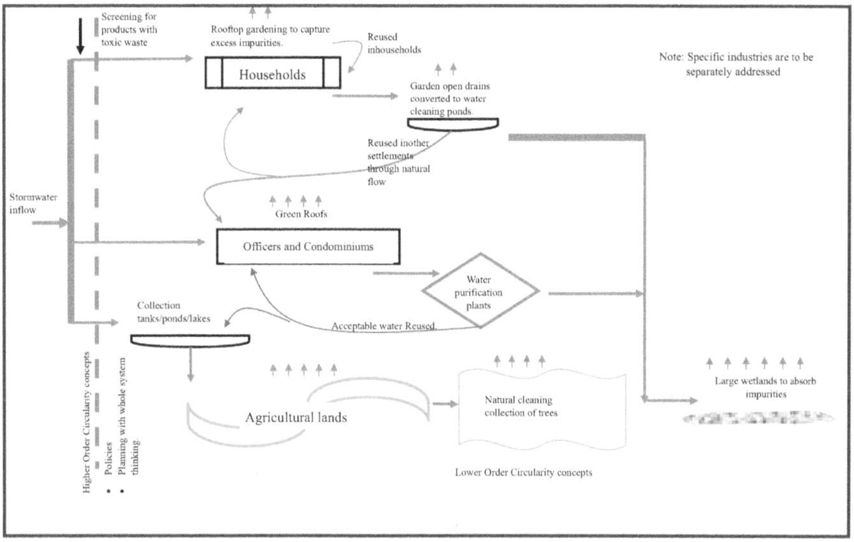

Fig. 4 Cascade system—a easy to read roadmap applied to present scenario

via a road map. It suggests that the cascade concept of the *ellangava* system can be applied to the present scenario by considering the subsystem as a household or an establishment or a factory or an agricultural land. The next step will be to identify the linkage patterns of these subsystems to the whole system through water bodies. Finally, the responsibility of each subsystem to the whole system must be identified and appropriate cost-effective, multi-objective, ecologically-comprehensive, whole system designs should be adopted.

References

1. Malhi Y, Franklin J, Seddon N, Solan M, Turner MG, Field CB, Knowlton N (2020) Climate change and ecosystems: threats, opportunities and solutions. Phil Trans R Soc B 375:20190104. https://doi.org/10.1098/rstb.2019.010
2. Borucke M, Moore D, Cranston G, Gracey K, Iha K, Larson J, Lazarus E, Morales JC, Wackernagel M, Galli A (2013) Accounting for demand and supply of the biosphere. Ecolo Ind 24:518–533. https://doi.org/10.1016/j.ecolind.2012.08.005
3. Sylva K, Iyer-Raniga U (2023) Partnership in the built environment for realizing the 2030 agenda: a soft systems model incorporating systems theory and the circular economy. In: Walker T, Cucuzzella C, Goubran S, Geith R (ed) The role of design, construction, and real estate in advancing the sustainable development goals. Sustainable development goals series. Palgrave Macmillan, Cham. https://doi.org/10.1007/978-3-031-28739-8_9
4. Inglezakis V, Poulopoulos S, Arkhangelsky E, Zorpas A, Menegaki A (2016) Aquatic environment. Environ Dev 137–212.https://doi.org/10.1016/b978-0-444-62733-9.00003-4
5. Narasimhan TN (2009) Encyclopedia of inland waters, hydrological cycle and water budgets. 714–720.https://doi.org/10.1016/B978-012370626-3.00010-7

6. Wear SL, Acuña V, McDonald R, Font C (2021) Sewage pollution, declining ecosystem health, and cross-sector collaboration. Biol Conserv. https://doi.org/10.1016/j.biocon.2021.109010
7. Hughes J, Cowper-Heays K, Olesson E, Bell R, Stroombergen A (2021) Impacts and implications of climate change on wastewater systems: a New Zealand perspective. Clim Risk Manage 31:100262. https://doi.org/10.1016/j.crm.2020.100262
8. Schlör H, Hake J-Fr (2015) Sustain assess circle. Energ Procedia 75:2641–2648. https://doi.org/10.1016/j.egypro.2015.07.361
9. Tchobanoglous G, Burton FL (1991) Wastewater engineering, treatment, disposal, and reuse, 3rd edn. McGraw Hill, Pennsylvania PA
10. Tchobanoglous G, Burton FL, Stensel HD (2003) Wastewater engineering, treatment and reuse, 4th edn. McGraw-Hill, Boston
11. Franceys R, Pickford J, Reed R (1992) Guide to the development of onsite sanitation. World Health Organization, Geneva
12. Chaffee K (2008) A new and improved septic tank. In: Proceedings of the water environment federation. https://doi.org/10.2175/193864708788809013
13. Hendrasarie N, Nugraha MYC, Fadilah K (2021) Restaurant wastewater treatment with a two-chamber septic tank and a sequencing batch reactor. E3S Web Conf 328:01011. https://doi.org/10.1051/e3sconf/202132801011
14. von Sperling M (2007a) Basic principles of wastewater treatment. In: Biological wastewater treatment series, vol 2. IWA Publishing
15. von Sperling M (2007b) Stabilization ponds. In: Biological wastewater treatment series waste, vol 3. IWA Publishing
16. von Sperling M (2007c) Activate sludge and aerobic biofilm reactors. In: Biological wastewater treatment series, vol 5. IWA Publishing
17. de Lemos Chernicharo CA (2007) Biological wastewater treatment series, vol 4—anaerobic reactors. IWA Publishing
18. EPA (1999) Wastewater technology fact sheet: sequencing batch reactors. Office of water, Washington, DC. Accessed on 2 Oct 2023 at https://www3.epa.gov/npdes/pubs/sbr_new.pdf
19. Andreoli CV, von Sperling M, Fernandes F (Eds) 2007 Biological wastewater treatment series, sludge treatment and disposal, vol 6. IWA Publishing
20. Drinan JE, Spellman FR (2012) Water and wastewater treatment: a guide for the non-engineering professional, 2nd edn. CRC Press
21. Rout PR, Shahid MK, Dash RR, Bhunia P, Liu D, Varjani S, Zhang TC, Surampalli RY (2021) Nutrient removal from domestic wastewater: a comprehensive review on conventional and advanced technologies. J Environ Manage 296:113246. https://doi.org/10.1016/j.jenvman.2021.113246
22. Vymazal J (2007) Removal of nutrients in various types of constructed wetlands. Science Total Environ 380:48–65
23. Stamm C, Räsänen K, Burdon FJ, Altermatt F, Jokela J, Joss A, Ackermann M, Eggen, RIL (2016) Unravelling the impacts of micropollutants in aquatic ecosystems: interdisciplinary studies at the interface of large-scale ecology, chap 4. In: Dumbrell, AJ, Kordas RL, Woodward G (eds) Advances in ecological research. Academic Press, vol 55. https://doi.org/10.1016/bs.aecr.2016.07.002
24. Chavoshani A, Hashemi M, Amin MM, Ameta SC (2020) Micropollutants and challenges: emerging in the aquatic environments and treatment processes. Elsevier Inc.
25. Margot J, Kienle C, Magnet A, Weil M, Rossi L, de Alencastro LF, Abegglen C, Thonney D, Chèvre N, Schärer M, Barry DA (2013) Treatment of micropollutants in municipal wastewater: ozone or powdered activated carbon. Sci Total Environ 461–462:480–498. https://doi.org/10.1016/j.scitotenv.2013.05.034
26. Virkutyte J, Varma RS, Jegatheesan V (2010)Treatment of micropollutants in water and wastewater. In: Virkutyte J, Varma RS, Jegatheesan V (eds). IWA Publishing, New York. ISBN: 9781843393160
27. Kröpfelová L, Vymazal J, Svehla J, Stíchová J (2009) Removal of trace elements in three horizontal sub-surface flow constructed wetlands in the Czech Republic. Environ Pollut 157(4):1186–1194

28. Reh L (2013) Process engineering in circular economy. Particology 11(2):119–133
29. Cramer J (2017) The raw material transition in the Amsterdam metropolitan area: add value for the economy, well-being and the environment. Environ Sci Policy Sustain Dev 501421. https://doi.org/10.1080/00139157.2017.1301167
30. Potting J, Hekkert M, Worrell E, Hanemaaijer A (2017) Circular economy: measuring innovation in the product chain. PBL Netherlands Environmental Assessment Agency: PBL Publishers
31. Iacovidou E, Hahladakis JN, Purnell P (2021) A systems thinking approach to understanding the challenges of achieving the circular economy. Environ Sci Pollut Res 28:24785–24806. https://doi.org/10.1007/s11356-020-11725-9
32. Churchman CW (1968) The systems approach. Dell Publishing Company
33. Khatri N, Tyagi S (2015) Influences of natural and anthropogenic factors on surface and groundwater quality in rural and urban areas. Front Life Sci 8(1):23–39. https://doi.org/10.1080/21553769.2014.933716
34. Tervahauta T, Rani S, Hernández LL, Buisman CJN, Zeeman G (2014) Black water sludge reuse in agriculture: are heavy metals a problem. J Hazard Mater 274:229–236. https://doi.org/10.1016/j.jhazmat.2014.04.018
35. Šuvalija S, Serdarević A, Džubur A, Lazović N (2022) Biocoagulants and bioflocculants in water and wastewater treatment technology. In: Karabegović I, Kovačević A, Mandžuka S (eds) New technologies, development and application V. NT 2022. Lecture notes in networks and systems, vol 472. Springer, Cham. https://doi.org/10.1007/978-3-031-05230-9_104
36. Kjellén M (2018) Wastewater governance and the local, regional and global environments. Water Altern 11(2):219–237
37. Sato T, Qadir M, Yamamoto S, Endo T, Zahoor A (2013) Global, regional, and country level need for data on wastewater generation, treatment, and use. Agric Water Manage 130:1–13. https://doi.org/10.1016/j.agwat.2013.08.007
38. Sarov A (2021) The use of waste sludge: benefits to the regenerative economy in Bulgaria. In: Andreucci MB, Marvuglia A, Baltov M, Hansen P (eds) Rethinking sustainability towards a regenerative economy. Future City, vol 15. Springer, Cham. https://doi.org/10.1007/978-3-030-71819-0_17
39. Lakshmikantha V, Hiriyannagowda A, Manjunath A, Patted A, Basavaiah J, Anthony AA (2021) IoT based smart water quality monitoring system. Global transitions proceedings.https://doi.org/10.1016/j.gltp.2021.08.062
40. Niazi MA (2021) Automation systems used in wastewater treatment facilities. December 10, 2021. Accessed 25 Sept 2023 at: Automation systems used in wastewater treatment facilities—technical articles (control.com)
41. Ahmed SF, Mehejabin F, Momtahin A, Tasannum N, Faria NT, Mofijur M, Hoang AT, Vo DVN, Mahlia TMI (2022) Strategies to improve membrane performance in wastewater treatment. Chemosphere 306. ISSN 0045-6535. https://doi.org/10.1016/j.chemosphere.2022.135527
42. Paulo PL, Azevedo C, Begosso L, Galbiati AF, Boncz MA (2013) Natural systems treating greywater and blackwater onsite: integrating treatment, reuse and landscaping. Ecol Eng 50:95–100. https://doi.org/10.1016/j.ecoleng.2012.03.022
43. Rufí-Salís M, Petit-Boix A, Leipold S, Villalba, G, Rieradevall J, Moliné E, Gabarrell X, Carrera J, Suárez-Ojeda ME (2022) Increasing resource circularity in wastewater treatment: environmental implications of technological upgrades. Sci Total Environ 838(3). ISSN 0048-9697. https://doi.org/10.1016/j.scitotenv.2022.156422
44. Ahlawat S, Nandal M (2020) Energy recovery from anaerobic digestion of wastewater. In: Kumar M, Munoz-Arriola F, Furumai H, Chaminda T (eds) Resilience, response, and risk in water systems. Springer transactions in civil and environmental engineering. Springer, Singapore. https://doi.org/10.1007/978-981-15-4668-6_20
45. Huyben D, Bevan D, Stevenson R, Zhou H, Moccia R (2018) Evaluation of membrane filtration and UV irradiation to control bacterial loads in recirculation aquaculture systems. Aqua Int. https://doi.org/10.1007/s10499-018-0301-z

46. Madduma Bandara CM (1985) Catchment ecosystems and village tank cascades in the dry zone of Sri Lanka, a time-tested system of land and water resource management. In: Lundqvist J, Lohm U, Falkenmark M (eds) Strategies for river basin management: environmental integration of land and water in a river basin. Springer Dordrecht, pp 99–113. https://doi.org/10.1007/978-94-009-5458-8
47. Fernando S (2019) Tank cascade system (Ellanga System). culture & heritage, ceylon Digest. Available at: tank cascade system (Ellanga System)—Ceylon Digest
48. Sylva KKK, Sylva KKLA (2021) Sustainable city planning and management strategies in vernacular settlement patterns in Sri Lanka. In: Iyer-Raniga U (ed), Sustainability in the built environment in the 21st century: lessons learned from India and the region. SDRAP 2019, Environmental Science and Engineering, Springer. https://doi.org/10.1007/978-3-030-61891-9_9
49. Tennakoon MUA (2015) Traditional dry zone irrigation—ecosystematic concepts revisited and reconceptualized. South Asia partnership Sri Lanka: Sri Jayawardenepura Kotte, Sri Lanka. Accessed on October 02 2023 at Traditional dry zone irrigation—ecosystematic concepts revisited and reconceptualised—South Asia partnership—Sri Lanka (sapsri.lk)
50. FAO in Sri Lanka (2018) Sri Lanka among globally important agricultural heritage systems. Accessed on 01 Oct 2023 at https://www.fao.org/srilanka/news/detail-events/es/c/1118377/
51. Perera MP (2017) Evolution of tank cascade studies of Sri Lanka. Saudi J Humanit Soc Sci 2(7):597–610
52. Geekiyanage N, Pushpakumara DKNG (2013) Ecology of ancient tank cascade systems in Island Sri Lanka. Mar Island Cultures 2(2):93–101
53. Sakalasooriya N (2021) Sustainability of the minor irrigation systems of the cascades in Dry zone of Sri Lanka: challenges and potentials—a cultural ecological analysis. Inter J Sci Arts Commer 6(1):70–93
54. FAO in Sri Lanka (2016) A proposal for declaration as a GIAHS, the cascaded tank-village system (CTVS) in the dry zone of Sri Lanka. Accessed on 01 Oct 2023, at http://www.fao.org/3/a-bp892e.pdf
55. Rathnayaka RMCY, Amarathunga AAD, Bellanthudawa BKA, Rathnasri PASA, Novak AT, Amarasiri C (2022) The effectiveness of a plant integrated lowcost treatment system to treat well water in coastal regions. Water Pract Technol 17(11):2242–2258. https://doi.org/10.2166/wpt.2022.111
56. Dharmasena PB (1995) Kattakaduwa a potential land for agro-forestry system development in Sri Lanka. In: Gunasena HPM (ed) Proceedings of the 6th regional workshop on multipurpose trees. University of Peradeniya, Peradeniya, Sri Lanka, 17–19th Aug 1995, pp 96–104
57. Rekha Nianthi KWG, Jayakumara MAS (2010) Progress of research on cascade irrigation systems in the dry zones of Sri Lanka. Water communities (community, environment and disaster risk management, vol 2. In: Shaw R, Thaitakoo D (eds). Emerald Group Publishing Limited. Bingley, pp 109–137. https://doi.org/10.1108/S2040-7262(2010)0000002010
58. Mendis DLO (2016) Water heritage of Sri Lanka, 3rd edn. Colombo, Sri Lanka. ISBN 955-647-003-3
59. Priyamala K, Perera MP (2021) Spatial differences of soil salinity in tank cascade Systems: a case study of Unaleeya tank cascade system in Daduru Oya basin in Puttalm district. In: National geography conference—2021. University of Peradeniya, Sri Lanka. Available at: (2) (PDF) Spatial differences of soil salinity in tank cascade systems: a case study of Unaleeya Tank cascade system in Daduru Oya Basin in Puttalm District (researchgate.net)
60. Reichenauer TG, Germida JJ (2008) Phytoremediation of organic contaminants in soil and groundwater. Chemsuschem 1(8–9):708–717. https://doi.org/10.1002/cssc.200800125
61. Dalupota MK, Watapila S (2003) Godage Brothers, Colombo
62. Leach ER (1961) Pul Eliya a village in Ceylon. Cambridge University Press, Cambridge
63. Rajasulochana P, Vijay P (2016) Comparison on efficiency of various techniques in treatment of waste and sewage water—a comprehensive review. Resour-Efficient Technol
64. Ragossing AM, Vujic G (2015) Challenges in technology transfer from developed to developing countries. Waste Manag Res 33(2):93–95

Environmental Assessment of Household Greywater in Chattogram City, Bangladesh: A Scoping Study

Samiha Mannan, Sayed Mohammad Nazim Uddin, Samiha Sabur, Muhammad Abu Bakar, and Saiful Islam

Abstract Untreated greywater that negatively impacts our environment is directly discharged even though it has the potential to be treated. Unlike traditional wastewater which requires a large-scale centralized treatment plants with huge infrastructural costs, greywater can be treated in a small-scale manner for use in irrigation. Hence, this chapter examines the primary quality of greywater present in Chattogram city by conducting an environmental assessment. 10 greywater samples were collected for chemical analysis and 5 for microbial examination from 5 different locations in Chattogram City. Water parameters like pH, Total Dissolved solids (TDS), Electrical conductivity (EC), Total hardness, Dissolved Oxygen (DO), Oxidation–Reduction potential, and other chemical analysis such as fluoride, chloride, phosphates, Sulphate, and nitrate levels were tested. Total coliforms and fecal coliforms were also evaluated through MPN tests and again cultured on EMB agar. Water parameters like Total hardness, Fluoride, Nitrates, and Sulfates conformed international standards which makes the water suitable for reuse. However, due to the exceeded turbidity and chloride levels found in the greywater, targeted treatment procedures are necessary for its reuse in irrigation. All samples were found to be contaminated with fecal matter (10^4 CFU/100 ml) and total coliform (1100 CFU/100 ml) that could cause diseases. In this regard, the chapter further reviews low-cost treatment systems from Kenya, India and Egypt which could be used in a developing country like Bangladesh. Moreover, this chapter gives recommendations for proper management approaches towards greywater reuse.

S. Mannan · S. M. N. Uddin (✉) · S. Sabur
Center for Climate Change and Environmental Health (3CEH), Environmental Sciences Program,, Asian University for Women, Chattogram, Bangladesh
e-mail: sayed.uddin@auw.edu.bd

S. Sabur
e-mail: samiha.sabur@post.auw.edu.bd

M. A. Bakar · S. Islam
Bangladesh Council of Scientific and Industrial Research (BCSIR), Chattogram, Bangladesh

S. Yadav et al. (eds.), *Emerging Technologies in Biological and Hybrid Wastewater Treatment: Lessons from Developed to Enhancing Practices in Developing Countries*, Earth and Environmental Sciences Library,
https://doi.org/10.1007/978-3-031-74515-7_15

Keywords Greywater · Water quality · Water conservation · Water reuse · Wastewater treatment

1 Introduction

In recent years, water scarcity stands as an important issue to tackle globally. Hence, water conservation and wastewater reuse for agricultural purposes has been a topic of attention internationally [20]. Reuse of wastewater can be safe only when the water undergoes treatment procedures as it contains large amounts of chemical and biological pollutants [31, 4]. The chemical effluents are nitrogen, phosphorus, heavy metals, detergents, pesticides and hydrocarbons and the biological contaminants are the microorganisms which come from human fecal matter consisting of different viruses, bacteria, and protozoa [4]. In developing nations, wastewater is directly discharged into the environment and surface water supplies [12]. Such negligence can have the potential to induce huge public health issues and harm aquatic ecosystems because of the degrading water quality [12]. In urban areas of low-income countries, large centralized treatment plants are expensive to develop [4]. Thus, reuse of greywater has gained interest as an alternative water source for irrigation to minimize such technological investments. Greywater is any wastewater that comes from bathtubs, showers, hand basins, kitchen sinks, dishwashers and laundry machines except water from toilets which is called blackwater [7]. Hence, greywater contains fewer organic pollutants and pathogens than wastewater making it more feasible for small-scale reuse.

Yet, greywater should not be used directly for irrigation as it has environmental effects and health implications for the people consuming the food [30]. Siggins et al. [25] proved that long-term on-site disposal of untreated greywater in irrigation has a moderately negative impact on the soil exposed to the greywater compared to the soil unexposed. Risks were determined in the study by comparison of biological and chemical indicators of the soil [25]. However, the treatment of greywater for irrigation involves only primary treatment like filtration, an optional secondary treatment stage such as use of bioreactors, and disinfection prior to irrigation [19]. This can be ensured by a greywater decentralized reuse system as an extensive treatment process to manage blackwater is not needed [16].

The greywater in a household is supposed to have a way less contaminant capacity as industrial effluents do not pollute this wastewater thus, it shows that greywater has a huge potential for reuse [16]. This is rather affordable as this irrigation as expensive fertilizers can be avoided [16]. The concept also increases the variation of technological availability as there could be many options to treat the water at low cost which can also reduce the risks of huge system failures [16]. This shows how greywater treatment is of a huge advantage in terms of economic and technological aspects when used for agriculture.

In Bangladesh, the prospect of greywater reuse is still a much disregarded topic. In a study conducted by Tamanna et al. [29], it was seen that greywater comprises almost 60–70% of the total household wastewater generated in Dhaka city. It is estimated that the reuse of this greywater can potentially save upto 59 million Taka or 786,000 USD per year in Dhaka city [3]. Furthermore, in another case study by Dakua et al. [10], it was mentioned that the use of greywater can have environmental benefits such as reduction in groundwater extraction and surface water pollution due to the water dependency change to recycled greywater. The study was conducted in water-scarce urban areas of Dhaka city which proved that 15–20% of the total water consumption can be saved by reusing greywater in residential halls and hospitals for purposes like gardening, cleaning, toilet flushing, etc. [10]. However, in Bangladesh such potential is only considered for research and not for long-term applications which is why it is important to bring attention to this prospect.

Though several studies have been conducted on greywater reuse in Bangladesh, the commercial capital of Bangladesh, Chattogram, has received less attention in terms of assessing greywater and adapting to greywater treatment processes. With the growth of the textile industries in the city, several studies have focused on assessing the industrial wastewater quality by analyzing the contamination levels. Analytical Methods have been implemented to check the deteriorating conditions of the surface water quality [5]. So, there has been no such monitoring of the untreated household greywater in Chattogram city, which is why this research is being implemented to add to the existing knowledge of wastewater in the locality.

In Chattogram city, there is no wastewater treatment plant available for the treatment of wastewater. Laws that are implemented in the zone for minimum wastewater standards for discharge are defied due to corruption of governing bodies. However, potential treatment policies are not the primary concern to the policymakers of Chittagong as such possibilities are yet to be analyzed in a long-term scenario. Thus, decentralized treatment systems at household level in Chattogram will be a start to reducing the environmental and social problems associated with water. Also, these studies were conducted to assess the industrial wastewater only. It has been seen that very limited research has been done to find out the quality of household greywater in Chattogram and to see the potential treatment procedures that can be applied in that region. So, the purpose of this chapter is to assess the quality of greywater and identify potential greywater treatment methods for households. In this respect, this chapter also aims to find out the existing practices regarding the reuse of treated greywater for irrigation purposes such as gardening through review.

2 Methodology

The study is a quantitative analysis based on primary sample collection and laboratory research. The parameters were determined based on experimental results from greywater samples. 10 grey water samples were tested, and the results were analyzed using secondary data review. The study is based on the Chattogram Metropolitan Area. Water samples were taken from households across the city from 5 different locations: Chandgaon R/A, Panchlaish, Amirbag R/A, Agrabad, Lalkhan Bazar. The locations were chosen approximately being 2 km apart from each other. Samples were collected from separate pipe outlets in the households which only discharged greywater using enclosed plastic bottles. Grab sampling method was used from households and the samples were directly taken in containers that were transported to the lab [26]. The sample did not undergo secondary transfer to any other container to minimize the contamination from external sources [26] (Fig. 1).

The collected samples were analyzed for pH, Total Dissolved solids (TDS), Electrical conductivity (EC), Oxidation-Reduction potential (ORP), Practical Salinity Unit (PSU) using a Multiparameter probe. The dissolved oxygen (DO) levels were measured using a dissolved oxygen meter immediately after it was taken to the lab so that no inter-mixing of the samples occurred. The levels of turbidity of the samples were measured by a turbidity meter. Ethylene-Diamine-Tetraacetic Acid (EDTA) titration was used to determine the total hardness of the greywater samples. The total hardness was then determined using the theory that 1 ml of 0.01 M Na_2 EDTA Solution ≡ 1 mg $CaCO_3$. For other chemical parameter analysis of chloride, phosphate, Sulphate, and nitrate levels, Ion Chromatography was used.

The Microbial tests for coliforms and fecal coliforms were done for only one sample from each location due to time constraints and limitations discussed later in the chapter. Firstly, to find out the coliforms present in the samples, an MPN test was conducted on each. For the presumptive test, MacConkey broth was prepared and 10 ml of the broth was taken in each of the 45 test tubes inserting Durham's tubes, 9 for each sample to be autoclaved. For each sample, 3 sets of 3 test tubes were taken

Fig. 1 Household greywater sample collection from pipe outlets

all containing 10 ml MacConkey broth and were inoculated with 10, 1, and 0.1 ml of sample respectively. The samples were then incubated for 24 h at 37 °C. The process was then repeated the next day to identify fecal coliforms with a different inoculation technique, this time inoculating each set with 1 ml from the positive tubes containing the sample. All the tubes were then kept in 44.5 °C for another 24 h. For the confirmative test, selective media EMB agar was prepared. 6 EMB agar plates were made, one for each sample and a control. The agar plate was divided into 3 parts to be streaked from 1 of the positive tubes from each set to identify for fecal coliform. The plates were then incubated for another 24 h at 35 °C to allow growth for the bacterial colonies. The results were then recorded. However, the completed test in this MPN test was not performed because the EMB agar restricts the growth of gram positive bacteria.

For system analysis, treatment systems for the greywater were reviewed from secondary data on the basis of the Laboratory analysis. So, extensive literature reviews on articles related to treatment systems that have been already implemented were done. Also, the existing practices are reviewed in this chapter to understand how the treated water can be used for irrigation in a small-scale manner by the help of online databases. No prototype is made.

3 Results and Discussion

3.1 Analysis of the Quality of Household Greywater in Chattogram City

3.1.1 Physical Properties

Table 1 shows the physical properties of the greywater samples collected:

In Chattogram City, it was found that the temperature of greywater samples ranges from 23.0 to 25.5 °C which is an ideal temperature to slow the growth of microbes as the optimum temperature for most microorganisms to grow is at 37 °C [13]. According to Stoner et al. [28], on average the treated greywater should have turbidity below 2 NTU for reuse in irrigation for food crops. However, the turbidity found for the samples ranged from 21.0 NTU to 239 NTU which is higher than the standard turbidity level [28]. According to the measurements, it was found that samples 4 and 7 had turbidity levels greater than 100 NTU which could be the result of higher levels of organic matter present in the water. Having a higher level of turbidity can affect the soil texture by reducing the hydraulic conductivity of the soil and also by contaminating soil surface [17]. Therefore, the greywater must be treated to lower its turbidity level which is possible through treatment processes such as coagulation-flocculation or other low-cost technology discussed later in the chapter [27] (Fig. 2).

Table 1 Results for physical parameters

Sample No.	Sample location	Temperature (°C)	Turbidity (NTU)	Electrical conductivity (µS/cm)	Total dissolved solids (ppm)
1	Chandgaon R/A	23.0	24.7	651	322
2	Chandgaon R/A	23.5	23.1	304	152
3	Panchlaish	23.5	43.8	233	111
4	Panchlaish	24.0	239	1792	935
5	Amirbag R/A	24.0	73.9	237	120
6	Amirbag R/A	23.5	28.9	120	60
7	Agrabad	25.5	119	1124	590
8	Agrabad	23.5	30.1	1299	648
9	Lalkhan Bazar	23.0	21.0	181	90
10	Lalkhan Bazar	23.5	39.2	303	150

Moving on to the Electrical conductivity of greywater the values found range from 120 to 1792 µS/cm which falls under the range of greywater composition suggested by Oteng-Peprah et al. [21] to be 14–3000 µS/cm as mentioned earlier in the chapter. The electrical conductivity standard in South Korea for food crops is set to < 700 µS/cm and < 2000 µS/cm for processed food crops, whereas it the standard in Israel, Italy and Portugal it is set to 1400, 3000 and 1000 µS/cm respectively for irrigation of all crop types [17]. According to the findings, it has been measured that the sample 4, 7 and 8 are not suited to be used for food crops to be eaten raw according to South Korean standards but they can be used for other food crops which are cooked/processed. However, ECs of all samples were found to be below < 2700 µS/cm which maintained the standards for reuse in irrigation according to WHO/FAO Guidelines [18].

3.1.2 Chemical Analysis of Greywater

According to Grumbles et al. [14], the EPA suggests the pH level of greywater for reuse to be within the 6.9–9.0 range. Hence, the results found for pH ranges 6.91–7.51 which aligns with the standard pH level for irrigation. Sample 3 has the highest pH which makes it slightly alkaline in nature which is expected as discussed in the literature review that because of the laundry detergents, greywater is supposed to be alkaline. However, Sample 6 has the lowest pH of 6.91 which is almost neutral. It was found that the Dissolved Oxygen level for use in irrigation has to maintain a concentration of more than 3.5 ppm or it can negatively impact crop growth [15].

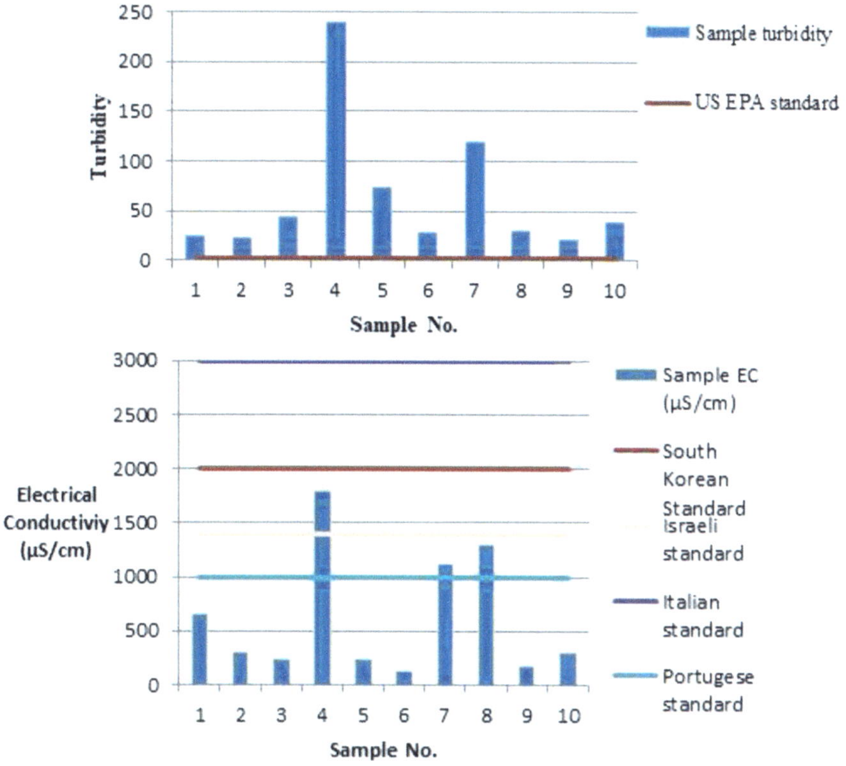

Fig. 2 Comparison of physical parameters to international standards

Therefore, all the samples of greywater had a dissolved oxygen level greater than 3.5 ppm which is why they are suitable for use in irrigation. Nevertheless, the best suited water for irrigation must have DO levels greater than 5.0 ppm which is not the case for sample 4, 5, 7, and 10 (Table 2; Fig. 3).

According to the Center for Agriculture, Food, and the Environment, the hardness of water should be between 0 and 100 ppm for irrigation [9]. All the results of the samples had $CaCO_3 < 1.6$ mg/L or ppm which is almost negligible and means that this range is suitable. Therefore greywater does not need to be treated for $CaCO_3$ before reused. Also, it can be deducted from the findings that the Fluoride ion levels were from Nil to 3 mg/L which is very low and does not have almost any effect on the quality degradation of greywater. The Chloride levels on the other hand, was found to be very high, ranging from 5.07 mg/L to 100.7 mg/L which is why Chloride levels need to be treated. This is because high levels of Chloride in Irrigation cause leaf burns and damages plant tissues when used for irrigation [8]. Also, Sulphate ions were present in all of the samples ranging from 0.452 to 20.71 mg/L which is supposed to come from the use of detergents [21].

Table 2 Results for chemical parameters of the samples

Sample No.	pH	ORP (mVpH)	DO (ppm)	PSU	$CaCO_3$ (mg/L)	Fl^- (mg/L)	Cl^- (mg/L)	Br^- (mg/L)	NO_3^- (mg/L)	PO_4^{3-} (mg/L)	SO_4^{2-} (mg/L)
1	6.98	− 14.6	5.28	0.31	1.0	Nil	100.7	Nil	Nil	Nil	18.44
2	7.42	− 36.6	5.76	0.14	0.8	Nil	30.68	Nil	Nil	Nil	20.53
3	7.51	− 40.4	7.30	0.11	1.2	Nil	24	Nil	1.6	Nil	20.1
4	7.42	− 36.6	3.54	0.88	1.2	0.792	25.235	Nil	2.081	5.313	2.832
5	7.04	− 17.7	3.61	0.11	1.1	3.106	16.17	Nil	1.12	Nil	9.89
6	6.91	− 10.4	6.46	0.06	1.1	0.314	5.07	Nil	0.611	0.157	6.114
7	7.27	− 28.7	3.79	0.57	1.1	0.111	22.24	Nil	Nil	4.71	3.073
8	7.25	− 28.6	5.11	0.65	1.6	Nil	77.62	Nil	Nil	Nil	0.452
9	6.98	− 14.4	5.90	0.08	1.6	0.730	13.00	Nil	Nil	Nil	10.9
10	7.19	− 25.3	3.93	0.14	1.4	0.604	14.17	Nil	0.774	Nil	20.71

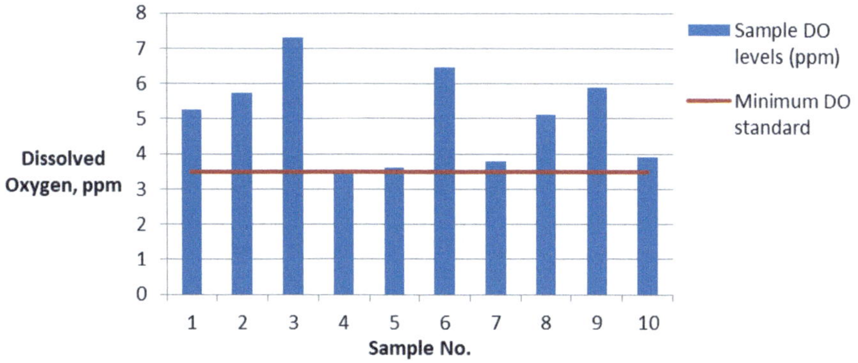

Fig. 3 Disolved oxygen levels

3.1.3 Biological Characteristics

Result of MPN test (presumptive test) after inoculation (Table 3):

Result of MPN test (confirmative test) on EMB agar (Table 4):

According to the WHO standards, it is said that the fecal coliforms for the reuse of greywater should be less than 1000 CFU/100 ml for restricted agriculture, which means for the water reused for crops for which food are eaten raw and 10,000 CFU/100 ml for food that are processed or cooked [32]. The total coliforms in the greywater were all found to be more than 1100 CFU/100 ml which meant that there were a lot of biological impurities in the water. The fecal coliforms found were likely to be *E. coli* and *Klebsiella pneumoniae,* depending on the colonies that formed in EMB agar. According to the WHO guidelines, fecal coliforms for the reuse of greywater should be less than 1000 CFU/100 ml for restricted agriculture and 10,000 CFU/100 ml for food that is processed or cooked [32].

Table 3 Result for presumptive test

Sample No.	Sample location	Results for total coliform (at 37° C)				Results for fecal coliform (at 44.5 °C)			
		10 ml	1 ml	0.1 ml	MPN per 100 ml	10 ml	1 ml	0.1 ml	MPN per 100 ml
1	Chandgaon R/A	3	3	3	> 1100	2	2	3	42
2	Panchlaish	3	3	3	> 1100	3	2	3	290
3	Amirbag R/A	3	3	3	> 1100	3	3	2	1100
4	Agrabad	3	3	3	> 1100	3	2	2	210
5	Lalkhan Bazar	3	3	3	> 1100	3	2	3	290

Table 4 Observations for fecal coliform colonies in EMB agar

Sample No.	Sample location	Set no	Bacterial colony color on EMB agar	Possible type of bacteria present
1	Chandgaon R/A	1	Green sheen colony	*E. coli*
		2	Mixture of Green sheen colonies + Purple colonies	*E. coli* *Klebsiella pneumoniae*
		3	Mostly Purple colonies + Green sheen colony	*E. coli* *Klebsiella pneumoniae*
2	Panchlaish	1	Mixture of Green sheen colonies + Purple colonies	*E. coli* *Klebsiella pneumoniae*
		2	Mixture of mostly Green sheen colonies + Purple colonies	*E. coli* *Klebsiella pneumoniae*
		3	Mixture of Green sheen colonies + Purple colonies	*E. coli* *Klebsiella pneumoniae*
3	Amirbag R/A	1	Green colonies	*E. coli*
		2	Green colonies	*E.Coli*
		3	Mixture of Green sheen colonies + Purple colonies	*E. coli* *Klebsiella pneumoniae*
4	Agrabad	1	Purple Mucoid Colony + Light green Sheen	*E. coli* *Klebsiella pneumoniae*
		2	Mixture of mostly Green sheen colonies + Purple colonies	*E. coli* *Klebsiella pneumoniae*
		3	Mixture of mostly Green sheen colonies + Purple colonies	*E. coli* *Klebsiella pneumoniae*
5	Lalkhan Bazar	1	Mixture of Green Colony + Purple Colonies	*E. coli* *Klebsiella pneumoniae*
		2	Green colonies + Purple colonies	*E. coli* *Klebsiella pneumoniae*
		3	Green colonies + Purple colonies	*E. coli* *Klebsiella pneumoniae*

As fecal coliforms were present in all the samples, it can be deducted that the greywater was contaminated by fecal coliforms due to poor hygiene practices and also because of behavioural aspects. According to the World Health Organization [33], these fecal contaminants are likely to come from household activities like from the laundry and washing of the fecal contaminated clothes, from childcare as many parents use wash basin for cleaning child excreta, and also from showering. In a study for the analysis of greywater in Dhaka city it also showed that greywater from kitchen waste also contained fecal coliforms [2]. Even though all samples could be used for unrestricted irrigation even without treatment of these fecal coliforms, the total coliforms were greater than 1100 CFU/100 ml which is why biological impurities need to be treated. This is because these impurities can result in bacterial and viral infections such as cholera, typhoid, viral enteritis etc. which cause gastrointestinal issues like diarrhoea [11].

3.2 A Review for the Selection of Suitable Treatment Systems to Reuse Greywater in Chattogram City

As discussed earlier in the chapter, the decentralized greywater treatment systems do not have as many steps involved for the treatment of water reuse like the centralized wastewater treatment plants. So, treatment of the greywater can be done through Primary treatment such as Filtration, Secondary treatment such as use of low-cost bioreactors and Disinfection after which the water can be used for Irrigation purposes. According to the findings and all the procedures that done in this study for sample collection, it can be seen that in Bangladesh there is no particular system to collect greywater separately from the laundry, washrooms, kitchen sinks, and wash basins. Therefore, treatment systems must be implemented accordingly. Other than that, as a developing country like Bangladesh where people usually do not care about such reuse and the concept of greywater, using expensive technologies for reuse in agriculture seems unrealistic which is why low-cost technologies are required for societal acceptance (Table 5).

Kariuki et al. [18] discussed in their paper the potential of low-cost greywater treatment technologies which could reduce the direct environmental impact of using untreated greywater. This treatment system designed by Kariuki et al. [18] had 5 recycled plastic barrels which were adjoined by PVC pipes where the flow of the greywater was dependent upon the gravity. The system was first made in a treatment site for laboratory pilot testing and then it was implemented at household level for field testing in Nairobi, Kenya. Each barrel consisted of sieves which collected oil, grease and food particles from the greywater. The first barrel was 100 L where initial filtration occurs after which the water moves onto the second barrel which was 200 L where rock alum was kept for flocculation purposes. The process in the second barrel required 8 h for proper sedimentation after which the greywater moved onto the third 200 L barrel where disinfection took place. For the disinfection, sodium

Table 5 Treatment system review

Description	Location	Potential to treat physical	Potential to treat chemical	Potential to treat biological	Potential for use in Chattogram City	Reference
A 5-barrel greywater treatment (GWT) system	Nairobi, Kenya	Rock alum was kept for flocculation purposes improves turbidity	Filtration	8 h for proper sedimentation and Disinfection with sodium hypochlorite (NaOCl) removes most of the pathogens	Can be used as the system is low cost and also helps in treating all three properties of the greywater	Kariuki et al. [18]
Gravel filter followed by sand filter down flow (GFSF) and horizontal flow sand filter (HFSF)	Cairo, Egypt	No effect	Removes Chemicals to great extent from Sand filtration	Sedimentation helps in removing biological impurities	Can be used to remove organic matter from greywater	Abdel-Shafy et al. [1]
Vegetated vermifilter	India	60–75% removal of turbidity with the help of pre-treatment; removal of suspended solids in a settling tank and Sand gravel Filter	Removes 60 to 70% of Chloride Concentrations	Vermi-Filter including earthworms reduces pathogens and biological impurities	Best suited for the greywater in Chattogram	Patil et al. [22]

hypochlorite (NaOCl) was used where the greywater was kept for 30 min after which the greywater was kept for storage and reuse [18].

The treatment system resulted in the reduction of the fecal coliforms in the greywater to under the count of 2.2×10^2 CFU/100 ml, respectively which is less than the standard set by WHO [18]. The treatment system was also able to reduce the total coliforms to under a count of 2.5×10^6 CFU/100 ml which was compared to the count of total coliform of 2.4×10^6 when treated with an expensive biological aerated filter [18]. However, the treatment system had no influence on improving the pH values, EC and salinity of the water which is a limitation to this treatment system but could be used for the reuse of greywater in Chattogram city as these values maintain the international standards according to the results found.

Abdel-Shafy et al. [1] discussed different sand filters and deducted that gravel filter followed by sand filter down flow (GFSF) and horizontal flow sand filter (HFSF) were technologies affordable for use in greywater treatment. However, before the use of such sand filters, pre-treatment of the water needs to be done in a sedimentation tank. Sand filters are usually filled with granular substances like sand and the remainder of the filtered water is collected at the bottom of the tanks. Sand filters have three stages for treatment—filtration, where only the water is drained out. Chemical sorption where the pollutants join to the surface of the sand particles and the biological microbes form a biofilm on the surface, and assimilation, where the aerobic microbes feeds on nutrients from the water and changes them to purify the water. The paper compares different sand filters as they are low-cost and reliable way to treat water which showed that the GFSF and HFSF could be used as they result in water having a COD, BOD, and TSS of 43, 16, and 7.5 and 40, 17, and 9 mg/l, for GFSF and HFSF respectively [1].

Another low-cost technology implemented by Patil et al. [22] in Rural India for the treatment of greywater was vegetated vermifilter which was seen to be very useful for agriculture. The system combines four components which are Settling tank, Sand Gravel filter, Laterite soil vegetated filter (LSVF) and laterite soil vegetated vermifilter (LSVVF). Diluted greywater was added into the system after which the water was allowed to settle for 20 h in the settling tank as raw greywater was polluted by in nature. The PVC pipes and water flow control valves in the tanks were added so that they could be open or closed when needed for water to pass to the next tank. The valve in the settling tank was kept above the sludge zone to restrict contaminants to move onto the other tank. The water then moved onto the up flow sand gravel filter which consisted of the bottom layer with 20 mm sized particles of gravel, second layer with 12.5 mm gravel, third/upper layer with 4.75 mm sized sand and topmost layer with 1 mm sand. The third tank the water passed to was the LSVF which consisted of four layers. "The bottom (first) layer was filled with 25–50 mm uni-sized laterite gravel (depth 0.15 m), second layer 12.5–25 mm laterite gravel (depth 0.15 m), third layer 6–12 mm laterite gravel (depth 0.15 m) and topmost layer laterite soil (depth 0.15 m). The total vertical depth of flow filter was 0.6 m with a freeboard of 0.20 m." This tank enhanced the filtration level of the greywater again after which the water reached the last filter which was the LSVVF which was identical to the third LSVF tank but allowed more filtration of the water. Moreover, the system was made more efficient by growing crops in the LSVF and LSVVF tanks such as by planting Canna indicia there was a healthy growth in the crops. As this system also works on gravity, there was no operative cost to maintain any mechanized system as it was a decentralized way for treatment of greywater at a low cost [22].

3.3 Environmental Impacts and Benefits from the Study

Greywater has environmental impacts that can directly concern public health when discharged untreated. In Chattogram city there is almost no plans or management system for these contaminated greywater and they are discharged into the environment without any treatment. The greywater is directly suspended to the drains which are left open and stagnant. These open drains contribute to bad smells, breeding places for mosquitoes which are also disease causing insects which needs to be avoided [24]. Other than that, these drains adjoin into rivers which are then get contaminated with the greywater with excess nutrients and disease-causing pathogens. This could result in eutrophication and outbreaks of waterborne diseases are very likely due to such reasons [24]. Therefore, the mismanagement of wastewater needs to be addressed and it can be directed for reuse purposes as mentioned in this research. Greywater has huge potential in terms of treatment in a low cost efficient way where decentralized systems can be used. It is clear after the analysis of the chapter that in Chattogram city the water is highly contaminated with biological impurities, Chloride, and has a high level of turbidity but these can be treated to an acceptable range of values using low cost technologies discussed above. The benefit of this study is that it gives an outlook of the contamination levels of the greywater in Chattogram city as there has been no such assessment conducted here. Moreover, the study shows that such contaminated water can be easily treated by the help of low cost technologies and used up for reuse in irrigation. This reused water is useful for helping the society to be more efficient and to not waste water in general. So when greywater is reused, it will not pollute the environment as a whole.

3.4 Social Implications of the Reuse of Treated Greywater

It is clear that the greywater has great potential for reuse. However, direct reuses of greywater in agriculture which are done by many families in developing and underdeveloped countries impose health concerns as well as have negative impacts on crop developments [6]. According to a case study conducted in Australia, Pinto and Maheshwari [23] saw as per their survey results that the main interest of the society for reuse of greywater in irrigation was that they wanted to know more about reuse strategies of greywater and the effects that this water had on the soil and plant health. For most participants, this was a new idea which made them keen to learn about this particular subject. Thinking about the same scenario in the context of a developing country like Bangladesh, people here have very limited knowledge on greywater which is why its treatment and reuse is not something that they are interested in. The reluctance of use of such systems might be generated from reasons such as instalment costs, soil nutrition, and also because users might not be environmentally conscious. However, the drivers affecting the applications of greywater reuse still need to be studied in the context of Bangladesh and internationally.

Limitations

The major limitation to this research was the coronavirus pandemic which caused the sample size to be very limited as there were restrictions in movement and the risks involved with fieldwork activities. Lack of proper funding to this research reduced the scope of study. The concentration of Surfactants, BOD_5, COD and TSS, which are some of the defining parameters for the quality of the greywater, could not be studied due to laboratory limitations. During the microbial examination, only 5 samples were taken from 5 locations for research because of limited equipment available and lack of enough test tubes. Even though the sampling during the microbial experiment was handled very carefully, the contamination of air microbes is somewhat possible. Also, if the treatment procedures were recommended on the basis of trial other than review, the study could have been more extensive research. However, due to budget constraints it was not possible. Lastly, due to the pandemic and lack of human interaction, the social influences on the greywater quality and the household practices could not be assessed which was integral to this research.

4 Conclusion

The management of greywater is a much-disregarded topic in Bangladesh mainly because of the lack of awareness and knowledge. This problem needs to be solved because it affects environmental health as a whole and creates public health concerns. This is why greywater requires attention and instead of thinking about wasting such a resource, it can be simply used to the benefit of society by reusing. Therefore, this paper showed the existing quality of greywater and how users can reuse such greywater with simple technologies which are of low cost. The paper is evidence of the potential of greywater and would like to facilitate more research in the future to add to this knowledge. This is because there is very limited knowledge on this topic in Chattogram as well as Bangladesh in general. Hence, it is necessary to look into the management aspect of this water and take necessary actions and provide investments for the treatment of water to stop the pollution from greywater/wastewater in general.

5 Recommendations

- National/Local standards for wastewater reuse in irrigation need to be set to bring the topic into attention.
- The social aspect of the research needs to be studied by conducting more research in the future to add to this knowledge of greywater reuse. User practices related to greywater and their perceptions regarding reuse of this resource needs to be understood in the future for social implications. Surveys could be implemented to see users' perception.

- Proper regulations and policies are necessary to forbid the negligent discharge of greywater into open drains as the water is fecal contaminated and can be disease causing as discussed earlier in the paper.
- More research is also necessary for the implementation of new technologies for reuse of greywater in a decentralized way which are also cheap for use in irrigation.

Acknowledgements Special thanks go to the Bangladesh Council of Scientific and Industrial Research (BCSIR) for funding the laboratory investigation. Approval from the Department of Environmental sciences at Asian University for Women to conduct this research is highly acknowledged. Tremendous thanks to Ms. Jannatul Ferdous, Research Fellow at BCSIR and to Ms. Sabekun Nahar for supporting the handling of the Microbiology laboratory equipment for data collection. Heartfelt appreciation is given to Dr. Mukesh Gupta, Farzana Sultana, Nazifa Rafa and Ismat Ara Sharin. Respective families of the authors are sincerely acknowledged for their support.

References

1. Abdel-Shafy H, El-Khateeb M, Shehata M (2013) Greywater treatment using different designs of sand filters. Desalin Water Treat 52(28–30):5237–5242
2. Abedin SB, Rakib ZB (2013) Generation and quality analysis of greywater at Dhaka city. Environ Res Eng Manage 64(2):29–41
3. Ahmed MA, Saiek ZM, Chowdhury OR, Tamanna M, Tewari S (2018) Sustainable wastewater management for underdeveloped communities-a hands-on method forqualitative and quantitative analysis of greywater. Int J Environ Sustain Dev 17(1):93–109
4. Akpor OB (2011) Wastewater effluent discharge: effects and treatment processes. Int Proc Chem Biol Environ Eng 20:85–91
5. Ali K, Sarraz A, Roy P, Nahar A, Jubae A, Das S, Mallick R (2017) Environmental assessment of industrial wastewater in Chittagong city, Bangladesh. Int J Sci Technol 6(3):733–741
6. Al-Jayyousi O (2002) Focused environmental assessment of greywater reuse in Jordan. Environ Eng Policy 3(1):67–73
7. Andreadakis A, Noutsopoulos C, Mantziaras I, Kouris N (2015) Grey water characterization and treatment. In: IWA Balkan Young Water Professionals conference, Hellenic Water Association, Athens, Greece, pp 328–334
8. Ayers R, Westcot D (1985) FAO Irrigation and drainage paper. In: Water quality for agriculture. Food and Agriculture Organization of the United Nations, Rome, Italy
9. Center for Agriculture, Food, and the Environment (2019) Water quality for crop production. Report, The University of Massachusetts Amherst. https://ag.umass.edu/greenhouse-floriculture/greenhouse-best-management-practices-bmp-manual/water-quality-for-crop. Accessed 18 Apr 2021
10. Dakua M, Mahmood M, Bhowmik S, Khaled F (2016) Potential of grey water recycling in water scarce urban areas in Bangladesh. Int J Environ Sci Dev 7(8):563
11. Drechsel P, Keraita B, Seidu R, Abaidoo RC (2014) Human health risks from wastewater-irrigated vegetable farming. In: Drechsel P, Keraita B (eds) Irrigated urban vegetable production in Ghana: characteristics, benefits and risk mitigation. International Water Management Institute (IWMI), Colombo, Sri Lanka, pp 104–115
12. Edokpayi JN, Odiyo JO, Durowoju OS (2017) Impact of wastewater on surface water quality in developing countries: a case study of South Africa. Water Quality, pp 401–416
13. Farrell J, Rose A (1967) Temperature effects on microorganisms. Annu Rev Microbiol 21(1):101–120

14. Grumbles BH, Gilman P, Schafer JE (2004) Guidelines for water reuse, Report EPA/625/R-04/108 U.S. Environmental Protection Agency, U.S. Agency for international development, Washington DC, USA
15. Holtman W, Van Duijn B, Blaakmeer A, Blok C (2005) Optimalization of oxygen levels in root systems as effective cultivation tool. In: Gavilán MU (ed) International symposium on soilless culture and hydroponics. Acta Horticulturae, Almeria, Spain, vol 697, pp 57–64
16. Imhof B, Mühlemann J (2005) Greywater treatment on household level in developing countries—a state of the art review. Report, Swiss Federal Institute of Technology, Zurich, Switzerland
17. Jeong H, Kim H, Jang T (2016) Irrigation water quality standards for indirect wastewater reuse in agriculture: a contribution toward sustainable wastewater reuse in South Korea. Water 8(169):1–18
18. Kariuki F, Kotut K, Ngángá V (2011) The Potential of a low cost technology for the greywater treatment. Open Environ Eng J 4:32–39
19. Matos C, Pereira S, Amorim EV, Bentes I, Briga-Sá A (2014) Wastewater and greywater reuse on irrigation in centralized and decentralized systems—an integrated approach on water quality, energy consumption and CO_2 emissions. Sci Total Environ 493:463–471
20. Meda A, Cornel P (2010) Aerated bio-filter with seasonally varied operation modes for the production of irrigation water. Water Sci Technol 61(5):1173–1181
21. Oteng-Peprah M, Acheampong MA, Devries NK (2018) Greywater characteristics, treatment systems, reuse strategies and user perception-a review. Water Air Soil Pollut 229(8):255
22. Patil YM, Patki VK, Jahagirdar S, Karale R, Angathekar V (2021) Greywater treatment by vegetated vermifilter: a low cost material for rural sanitation in India. Mater Today: Proc 1–5
23. Pinto U, Maheshwari BL (2010) Reuse of greywater for irrigation around homes in Australia: understanding community views, issues and practices. Urban Water J 7(2):141–153
24. Ridderstolpe P (2004) Introduction to greywater management. Report 2004/4, Stockholm Environment Institute, EcoSanRes Publication Series, Stockholm, Sweden
25. Siggins A, Burton V, Ross C, Lowe H, Horswell J (2016) Effects of long-term greywater disposal on soil: a case study. Sci Total Environ 557:627–635
26. Simpson B, France D, Lewis B (2013) Wastewater sampling. U.S Environmental Protection Agency, Science and Ecosystem Support Division, Athens, Georgia, USA
27. Sivakumar M, Ramezanianpour M (2012) Green technology options for greywater treatment. National Convention of Environmental Engineers Mangalore Local Centre, Karnataka
28. Stoner N, Postel E, Kadeli L (2012) Guidelines for water reuse. Report EPA/600/R-12/618 U.S. Environmental Protection Agency, U.S. Agency for International Development, Washington DC, USA
29. Tamanna M, Jalil MA, Islam S, Anam S (2011) Greywater generation and quality measurement at a specific site in Dhaka city. Undergraduate thesis, Bangladesh University of Engineering and Technology, Bangladesh
30. Uddin SMN, Li Z, Ulbrich T, Mang H, Adamowski JF, Ryndin R (2016) Household greywater treatment in water-stressed regions in cold climates using an 'Ice-Block Unit': perspective from the coldest capital in the world. J Cleaner Prod 133:1312–1317
31. Uitto JI (2001) Global freshwater resources. In: Palo M, Uusivuori J, Mery G (eds) World forests, markets and policies, world forests, vol 3. Springer, Dordrecht, pp 47–58
32. WHO (2006) Guidelines for the safe use of wastewater, excreta and greywater. In: Policy and regulatory aspects. World Health Organization, United Nations Environment Programme, France
33. World Health Organization (2006) Overview of greywater management health considerations. Report WHO-EM/CEH/125/E. Centre for Environmental Health Activities, Amman, Jordan

Experiences from Developed Countries

Advancements in Nutrient Removal Technologies for Wastewater

Matthew C. Cotton, Farhad Majdzadehmoghadam, and Catherine N. Mulligan

Abstract Nutrient (nitrogen and phosphorus) removal from wastewater is becoming even more important due to climate change. The basic principles of both aerobic and anaerobic processes, including microbiological characteristics for industrial, stormwater and municipal wastewater systems are described. Nitrogen removal is often a two-step process including nitrification and denitrification. More recent attention has been given to the annamox process, an anaerobic ammonium oxidation process that converts ammonia and nitrite to nitrogen gas without the addition of carbon containing compounds. A combination of aerobic and anaerobic conditions is needed for phosphorus removal. In addition, combined nitrogen and phosphorus systems can also be considered. Various reactor configurations are described for nutrient (nitrogen and/or phosphorus) removal including aerated tanks, sequencing batch reactors, biofilters, fixed film, UASB, airlift, and membrane bioreactors, among others. Applicability, design parameters, monitoring, maintenance, and operational requirements for each reactor type are discussed. In addition to conventional nutrient removal processes, nature-based solutions such as wetlands will be described. Challenges and future research directions are identified.

Keywords Nitrogen · Phosphorus · Wastewater treatment · Aerobic · Anaerobic

1 Introduction

Nitrogen and phosphorus containing compounds are of vital importance to the functioning of aquatic and terrestrial ecosystems. These nutrients, in excess, can also cause issues in those systems, such as eutrophication of water systems, and the resultant break down of the normal ecological processes the ecosystem requires to

M. C. Cotton · F. Majdzadehmoghadam · C. N. Mulligan (✉)
Department of Building, Civil, Environmental Engineering, Concordia Institute of Water, Energy and Sustainable Systems, Concordia University, Montreal, Canada
e-mail: catherine.mulligan@concordia.ca

S. Yadav et al. (eds.), *Emerging Technologies in Biological and Hybrid Wastewater Treatment: Lessons from Developed to Enhancing Practices in Developing Countries*, Earth and Environmental Sciences Library,
https://doi.org/10.1007/978-3-031-74515-7_16

maintain homeostasis. This can lead to issues impacting all water uses we rely on from drinking through to industrial usage.

Beginning in 1915, the method for biological treatment of wastewater has been via aeration and activated sludge processes in one form or [53]. Today the processes are still used extensively to remove COD and nutrients. In the activated sludge process an effluent stream is mixed with recycled activated sludge from a later process settlement stage. This mixture, referred to as mixed liquor suspended sludge (MLSS) is then fed into an aeration chamber where microorganisms begin the removal of COD and nutrients. The outflow proceeds to a clarification tank where the sludge (activated sludge) settles out. Part of this activated sludge is removed and sent back to the beginning of the process (recycled sludge) while the remainder is sent for a sludge treatment process. The outflow of the clarification process can then proceed to an anerobic treatment tank where the process is repeated with anerobic microorganisms to further treat the water for release followed by another clarification and sludge recycling process for the anerobic step.

Most biological treatments systems, both those in use and those being researched, use some semblance of this process. As will be seen in this chapter much of the refinements, and research, being done to the process are for the reduction in energy use and space requirements, improvements in overall removal efficiencies, reduction in hydraulic residence time (HRT), or removal/reduction of sludge recycling/ production, or some combination of them all. The definitions of HRT and solids retention time (SRT) are as follows:

$$HRT = \frac{Reactor\,volume(\mathrm{m}^3)}{Flow\,rate(\mathrm{m}^3/\mathrm{h})} \tag{1}$$

$$SRT = \frac{Total\,sludge\,present\,in\,reactor(\mathrm{kg})}{Sludge\,withdrawn\,per\,day(\mathrm{kg/day})} \tag{2}$$

The intent is not just to improve new plants being designed for future use but also to improve those already in operation. By improving operational plant efficiencies without increasing over all footprints, their operational life spans can be increased avoiding the necessity of, and costs, of replacement. In addition, reducing energy and space requirements for high load waste waters from industrial sources means those industries may be able to assist by pre-treating of their effluent streams before release into municipal systems. This in turn will assist in meeting regulatory requirements and help save our water systems for future uses.

Recent studies have shown that where organic and nutrient removal is of paramount importance but the concern for other contaminants is minimal, such as in dairy/pig farming or brewery waste, microalgae cultivation can serve as a single pot WWT while also providing the secondary benefit of resource extraction and value generation through biomass generation [22, 73].

Other recent research has looked into the benefits of high-rate activated sludge (HRAS) treatment [10] or how aeration methods factor into the generation of greenhouse gasses (GHGs) [44]. The HRAS study concluded it could be a promising method of primary clarifier at low (0.5 mg L^{-1}) showing average removal efficiencies: 23% for total Kjeldahl Nitrogen (TKN), 7.3% for N–NH_4^+, 56% for total phosphorus (TP), and 11% for P–PO_4^{3-} as well as COD removal but that it may not work as a secondary or tertiary treatment option. The intermittent aeration (IA) study looked at varying aeration in the oxic-settling-anaerobic process (OSA) and compared it to GHG emissions of the conventional activated sludge (CAS) method. OSA-IA showed better PO_4–P (ortho-phosphate) and COD (chemical oxygen demand) removal whereas CAS performed better for removing TSS (total suspended. However, the overall results showed that OSA-IA released more GHGs than CAS mostly due to more energy being required to run the system, mixers and sludge recirculation pumps.

2 Conventional Nitrogen Removal Systems

Conventional nitrogen removal systems rely on a two-stage process to convert ammonium (NH_4^+) to nitric oxide (NO), nitrous oxide (N_2O) and to nitrogen gas (N_2) and released to the atmosphere. This is done in nature as part of the nitrogen cycle and relies on microbial communities for the conversion.

Stage 1: Nitrification (oxic conditions, chemolithoautotrophic bacteria)

In the first step of this stage, a group of ammonia oxidizing bacteria (AOB) converts ammonia to nitrite (NO_2^-):

$$NH_4^+ + 1.5\,O_2 \rightarrow NO_2^- + 2H^+ + H_2O \tag{3}$$

In the second step, the oxidation of the nitrite to nitrate (NO_3^-) is carried out by a group of nitrite-oxidizing bacteria (NOB):

$$NO_2^- + 0.5O_2 \rightarrow NO_3^- \tag{4}$$

The optimum pH range for nitrification is between 7 and 8 which can quickly change due to the creation of hydrogen ions during the nitrification process. At the same time free and saline ammonia (FSA) (NH_3) has been shown to have an inhibitory effect on the process at higher concentrations so operating the system at a higher pH will also negatively effect nitrification. To maintain the system within the desired range it is recommended that pH be monitored and controlled by the addition of alkalinity via the addition of carbonate (CO_3) at a ratio of 7.1 CO_3: NH_4^+ [5, 15, 59].

Stage 2: Denitrification (anaerobic conditions, heterotrophic bacteria)

Denitrification requires anaerobic conditions and is carried out by heterotrophic bacteria, usually using bacteria from the *Paracoccus* and *Pseudomonas* genera. First the nitrate is reduced to nitrite then further to nitric oxide (NO) and nitrous oxide

(N_2O). Finally in the nitrous oxide reduction N_2O is further reduced to nitrogen gas (N_2) and released into the atmosphere [15, 28, 33, 36]. Because the bacteria used is heterotrophic, a carbon source is required for the reduction reactions to occur. This means the stoichiometric equations for the reactions will vary depending on the carbon source used. In general, though the four reduction reactions can be expressed as seen in Eqs. 5–8 using acetate (CH_3COO-).

Nitrate Reductase

$$\frac{1}{2}NO_3^- + \frac{1}{8}CH_3COO^- + \frac{1}{8}H^+ \rightarrow \frac{1}{2}NO_2^- + \frac{1}{4}H_2O + \frac{1}{4}CO_2 \tag{5}$$

Nitrite Reductase

$$NO_2^- + \frac{1}{8}CH_3COO^- + \frac{9}{8}H^+ \rightarrow NO + \frac{3}{4}H_2O + \frac{1}{4}CO_2 \tag{6}$$

Nitric-oxide Reductase

$$NO + \frac{1}{8}CH_3COO^- + \frac{1}{8}H^+ \rightarrow \frac{1}{2}NO_2^- + \frac{1}{4}H_2O + \frac{1}{4}CO_2 \tag{7}$$

Nitrous-oxide Reductase

$$\frac{1}{2}N_2O + \frac{1}{8}CH_3COO^- + \frac{1}{8}H^+ \rightarrow \frac{1}{2}N_2 + \frac{1}{4}H_2O + \frac{1}{4}CO_2 \tag{8}$$

The Modified Ludzack-Effinger (MLE) is probably one of the most used methods for nitrogen removal worldwide. As discussed earlier, the denitrification process requires a carbon source, and the products form the nitrification step to denitrify the wastewater. The nitrification process, however, requires the system to be aerated to function, but the presence of COD would greatly increase the cost associated with the aeration step while also removing the carbon that would be used by the anerobic step. To get around this, the MLE directly feeds the COD rich effluent to an anerobic tank (Fig. 1). The anerobic tank then flows into an aerobic chamber where the nitrification takes place. In the aerobic step part of the mixed liquor and suspended solids is removed and fed directly back into the anerobic chamber. The aerobic chamber then empties into a clarifier where the sludge settles out, is collected, and is fed back into the incoming effluent and the anerobic chamber. This method allows for a more complete removal of the nitrogen processes while lowering the cost of aeration.

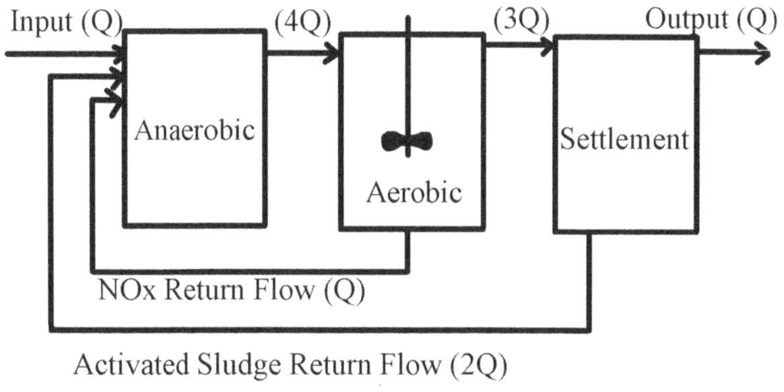

Fig. 1 Modified Ludzack-Effinger (MLE) showing idealized flows (Q) between tanks

3 Advanced Nitrogen Removal System and Anammox

While conventional nitrogen removal technologies have been employed in wastewater treatment since the early 1900s they do have their drawbacks. Add to this the requirements imposed by improved regulatory standards and older plants are looking to improve or retrofit their current systems while newer plants are looking to save space and costs. As such, much effort and research has been done to find means of improving efficiency, lowering cost, and reducing the monitoring requirements while saving space.

Simultaneous nitrification and denitrification (SND)

Requiring a carbon to nitrogen ratio of 10, the entire denitrification process is done aerobically, in a single reactor, without the need for additional carbon and alkalinity. Efficiencies of 80–96% nitrogen removal at dissolved oxygen (DO) levels of just 0.3–0.7 mg/l [40] can lead to a substantial savings in operational cost. As in the conventional denitrification, the process under aerobic conditions is still done by using nitrate as an oxidizing agent (electron acceptor) and is reduced to nitrogen gas by heterotrophic bacteria [29]. This is done via the creation of large flocs containing both aerobic and anerobic condition and microbial communities. This granulation process will be discussed further in later sections.

Nitrite Shunt

In the nitric shunt method partial nitrification occurs by converting ammonia to nitrite by AOB with no further oxidation to nitrate taking place. Instead of the nitrite being oxidized to nitrate by NOB, low DO levels inhibit or prevent their growth, and instead heterotrophic bacteria convert the nitrite to nitrogen gas [40]. This method has been shown to reduce aeration cost by 25% while increasing nitrogen removal by 40%. The method has also increased efficiencies when used in a sequential batch reactor (SBR) treating landfill leachates heavily loaded with ammonia [17].

Anammox

Partial nitration (PN) takes place in a reduced DO aerobic environment. The resultant nitrite produced by AOB is then used to oxidize ammonium further in an anerobic environment by anerobic ammonium oxidizing bacteria (ANAMMOX) [40]. The entire process takes place in a single tank with less aeration and higher nitrogen removal [26]. However, it has longer start up times due to the longer time for biomass creation (7–22 day doubling time) and it suffers from decreased efficiencies at lower temperatures than 35–40 °C which are optimal [21]. It has been shown that the addition of 20 mg N/L nitrite addition had an over all improvement in the annamox process. This improved efficiency continued even after the nitrite loading was removed [80] with nitrogen removal rates approaching 99%.

Implementation

While the above methods have mostly been investigated in research labs some of their methods have been combined into systems for testing and use in wastewater treatment plants. Despite the benefits in the nitrogen removal process, their implementation is hampered by the necessary pretreatment (removal of COD and ammonium concentration) of the wastewater fed to them [29]. Hence, these systems are mostly focused on the removal of nitrogen from side stream processes taking place in the plant. Side stream processes refer to processes in the treatment plant where ammonium is concentrated due to various practices associated with treatment. They include, but are not limited to, supernatant fluid from sludge digestion, sludge dewatering, and reject streams from membranes [71]. They can also be used for the treatment of any high ammonium content wastewater. Two of the most common methods being implemented and tested are the Completely Autotrophic Nitrogen-removal Over Nitrite (CANON®) and the Single Reactor for High activity Ammonia Removal Over Nitrite (SHARON®) processes as follows.

CANON®

In a single reactor the partial nitrification (PN) of ammonium occurs by *Nitrosomonas*-like aerobic ammonium oxidizing bacteria (AOB) in a carefully controlled low DO environment leading to the production of nitrites. These nitrites are then used by *Planctomycete*-like annamox bacteria to anaerobically oxidize the remaining ammonium to N_2 [70].

SHARON®

As the name suggests this method uses a single, well mixed, reactor to perform PN and denitrification of high ammonia wastewater using autotrophic nitrifiers. The process uses high temperatures (30–40 °C), a pH between 7–8, while operating without sludge loss [27, 48].

In addition to these research and methods, there is a novel approach for the denitrification process being explored. This process, solid-phase denitrification bioreactors (SPDRs), uses a solid carbon source to fill a reactor and the nitrified waste is fed through the selected medium. A heterotrophic denitrification is then possible with the bacterial communities growing on and using the carbon source for biomass generation and nitrates as electron donors. These carbon sources may be byproducts of agribusiness or biodegradable polymers. A recent study used wood chips as the carbon sources in two bioreactors with different filling patterns [84]. Using a uniform filling reactor and a gradient filling reactor they found that nitrate removal was enhanced in the gradient filling as compared to the uniform filling reactor with nitrate removal efficiencies of more than 97%. This was accomplished with extremely low nitrogen loading rates 7.78–47.82 g N/m^3 per day with flow rates of only 0.3–1.8 mL/min and in a column of only 25–35 cm with an HRT of 0.5–3 days. They also cautioned that while woodchip filled SPDRs can last for years before needing the chips replaced, the removal rates are limited by the rate the carbon can be dissolved from the woodchips which are apparently not that fast. It is, nonetheless, an interesting approach to the removal of nitrates from wastewater.

4 Phosphorus Removal Systems

Phosphorus is an essential element for life. However, excessive phosphorus can lead to a variety of environmental issues, including eutrophication of water systems. Factors such as phosphorus concentration, economic consideration and the volume of wastewater to be treated, primarily influence the selection of an appropriate method for removal of phosphorus from wastewater streams [74].

Chemical Precipitation

One of the most common methods used for phosphorus removal is chemical precipitation, in which metal salts are added to wastewater. Multivalent metal ions such as aluminum (III) and iron (III) react with water and form metal hydroxides ($Al(OH)_3(s)$ or $Fe(OH)_3(s)$). Phosphate makes strong bonds with the created metal hydroxyl complex and produces a metal-phosphate precipitant ($AlPO_4(s)$ or $FePO_4(s)$). The effectiveness of phosphate removal is influenced by the chemical dosage, pH levels, and temperature [23].

$$Metal\ Salt + Phosphorus \rightarrow Insoluble\ Compound + Water \quad (9)$$

Membrane Filtration

Membranes act as a physical barrier, filtering out the suspended solids, and accordingly pollutants such as phosphorus, in wastewater. Microfiltration, Ultrafiltration, Nanofiltration, Reverse Osmosis (RO) and Membrane Bioreactor (MBR) are the most common pressure-driven membrane filtration methods. Membrane filtration is basically a versatile unit operation, that means it can be introduced into the biological wastewater at various stages of processing. For example, RO systems are separately located downstream of the biological step, while membranes in MBR systems are integrated into a biological process.

Adsorption Technologies

In this method, a suitable adsorbent including activated carbon, ion oxides and synthesized or modified adsorbents such as aluminium-modified biochar [81] and nano-adsorbents [52] is selected based on the physicochemical properties of the adsorbent, the level of concentration and the conditions of operation. These materials have surfaces with high affinity for phosphorus, effectively binding and removing it from wastewater [82].

Highlights and challenges

Chemical precipitation is operationally a simple and reliable method. Membrane filtration provides a flexible and environmentally friendly technology and adsorption is an effective method for removal of phosphorus at low concentrations. However, chemical precipitation requires a careful disposal of produced sludge and membranes filters need high initial investment and energy consumption. Therefore, prior to selecting the appropriate method, a careful evaluation needs to be performed to increase the amount of phosphorus that can be recycled in the treatment systems, while minimize the environmental footprints.

5 Biological Phosphorus Removal System

Enhanced biological phosphorus removal (EBPR) is an eco-friendly process that occurs in anoxic or anaerobic conditions. Facilitated by the metabolic activity of phosphorus-accumulating and polyphosphate-accumulating organisms (PAOs). Due to sludge wasting, these bacteria successfully remove soluble orthophosphates from wastewater by assimilation and storage as non-soluble intracellular polyphosphate. PAOs can then use these polyphosphates as an energy source for the creation of biomass and energy storage molecules. The process is cyclic in nature with the uptake of orthophosphates and their conversion to polyphosphates occurring under aerobic conditions, that coincides with biomass generation, and the breakdown of the polyphosphates and the release of orthophosphates occurring under anaerobic conditions. While other methods, such as chemical precipitation, have higher operating costs due to the requirement for chemical purchasing and sludge processing biological phosphorus removal, particularly EBPR, is considered to be both an extremely sustainable and cost-effective method of treating wastewater. One of the best system configurations for EBPR can be seen in Fig. 2 [66].

The first stage of the processes is the anerobic reactor where to ensure proper mixing the HRT is maintained between 30 min and 2 h [31]. Through this interaction, PAOs can consume carbon in the form of volatile fatty acids, most often acetate, and produce molecules that store carbon. Known as polyhydroxyalkanoates (PHA), the most common being poly-hydroxy-butyrate (PHB) these molecules will serve as the primary energy source for cell growth and biomass generation in the aerobic phase.

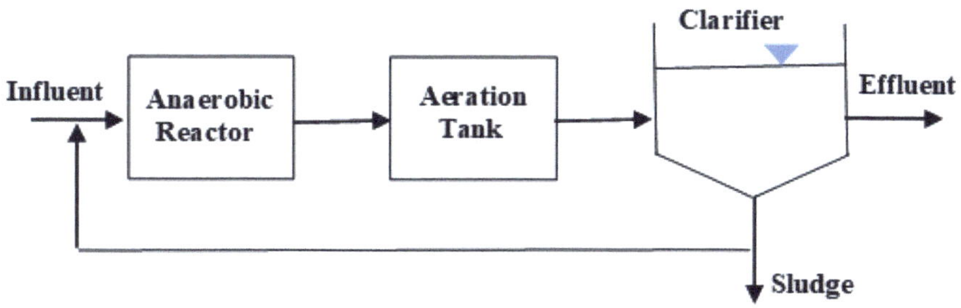

Fig. 2 Biological suspended growth process configuration

This is done by hydrolysis of polyphosphate by ADP and the enzyme phosphotransferase or through sugar phosphorylation using the enzyme polyphosphate glucokinase and results in the polyphosphate being converted back to an orthophosphate and released from the cell back to the surrounding medium.

The second stage of the EBPR is the aerobic zone. In this section the presence of dissolved oxygen allows for the oxidising of the stored PHAs releasing the stored energy and carbon they contain. The energy released is used for the creation of glycogen as well as the uptake and conversion of environmental orthophosphates to intracellular polyphosphates. Released carbon on the other hand is used for cellular growth and biomass generation [30, 39].

In the clarifier section part of the previously generated biomass, complete with stored polyphosphates is removed from the system while another part is recycled back into the system to begin the process all over again in the anerobic tank. It is through this removal of the generated biomass that phosphorus is removed from the wastewater being treated [30, 39].

6 High-Rate Full-Scale Nitrogen and Phosphorus Removal Systems

As populations rise, and demand increases, wastewater streams from both industry and the public are becoming increasingly nutrient rich. At the same time, the understanding of the effects these nutrients have on the surrounding environment is also improving. This increase in waste streams with high nutrient loading has led to research into and, in some cases, adoption of high-rate full scale nitrogen and phosphorus removal systems. These are systems such as granular biomass, sequential batch reactors (SBR), microalgae, biofilm-based treatment, and integrated up-flow anaerobic sludge blanket, (UASB) along with natural systems, such as microalgae and constructed wetlands, that have been shown to be able to handle the increased nutrient loading. These systems show great promise for lowering nutrient content in post treatment waste waters. Due to their ability to handle high loading rates while occupying a small footprint they are ideal for integrating into industrial processes or for adding additional processing power to older more traditional wastewater treatment plants via integration into to side stream processing.

7 Granular Biomass System

One of the systems researchers and WWTP operators have been investigating over the last few decades is the granular biomass system. Consisting of mixed microbial cultures these aerobic granules may be of great importance as a possible replacement for current activated sludge methods. [62]. Other benefits of the system are the dense microbial communities that appear to isolate the system to load shocking and the fact that the floccular size (0.5–3 mm) lends itself to quick settling [58, 62, 77]. These granules are essentially an outer shell of aerobic nitrogen oxidizing organisms surrounding, and protecting, an anerobic core composed of denitrifying phosphate accumulating organisms and those capable of removing COD. This colony of organisms allows for the simultaneous removal of COD, phosphorus, and nitrogen containing compounds from the system in a single reactor [57].

Although the method of granulation is still being researched, certain factors have been found to influence their development. Some of the factors found so far include the feast/famine and starvation time, HRT, hydrodynamic selective pressure, and shear force (intensity of aeration), waste characteristics (organic loading rates (OLR), COD to nitrogen ratio (C/N), carbon source, etc.), seed sludge, and, of course, reactor design [1]. It has also been shown that favouring slow growing bacteria, such as polyphosphate accumulated organism (PAOs), may enhance granulation [56]. In other words, on a micro scale granulation may be encouraged via the following four factors [78]:

1. Cell-to-cell attraction: the attraction of microbial cells via various forces
   - Physical as in thermodynamic, van der Waals, or opposite charge.
   - Chemical such as ionic pairing, triplet ionic pairing, or inter-particulate bridging.
   - Biochemical actions like cell receptor attraction, cell surface dehydration, and cell membrane fusion.
2. Microbe-to-microbe attachment: aggregate formation is influenced via hydrodynamic, diffusional, and/or thermodynamic actions.
3. Microbial adhesion enhancement: Some microbes secret an extracellular polymeric substance (EPS) which helps to cement the granules together.
4. Hydrodynamic shear force: the tiny differences in fluid layer velocity caused by aeration helps to stabilize the granules in 3 dimensions.

Operating under the brand Nerada®, the system has been used to treat industrial (2006) and domestic (2009) waste-water flows. Demonstration (full scale) plants relying on sequential batch reactor systems have been built in Gansbaai, South Africa and Frielas, Portugal. The data from these systems was then used to construct a system in Epe, the Netherlands [20, 57, 72]. Since then, plants have been built in Ryski and Vroomshop (Poland) with plans or construction proceeding on 40 others worldwide. Results from the Polish plants show a 25–75% reduction in system footprint and a 10–50% reduction in energy usage. This is attributed the elimination of secondary settlement tanks and the increase in biomass concentrations in the reactor [56].

8 Sequencing Batch Reactor (SBR)

The Sequencing Batch Reactor (SBR) can be considered as a fill-and-draw version of conventional activated sludge process (ASP). The SBR runs in batch cycles, in contrast to the continuous nature of ASP, providing an improved process control and flexibility that eventually results in compliance with strict effluent discharge criteria. The SBR is popular because of its ability to automate tasks effectively, which may reduce operational costs by over 60% when compared to traditional ASP [55]. Short aeration intervals allow for the generation of high-quality effluent, which makes it perfect for densely populated places like Europe and India, where the SBR is a preferred technology because to its compact footprint and lower workforce needs. Moreover, SBRs target on the removal of nutrients, the reduction of BOD, and/or the control of filamentous bacteria, while additionally consuming less energy [13].

This process is characterized by five main stages as illustrated in Fig. 3:

1. Fill Phase

In the fill phase, influent wastewater is introduced into the reactor.

2. React Phase

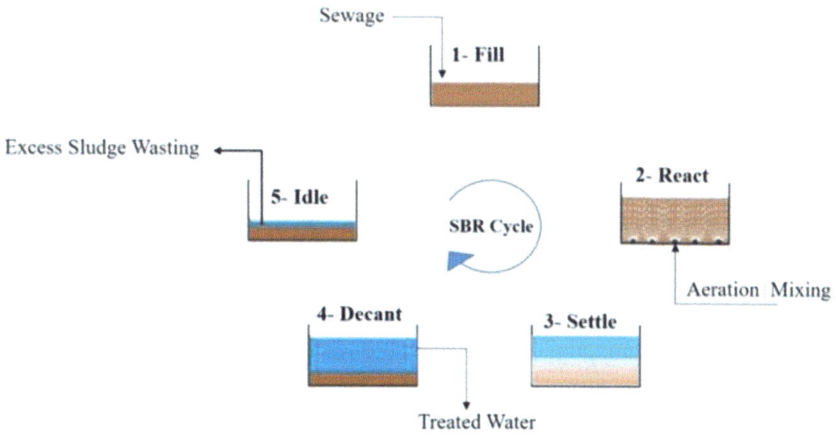

Fig. 3 Main stages of SBR process

Aeration and/or mixing allows the bacterial flocs to be circulated through the tank allowing for the biological treatment of the effluent to occur.

3. Settle Phase

Aeration and/or mixing is discontinued. The heavier biomass settles to the bottom of the tank leaving the treated effluent in the upper reaches of the tank.

4. Decant Phase

In this phase the clear treated effluent is removed from the tank.

5. Idle Phase

Allows the biomass to enter the anoxic or anaerobic stage off their cycle, if required, allowing for further nutrient removal or a resting period for the microorganisms.

The working or effective volume of the SBR can be calculated using Eq. (10) and is influenced by factors such as aeration duration and food to microorganism ratio (F/M) [69].

$$V_s = \frac{Q_s * C_s}{L_s * C_A * \varepsilon} \tag{10}$$

where

V_S effective volume (m^3)
Q_S flow rate (m^3/day)
C_S average concentration of BOD_5 (mg/L)
L_S F/M ratio (kg BOD/kg SS·day)
C_A average concentration MLSS (mg/L)
ε aeration time.

Aeration time depends on influent loading. Therefore, the settling and discharging stages need to be optimized for more flexibility in bio-reaction timing. This adaptability enables constant and effective nutrient removal by enabling responses to variations in influent loads.

9 Integrated Up-Flow Anaerobic Sludge Blanket (UASB) and Short Aerobic System

The UASB is, as the name suggests, an anaerobic system where incoming wastewater is processed via flowing upwards through an accumulated mass of a granular sludge blanket. This up-flow through the sludge bed allows for very thorough mixing of the incoming wastewater and the sludge granules. Due to this mixing, rapid decomposition of organic matter and nutrient removal is possible. The large size of the granules leads to quick settling of the flocs in the water column leading to increased solid retention and bed height as time goes on. At the top of the system, the water flows out into another tank where a short period of aeration is performed to eliminate pathogens [43].

The system, due to its anaerobic decomposition (Fig. 4), can be a great method for methane production which can be used to lower the energy requirements of the operating plant. Usually started with inoculated granules occupying 10–30% of the reactor volume with higher initial loading rates being associated with higher initial granule volumes, these granular sludges can operate under low flow conditions and high strength wastewater without the need for recirculation or external separation [35]. Similarly, studies have shown that by varying the substrate materials, inhibiting mutagenic bacteria through pre-treating the sludge thermally, headspace gas recirculation, or acid-base procedures, the system can instead be optimized for a variety of outputs such as bio-hydrogen or volatile fatty acids which can be sold to various industrial sectors [9, 16, 42, 43, 47].

As the microbial community make up of the reactors is dependent on the make up of the waste flowing through it and the waste that is leaving the bioreactor has a different makeup, it is possible to establish another bioreactor in sequence with a different microbial community being contained in the granule makeup. Indeed, by using a two-stage bioreactor, UASB as the second stage, it was demonstrated in laboratory experiments [60] that a 10–30% increase in methanogenesis could be achieved. However, this increase is also coupled with an increase in operational issues decreasing their reliability. Due to the increase in cost and decreased reliability, industry adoptability has been low.

Another option to increase methanogenesis, while also increasing COD removal, is through the addition of micro-nutrient supplementation (Fe, Co, Ni, Se, Mo) or via co-digestion with food waste (due to its C/N ratio) [11, 38]. This has been shown to both increase methanogenesis and decrease nutrient loads in the treated wastewater.

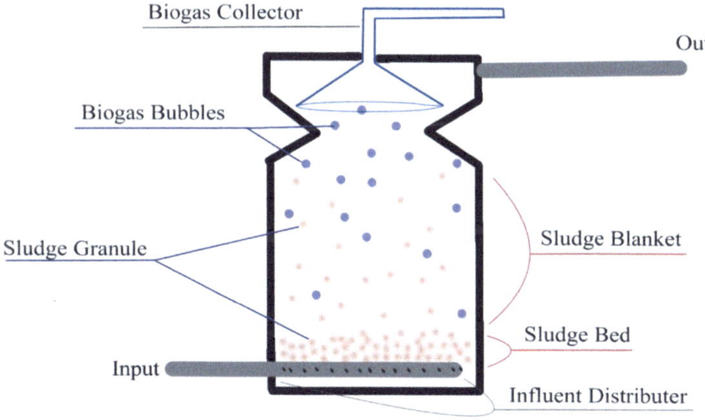

Fig. 4 Integrated up-flow anaerobic sludge blanket (UASB) showing simplified reactor design and constituents (granular sludge, sludge bed, sludge blanket, biogas)

10 Biofilm Based Treatment System

Complex communities of microorganisms, such as bacteria, protozoa, algae and fungi, that attach to surfaces and are embedded in a matrix of extracellular polymeric substances (EPS) are called biofilms. EPS, which is a mixture of proteins, lipids, carbohydrates and nucleic acids, supports the stability of and protection of biofilm [33].

These communities of organisms have been shown to efficiently remove both organic and inorganic contaminants from their environments including nutrients such as phosphorus and nitrogen. Despite being an integral part of the bioremediation and water treatment process these biofilms do come with their own hardships. These biofilms can be difficult to form and maintain long term due to issues such as film fouling and community growth and control [49, 63].

Biofilm formation is the result of a complex interaction between biotic and abiotic factors, including the diversity of the microbial community, the physical and chemical properties of the surface carrier, and environmental factors such as pH and temperature.

The first step of biofilm formation starts with initial attachment in which bacteria contacts with the surface, through van der Walls and electrostatic forces. Then, as Fig. 5 shows, cell-cell adhesion occurs on the surface. Surface-associated proteins, such as pili and flagella, mediate bacterial attachment and adhesion, and facilitate the transition of bacteria from a planktonic state to the surface.

Next is the proliferation stage, during this stage biomass is increased and a certain type of self sorting occurs. This sorting is the organization of the organisms into micro-colonies. These micro-colonies are arranged spatially and allow for the sharing of resources or the organizing of metabolic tasks in an almost factory like precession. It also allows for the creation of mini-environments where there may be bacteria requiring an anaerobic or anoxic environment being sheltered from oxygen contact by aerobic bacteria. As the biofilm matures, some extracellular chemical signalling molecules promote the production of EPS and the formation of a three-dimensional matrix.

As a result of bacterial division, mechanical pressures, or environmental conditions, the biofilm may eventually detach and spread, that it will allow the sessile cells to become mobile and potentially colonise other surfaces. Biofilm reformation is facilitated by the involvement of enzymes and modifications in gene expression during the detachment phase [7].

11 Wetland Process

In nature, wetlands provide key ecosystem filtration services removing access COD, phosphorus and nitrogen containing compounds as well as removing pathogens [51, 61]. New York City, when faced with worsening water supplies, looked into the

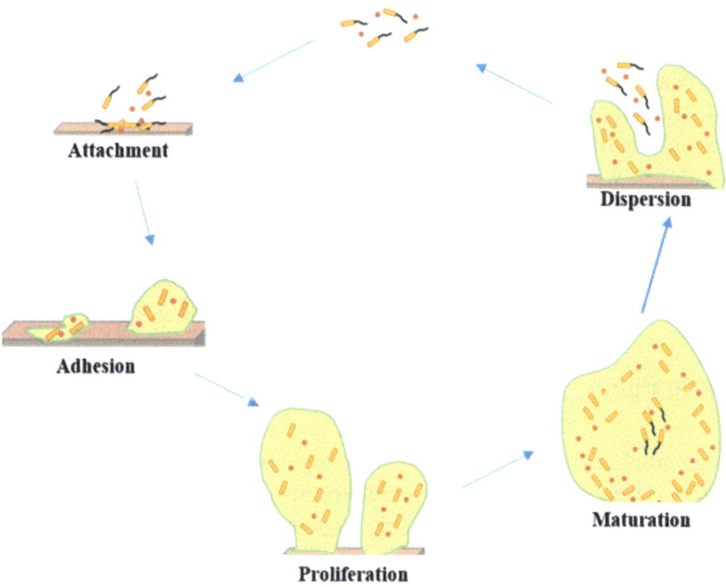

Fig. 5 Life cycle of a biofilm: adhesion, attachment of microbes, proliferation, maturation of biofilm, and dispersion

process of building and running a water treatment plant. Faced with a cost estimation of 8–10 billion US$ to build and 1 million US$ per day to operate and maintain [50] the city instead opted to improve and protect wetland functions in the watershed. This was done through a variety of means including entering partnership with the farmers to address nonpoint source pollutants lowering water quality as well as restoring, and integrating, natural streams into their storm management system. Between these initiatives, and others, New York City has some of the best unfiltered water in the United States and was at the forefront of ecosystem service thinking [6, 50].

Today, much research has been done on constructed wetlands (CW). Ideal at removing and containing pollutants from nonpoint sources, they are simple to operate, low cost, and capable of handling a variety of effluents [12]. Constructed wetlands can be designed to create surface and subsurface flows to take advantage of natural filtration, adsorption, and absorption, as well as anerobic and aerobic micro-organisms to remove nutrients and pollutants from the waste flows [18, 34, 51]. The addition of various plants, both rooted and floating has also been shown to improve nutrient removal as well as slowing surface flows through the system allowing for greater solid settling while also maintaining greater porosity in the soils of the wetlands through root action. Thus, allowing for subsurface flows to continue for longer periods with less maintenance. CW with plants also showed higher evapotranspiration and, therefore, longer hydraulic retention times, during summer months [46, 54, 68].

Constructed wetlands are free water surface flow (FWSCW), subsurface flow (SFCW), or hybrid flow. SFCW can further be broken down into horizontal, vertical, or french vertical flow (at least two vertical flow sections, with different filter media, placed sequentially) and have been in use since 1968 in Hungary [54]. Each have their own benefits and possibilities that can be tailored to fit or combine with a variety of situations and loading/flow requirements with space requirements being one of the only hurdles to overcome depending on volume of flow and HRT. With horizontal flow CWs, the area requirement is 10 m^2 per population equivalent (PE) whereas for vertical flow, the requirement ranges from 1 to 3 m^2/PE but requires more maintenance than the horizontal flow. However, with operation and maintenance cost being 1–2% of traditional treatment plants [54] is still a viable economic option.

12 Microalgae

Microalgae, single celled photosynthesizing organisms, existing in both fresh and saline waters. These organisms may be autotrophic, using atmospheric CO_2 for the creation of biomass, heterotrophic, using organic carbon sources for energy and biomass production, or mixotrophic, able to use both autotrophic and heterotrophic methods of energy extraction and carbon sources for biomass generation allowing for their implementation in municipal, industrial, and agricultural WWT [14, 65, 79].

Regardless of metabolic methods, all types of microalgae require the same types of nutrients as macroalgae ad vascular plants. This makes them extremely useful for the removal of nutrients from wastewater as well as the removal of BOD and COD sources if hetero/mixotrophic verities are used [3, 65, 79]. They also have the ability, through bio-absorption, of removing many types of heavy metal pollutants [41, 67]. For these reasons, the cultivation and use of microalgae for wastewater treatment may be introduced during secondary treatment, for nutrient and organic carbon removal, or as a tertiary treatment for removing the remaining nutrients and heavy metals prior to release [24, 67, 79].

In general, the types of microalgal systems in WWTP can be classified under two main categories suspended or immobilized systems. Suspended systems rely on microalgae being equally distributed throughout and suspended in the water column while immobilised systems rely on the organism being fixed in place and can be further broken down into fixed bed or fluidized bed systems [79].

There is a variety of suspended photobioreactor (PBR) systems available for introduction and use in WWTPs. While pond-based PBRs, such as those used in bacterial cultivation and WWT, are currently the most common for microalgae WWT and offer many advantages, such as low cost, simple construction, and easy maintenance, there are adaptations made to try and minimize some of the problems associated with these types of reactors. One of the main issues hindering suspended PBRs introduction during upgrading, or planning, of WWTPs is the optical capacity of water. If using autotrophic microalgae for the treatment process, the availability of light is one of the key limiting factors for the production of biomass in the PBR [2, 14, 79]. This means using large shallow ponds or the addition of a mixing method. This means either a large space or increase in energy requirement for the system. There is also the fact that for optimum biomass production and nutrient removal the C:N:P ratio must be maintained [79]. When using autotrophic bacteria this means the introduction of CO_2. The pumping of atmospheric CO_2, while providing a source of agitation for the ponds, means an increase in the set-up, operation, and maintenance costs for the pond PBR. While some methods, such as vertical tubes, flat-panels, horizontal tubes, and raceway ponds, have been developed to minimize the space requirements, and improve the light availability and mixing requirements they can still be rather large in comparison to other methods of nutrient removal and may still suffer from some of the other limiting factors such as temperature and pH of the waste waters [79]. However, there are a wide variety of extremophile microalgae that can operate in a wide range of conditions. In addition, the co-cultivation of other microbial organisms has been shown to increase removal efficiencies. Therefore, the proper selection of microbial constituents can lend itself to the establishment of microbial communities with synergistic effects in the selected PBR [24, 45, 79].

Regardless of the suspended PBR selected one of the hardest, and therefore most expensive, aspects of microalgal WWT is the down system extraction and processing of these microorganisms [79]. The use of an immobilized system helps reduce the cost of capture and processing by immobilizing the microalgae either on a type of growth matrix (fixed-bed) or on a floating substrate (fluidized bed). In both cases the selection of the growing medium is important to the development of the microalgal biofilm with high surface to volume ratios being one of the most important factors. Also, in both cases it allows for a thin layer of water to be trickled over, or to suspended, the growing medium allowing for maximum contact between biofilm and contaminants. The biofilm aspect also allows for easy extraction and processing of the attached biomass. As in the suspended PBRs the proper selection of microalgal species, and accompanying microorganisms, is key to the system functioning optimally under the required conditions [79].

The use of microalgae systems has an added benefit allowing for the extraction of biomass as resource. Microalgae, depending on type, has uses as an animal and human food additive, the creation of cosmetics, chemicals and fertilizers, as well as in the biofuel industry. All these benefits can be used to defray the system costs either through the sale of the biomass as resource or the use of the biomass in house for energy generation via biogas generation [8, 14, 24].

13 Challenges Associated with Nutrient Removal Technologies

While microalgae-based technologies show great promise, challenges such as variability in environmental conditions, competition with other microorganisms, and the need for cost-effective harvesting methods remain. Researchers are actively working to address these challenges through innovations in genetic engineering, optimization of cultivation conditions, and the development of novel harvesting techniques [3, 65].

Constructed wetlands are a seasonal system, especially in colder climates, with removal efficiencies increasing during summer and decreasing during winter due to the plant function in the removal process. In addition, these plants may take up and retain harmful chemicals from the effluent. This is fine as long as they stay growing but as the accumulation of biomass is larger in a CW than in other systems this requires regular harvesting and disposing of the contaminated biomass. There is also an issue, in FWSWCWs, of mosquito breeding and the fact some pollutants, such as heavy metals, may not be removed [54].

PN, ANAMAMOX, SNAP, and other improvements made to the traditional methods of nitrogen removal show much promise in the ability to improve energy efficiencies and space requirements. However, the micro-organisms slow development extends start up times and decreased efficiency in low temperatures is a hurdle that needs to be overcome for their widespread adoption in in industrial and public effluent treatment. The ANNAMOX process also may need an additional denitrification step due to a tendency of two much nitrate in the outflow. This additional denitrification step would also require the addition of a carbon source [74].

14 Conclusions and Future Prospects

In Table 1 some of the known advantages and disadvantages of nutrient removal methods are outlined. It also compares removal rates for the various methods discussed throughout this chapter. When comparing the removal rates, the variation in nutrient loading rates between the methods in the studies must be considered. Some were looking at smaller laboratory or pilot scale experiments while others were taken from full scale treatment plants, and all were dealing with differences in nutrient concentrations and flow rates. Keeping that in mind, the bottleneck in current methods and research seems to be the efficient removal of nutrients from low to medium strength waste waters. While conventional activated sludge methodology has been working quite well for treating these flows and removin organics, their larger footprints and energy requirements make them not feasible for increasing established plants processing capabilities or establishing newer smaller scale treatment facilities servicing isolated communities or industries. Solutions for these situations may be to combine methods such as SBR and aerobic granular sludge to increase nutrient removal and biomass elimination while also reducing operational footprints.

For industries, such as urban greenhouses or microbreweries, seeking to pretreat their water prior to release into municipal systems unable to handle the increased volume or loading, the solution may be the same as that for established treatment plants looking to scale up, while lacking space. One of the easiest ways to improve overall nutrient removal in these situations would be to implement some of the newer methods for handling of side streams. For established treatment plants these

Table 1 Comparison of nutrient removal technologies, nitrogen and phosphorus removal rates, advantages, and disadvantages

Method	Phosphate removal	Nitrogen removal	Advantages	Disadvantages	Sources
MLE	56–99%	75–99%	Proven technology Works with low nutrient loading	Comparatively high energy and space requirements High sludge production and treatment requirements	[15, 37, 83]
SND	–	43–96%	Low DO use Low space and energy requirements Minimal sludge creation	Slow biomass production Granulation is not fully understood	[5, 40]
Nitrite Shunt	–	36–95%	Low sludge generation Low DO No added carbon Low energy and space requirements	Slow biomass generation Difficult to troubleshoot High optimal temperature	[5, 40]
ANNAMOX	–	65–99%	Low sludge generation Low DO No added carbon Low energy and space requirements	Slow biomass generation Difficult to troubleshoot High optimal temperature	[5, 21]
Granular biomass	67–98%	80–97%	Low to no DO requirement Low energy and space requirements No sludge recycling	Granular sludge formation not fully understood	[4, 25]
SBR	> 90%	> 96%	Low energy and space requirements Tolerates shock loads	Higher maintenance and monitoring costs	[40, 76]

(continued)

Table 1 (continued)

Method	Phosphate removal	Nitrogen removal	Advantages	Disadvantages	Sources
UASB	20–90%%	20–98.3%	Biogas production for energy cost offset Low space and energy requirements	Granular sludge creation not fully understood Granule flotation may not occur at low flow rates and larger granule size Slow biomass development Sensitive to organic shock loads	[19, 43]
Biofilm	< 95.8%	52–84.6%	Multiple application methods Can be tailored to fit treatment needs	Prone to inhibitory actions of wastewater constituents such as ammonium, nitrites, heavy metals, etc	[32, 40]
Wetland processes	40–82.11%	40–84.32%	Low maintenance and operational costs Simple to plan and set up	Space requirements Function best in moderate to hot climates	[64, 75]
Microalgae	54–99%	54–99%	Can be paired with many other treatment protocols and systems Can also treat pollutants of emerging concern	Economic feasibility (energy intensive) Odor Algae growth can block light transmission and a reduction in effectiveness Low biomass productivity	[65]

would be optimal as they could be fit into existing processes and improve operational efficiencies with minimal cost or space requirements. The same holds true for the pretreatment of industry waste, the use of an anammox system or UASB and short aerobic system may be ideal for industries like those mentioned. If planned and implemented as part of the startup system, the slow build-up of biomass in the system may not be as much of a hurdle as these industries have a slow startup time.

Research wise, more resources need to be put toward the understanding of the establishment of granular sludges and improving the startup times and stability for systems that rely on them, and anammox. This is also true of the other more advanced form of nitrogen removal. While the technology is promising, the temperatures, and nitrogenous loading, at which they must function relegates them strictly to treating waste flows that have higher incoming temperatures and concentrated nitrogen such as certain industrial processes and wastewater treatment plant side streams. A cost-effective way of having them work at lower temperatures or lower loading rates would be an excellent avenue to explore. This would aid in wider industry adoption and overall nutrient removal with a smaller footprint.

In addition, new avenues of research need to be devised for low cost, smaller areal footprints, and treatment of low nutrient load wastewater. This all may be possible through a better understanding of the metabolic pathways whereby the microorganisms involved remove the nutrients or by better understanding how to improve biomass increase or the granulation of these organisms.

15 Recommendations

From a legislation or policy perspective a level of care must be followed with the assumption being release levels for treated wastewater should be as stringent as technologically possible with the allowable levels constantly being reassessed as technology improves. This holds true for all contaminants and pollutants found in wastewaters. The reason being many of the novel and emerging pollutants, as well as the known and treated for ones such as nutrients, have ecological ramifications that may not become apparent for years if not decades. There is a duty of care required at the policy level to ensure that the natural systems we rely on are, at a minimum, maintained for future generations at today's level of functioning.

At a research, planning, or upgrading of WWTPs and techniques level more emphasis needs to be placed on the minimization of costs, especially space and energy requirements, while increasing overall efficiencies. As our population continues to grow the availability of space and energy resources are going to deplete as our usable water requirements increases. This means all levels of WWT are going to have to do more with less. A possible method of doing so is by focussing on ways to improve natural and biological systems by making them more robust under a variety of conditions and with faster start up times. Natural wetlands show us how a wide variety of microbial communities may function as a purification system. The goal moving forward should be how to decrease the space requirements necessary to achieve those levels of purification.

At the same time all levels of stakeholders should be seeking ways of treating wastewater as a resource and its treatment as a method of resource extraction. Whether this means through the seeking of added value substances such as the creation of chemicals, fertilizers, food additives, etc. that can be sold or if it means the onsite reuse of those resources such as the biogas for heat or electricity generation. The purpose should be to minimize resource wastage while offsetting treatment costs.

References

1. Abbass JK, Yuan Q (2017) Long-term stability and nutrient removal efficiency of aerobic granules at low organic loads. Bioresour Technol 234:336–342. https://doi.org/10.1016/j.biortech.2017.03.057
2. Abdel-Raouf N, Al-Homaidan AA, Ibraheem IBM (2012) Microalgae and wastewater treatment. Saudi J Biol Sci 19:257–275. https://doi.org/10.1016/j.sjbs.2012.04.005
3. Abdelfattah A, Ali SS, Ramadan H et al (2023) Microalgae-based wastewater treatment: mechanisms, challenges, recent advances, and future prospects. Environ Sci Ecotechnology 13:100205. https://doi.org/10.1016/j.ese.2022.100205
4. Adav SS, Lee D-J, Show K-Y, Tay J-H (2008) Aerobic granular sludge: recent advances. Biotechnol Adv 26:411–423. https://doi.org/10.1016/j.biotechadv.2008.05.002
5. Ahn Y-H (2006) Sustainable nitrogen elimination biotechnologies: a review. Process Biochem 41:1709–1721. https://doi.org/10.1016/j.procbio.2006.03.033
6. Appleton AF (2002) How New York City used an ecosystem services strategy carried out through an urban-rural partnership to preserve the pristine quality of its drinking water and save billions of dollars and what lessons it teaches about using ecosystem services. Tokyo
7. Asri M, El Abed S, Saad I, El Ghachtouli N (2018) Biofilm-based systems for industrial wastewater treatment
8. Bora A, Thondi Rajan AS, Ponnuchamy K et al (2024) Microalgae to bioenergy production: recent advances, influencing parameters, utilization of wastewater—a critical review. Sci Total Environ 946:174230. https://doi.org/10.1016/j.scitotenv.2024.174230
9. Buitrón G, Muñoz-Páez KM, Quijano G et al (2020) Biohydrogen production from winery effluents: control of the homoacetogenesis through the headspace gas recirculation. J Chem Technol Biotechnol 95:544–552. https://doi.org/10.1002/jctb.6263
10. Canals J, Cabrera-Codony A, Carbó O et al (2024) Nutrients removal by high-rate activated sludge and its effects on the mainstream wastewater treatment. Chem Eng J 479:147871. https://doi.org/10.1016/j.cej.2023.147871
11. Chan PC, Lu Q, de Toledo RA et al (2019) Improved anaerobic co-digestion of food waste and domestic wastewater by copper supplementation—microbial community change and enhanced effluent quality. Sci Total Environ 670:337–344. https://doi.org/10.1016/j.scitotenv.2019.03.081
12. Cooke JG (1992) Phosphorus removal processes in a wetland after a decade of receiving a sewage effluent. J Environ Qual 21:733–739. https://doi.org/10.2134/jeq1992.00472425002100040032x
13. Dutta A, Sarkar S (2015) Sequencing batch reactor for wastewater treatment: recent advances. Curr Pollut Rep 1:177–190. https://doi.org/10.1007/s40726-015-0016-y
14. Díaz V, Leyva-Díaz JC, Almécija MC et al (2022) Microalgae bioreactor for nutrient removal and resource recovery from wastewater in the paradigm of circular economy. Bioresour Technol 363:127968. https://doi.org/10.1016/j.biortech.2022.127968
15. Ekama GA, Wentzel MC, Van Loosdrecht MC (2020) Nitrogen removal. In: Chen G, Ekama GA, Van Loosdrecht MCM, Brdjanovic D (eds) Biological wastewater treatment: principles, modeling and design, 2nd edn. IWA Publishing, pp 161–238
16. Eregowda T, Kokko ME, Rene ER et al (2021) Volatile fatty acid production from Kraft mill foul condensate in upflow anaerobic sludge blanket reactors. Environ Technol 42:2447–2460. https://doi.org/10.1080/09593330.2019.1703823
17. Fudala-Ksiazek S, Luczkiewicz A, Fitobor K, Olanczuk-Neyman K (2014) Nitrogen removal via the nitrite pathway during wastewater co-treatment with ammonia-rich landfill leachates in a sequencing batch reactor. Environ Sci Pollut Res 21:7307–7318. https://doi.org/10.1007/s11356-014-2641-1
18. GARCÍA J, ROUSSEAU DPL, MORATÓ J et al (2010) Contaminant removal processes in subsurface-flow constructed wetlands: a review. Crit Rev Environ Sci Technol 40:561–661. https://doi.org/10.1080/10643380802471076

19. Ge H, Batstone DJ, Keller J (2013) Operating aerobic wastewater treatment at very short sludge ages enables treatment and energy recovery through anaerobic sludge digestion. Water Res 47:6546–6557. https://doi.org/10.1016/j.watres.2013.08.017
20. Giesen A, de Bruin LMM, Niermans RP, van der Roest HF (2013) Advancements in the application of aerobic granular biomass technology for sustainable treatment of wastewater. Water Pract Technol 8:47–54. https://doi.org/10.2166/wpt.2013.007
21. Gonzalez-Martinez A, Muñoz-Palazon B, Rodriguez-Sanchez A, Gonzalez-Lopez J (2018) New concepts in anammox processes for wastewater nitrogen removal: recent advances and future prospects. FEMS Microbiol Lett 365. https://doi.org/10.1093/femsle/fny031
22. Gowd SC, Mehta K, Ganeshan P et al (2024) Microalgae as a single-pot system for nutrient removal and wastewater treatment: comparison of effluents and species performance. Clean Technol Environ Policy. https://doi.org/10.1007/s10098-024-02808-z
23. Gu AZ, Liu L, Onnis-Hayden A, et al (2014) Phosphorus fractionation and removal in wastewater treatment—implications for minimizing effluent phosphorus. IWA Publishing
24. Gupta SK, Bux F (eds) (2019) Application of microalgae in wastewater treatment: volume 2. biorefinery approaches of wastewater treatment. Springer International Publishing, Cham
25. Hamza R, Rabii A, Ezzahraoui F et al (2022) A review of the state of development of aerobic granular sludge technology over the last 20 years: full-scale applications and resource recovery. Case Stud Chem Environ Eng 5:100173. https://doi.org/10.1016/j.cscee.2021.100173
26. Hasan MN, Altaf MM, Khan NA et al (2021) Recent technologies for nutrient removal and recovery from wastewaters: a review. Chemosphere 277:130328. https://doi.org/10.1016/j.chemosphere.2021.130328
27. Hellinga C, Schellen AAJC, Mulder JW et al (1998) The sharon process: an innovative method for nitrogen removal from ammonium-rich waste water. Water Sci Technol 37:135–142. https://doi.org/10.1016/S0273-1223(98)00281-9
28. Hu Z, Wentzel MC, Ekama GA (2003) Modelling biological nutrient removal activated sludge systems—a review. Water Res 37:3430–3444. https://doi.org/10.1016/S0043-1354(03)00168-4
29. James SN, Vijayanandan A (2023) Recent advances in simultaneous nitrification and denitrification for nitrogen and micropollutant removal: a review. Biodegradation 34:103–123. https://doi.org/10.1007/s10532-023-10015-8
30. Janssen PMJ, Meinema K, van der Roest HF (2002) Biological phosphorus removal. IWA Publishing
31. Jolis D, Mitch AA, Marneri M, Ho C-F (2007) Effects of anaerobic selector hydraulic retention time on biological foam control and enhanced biological phosphorus removal in a pure-oxygen activated sludge system. Water Environ Res 79:472–478. https://doi.org/10.2175/106143006X115381
32. Kermani M, Bina B, Attar H et al (2008) Application of moving bed biofilm process for biological organics and nutrients removal from municipal wastewater. Am J Environ Sci 4. https://doi.org/10.3844/ajessp.2008.675.682
33. Kosgey K, Zungu PV, Bux F, Kumari S (2022) Biological nitrogen removal from low carbon wastewater. Front Microbiol 13
34. Land M, Granéli W, Grimvall A et al (2016) How effective are created or restored freshwater wetlands for nitrogen and phosphorus removal? A systematic review. Environ Evid 5:9. https://doi.org/10.1186/s13750-016-0060-0
35. Lim SJ, Kim T-H (2014) Applicability and trends of anaerobic granular sludge treatment processes. Biomass Bioenergy 60:189–202. https://doi.org/10.1016/j.biombioe.2013.11.011
36. Lin W, Lin W, Cheng X et al (2021) Microbially induced desaturation and carbonate precipitation through denitrification: a review. Appl Sci 11:7842. https://doi.org/10.3390/app11177842
37. Liu G, Wang J (2017) Enhanced removal of total nitrogen and total phosphorus by applying intermittent aeration to the Modified Ludzack-Ettinger (MLE) process. J Clean Prod 166:163–171. https://doi.org/10.1016/j.jclepro.2017.08.017

38. Loizia P, Neofytou N, Zorpas AA (2019) The concept of circular economy strategy in food waste management for the optimization of energy production through anaerobic digestion. Environ Sci Pollut Res 26:14766–14773. https://doi.org/10.1007/s11356-018-3519-4
39. Lopez CM, Wentzel MC, Comeau Y et al (2023) Chapter 6 Enhanced biological phosphorus removal. In: Biological wastewater treatment: principles, modelling and design, 2nd edn. IWA Publishing, pp 239–326
40. Magdum S, Kalyanraman V (2017) Existing biological nitrogen removal processes and current scope of advancement. Res J Chem Environ 21:43–53
41. Mahlangu D, Mphahlele K, De Paola F, Mthombeni NH (2024) Microalgae-Mediated biosorption for effective heavy metals removal from wastewater: a review. Water 16:718. https://doi.org/10.3390/w16050718
42. Mahmod SS, Azahar AM, Tan JP et al (2019) Operation performance of up-flow anaerobic sludge blanket (UASB) bioreactor for biohydrogen production by self-granulated sludge using pre-treated palm oil mill effluent (POME) as carbon source. Renew Energy 134:1262–1272. https://doi.org/10.1016/j.renene.2018.09.062
43. Mainardis M, Buttazzoni M, Goi D (2020) Up-Flow anaerobic sludge blanket (UASB) technology for energy recovery: a review on state-of-the-art and recent technological advances. Bioengineering 7:43. https://doi.org/10.3390/bioengineering7020043
44. Mannina G, Bosco Mofatto PM, Cosenza A et al (2024) The effect of aeration mode (intermittent vs. continuous) on nutrient removal and greenhouse gas emissions in the wastewater treatment plant of Corleone (Italy). Sci Total Environ 924:171420. https://doi.org/10.1016/j.scitotenv.2024.171420
45. Metting FB (1996) Biodiversity and application of microalgae. J Ind Microbiol 17:477–489. https://doi.org/10.1007/BF01574779
46. Miranda ST, de Matos AT, de Matos MP et al (2019) Influence of the substrate type and position of plant species on clogging and the hydrodynamics of constructed wetland systems. J Water Process Eng 31:100871. https://doi.org/10.1016/j.jwpe.2019.100871
47. Moreno Dávila IMM, Tamayo Ordoñez MC, Morales Martínez TK et al (2020) Effect of fermentation time/hydraulic retention time in a UASB reactor for hydrogen production using surface response methodology. Int J Hydrog Energy 45:13702–13706. https://doi.org/10.1016/j.ijhydene.2019.12.137
48. Mulder JW, Duin O, Goverde AJ et al (2006) Full-Scale experience with the sharon process through the eyes of the operators. Proc Water Environ Fed 2006:5256–5270. https://doi.org/10.2175/193864706783763444
49. Murshid S, Antonysamy A, Dhakshinamoorthy G et al (2023) A review on biofilm-based reactors for wastewater treatment: recent advancements in biofilm carriers, kinetics, reactors, economics, and future perspectives. Sci Total Environ 892:164796. https://doi.org/10.1016/j.scitotenv.2023.164796
50. NYS DEC New York City Water Supply—NYDEC. In: N. Y. City Water Supply. http://www.dec.ny.gov/nature/waterbodies/lakes-rivers/new-york-city-water-supply. Accessed 4 Nov 2023
51. Nichols DS (1983) Capacity of natural wetlands to remove nutrients from wastewater. J Water Pollut Control Fed 55:495–505
52. Nnaji CO, Jeevanandam J, Chan YS, Danquah MK, Pan S, Barhoum A (2018) Engineered nanomaterials for wastewater treatment: current and future trends. In: Barhoum A, Maklouf ASH (eds.), Fundamentals of Nanoparticles. Elsevier, Amsterdam, The Netherlands, pp 129–168
53. Orhon D (2015) Evolution of the activated sludge process: the first 50 years. J Chem Technol Biotechnol 90:608–640. https://doi.org/10.1002/jctb.4565
54. Parde D, Patwa A, Shukla A et al (2021) A review of constructed wetland on type, treatment and technology of wastewater. Environ Technol Innov 21:101261. https://doi.org/10.1016/j.eti.2020.101261
55. Piotrowski R, Sawicki H, Żuk K (2021) Novel hierarchical nonlinear control algorithm to improve dissolved oxygen control in biological WWTP. J Process Control 105:78–87. https://doi.org/10.1016/j.jprocont.2021.07.009

56. Pronk M, Giesen A, Thompson A et al (2017) Aerobic granular biomass technology: advancements in design, applications and further developments. Water Pract Technol 12:987–996. https://doi.org/10.2166/wpt.2017.101
57. Pronk M, de Kreuk MK, de Bruin B et al (2015) Full scale performance of the aerobic granular sludge process for sewage treatment. Water Res 84:207–217. https://doi.org/10.1016/j.watres.2015.07.011
58. Pronk M, van Dijk EJH, Van Loosdrecht MCM (2020) Aerobic granular sludge. In: Chen G, Ekama GA, Van Loosdrecht MCM, Brdjanovic D (eds) Biological wastewater treatment: principles, modeling and design, 2nd edn. IWA Publishing, pp 497–522
59. Rahimi S, Modin O, Mijakovic I (2020) Technologies for biological removal and recovery of nitrogen from wastewater. Biotechnol Adv 43:107570. https://doi.org/10.1016/j.biotechadv.2020.107570
60. Rajendran K, Mahapatra D, Venkatraman AV, Muthuswamy S, Pugazhendhi A (2020) Advancing anaerobic digestion through two-stage processes: current developments and future trends. Renew Sustain Energy Rev 123:109746. https://doi.org/10.1016/j.rser.2020.109746
61. Reinoso R, Torres LA, Bécares E (2008) Efficiency of natural systems for removal of bacteria and pathogenic parasites from wastewater. Sci Total Environ 395:80–86. https://doi.org/10.1016/j.scitotenv.2008.02.039
62. Sarma SJ, Tay JH (2018) Aerobic granulation for future wastewater treatment technology: challenges ahead. Environ Sci Water Res Technol 4:9–15. https://doi.org/10.1039/C7EW00148G
63. Shahot A, Idris A, Omar R, Yusoff HM (2014) Review on biofilm processes for wastewater treatment. Life Sci J 11:1–13. https://doi.org/10.7537/marslsj111114.01
64. Sim CH, Yusoff MK, Shutes B et al (2008) Nutrient removal in a pilot and full scale constructed wetland, Putrajaya city, Malaysia. J Environ Manage 88:307–317. https://doi.org/10.1016/j.jenvman.2007.03.011
65. Song Y, Wang L, Qiang X et al (2022) The promising way to treat wastewater by microalgae: approaches, mechanisms, applications and challenges. J Water Process Eng 49:103012. https://doi.org/10.1016/j.jwpe.2022.103012
66. Srivastava G, Kazmi AA (2023) Intricacies of enhanced biological phosphorus removal in full-scale sequencing batch reactors based on wastewater treatment plants of India. J Environ Eng 149:04022082. https://doi.org/10.1061/(ASCE)EE.1943-7870.0002075
67. Sukla LB, Subudhi E, Pradhan D (eds) (2019) The role of microalgae in wastewater treatment. Springer Singapore, Singapore
68. Tanner CC, Sukias JP (1995) Accumulation of organic solids in gravel-bed constructed wetlands. Water Sci Technol 32:229–239. https://doi.org/10.2166/wst.1995.0145
69. Tchobanoglous G, Burton FL, Stensel HD (2003) Wastewater engineering: treatment and reuse—ProQuest. Am Water Works Assoc 95
70. Third KA, Sliekers AO, Kuenen JG, Jetten MSM (2001) The CANON system (completely autotrophic nitrogen-removal over nitrite) under ammonium limitation: interaction and competition between three Groups of Bacteria. Syst Appl Microbiol 24:588–596. https://doi.org/10.1078/0723-2020-00077
71. USEPA, OW, SCIB et al (2019) Side stream nutrient removal fact sheet
72. van der Roest HF, de Bruin LMM, Gademan G, Coelho F (2011) Towards sustainable waste water treatment with Dutch Nereda® technology. Water Pract Technol 6:wpt2011059. https://doi.org/10.2166/wpt.2011.059
73. Vellaiyan S (2024) An integrated approach for wastewater treatment and algae cultivation: Nutrient removal analysis, biodiesel extraction, and energy and environmental metrics enhancement. J Environ Manage 354:120410. https://doi.org/10.1016/j.jenvman.2024.120410
74. Vu MT, Duong HC, Wang Q et al (2023) Recent technological developments and challenges for phosphorus removal and recovery toward a circular economy. Environ Technol Innov 30:103114. https://doi.org/10.1016/j.eti.2023.103114
75. Vymazal J (2007) Removal of nutrients in various types of constructed wetlands. Sci Total Environ 380:48–65. https://doi.org/10.1016/j.scitotenv.2006.09.014

76. Wang D, Li X, Yang Q et al (2008) Biological phosphorus removal in sequencing batch reactor with single-stage oxic process. Bioresour Technol 99:5466–5473. https://doi.org/10.1016/j.biortech.2007.11.007
77. Wei D, Si W, Zhang Y et al (2012) Aerobic granulation and nitrogen removal with the effluent of internal circulation reactor in start-up of a pilot-scale sequencing batch reactor. Bioprocess Biosyst Eng 35:1489–1496. https://doi.org/10.1007/s00449-012-0738-9
78. Winkler M-KH, Meunier C, Henriet O et al (2018) An integrative review of granular sludge for the biological removal of nutrients and recalcitrant organic matter from wastewater. Chem Eng J 336:489–502. https://doi.org/10.1016/j.cej.2017.12.026
79. Wollmann F, Dietze S, Ackermann J-U et al (2019) Microalgae wastewater treatment: biological and technological approaches. Eng Life Sci 19:860–871. https://doi.org/10.1002/elsc.201900071
80. Xiang Y, Zhou T, Deng S et al (2023) Nitrite improved nitrification efficiency and enriched ammonia-oxidizing archaea and bacteria in the simultaneous nitrification and denitrification process. Water Res X 21:100204. https://doi.org/10.1016/j.wroa.2023.100204
81. Yin Q, Ren H, Wang R, Zhao Z (2018) Evaluation of nitrate and phosphate adsorption on Al-modified biochar: influence of Al content. Sci Total Environ 631:895–903
82. Zhang A, Fang S, Xi H et al (2023) Highly efficient and selective removal of phosphate from wastewater of sea cucumber aquaculture for microalgae culture using a new adsorption-membrane separation-coordinated strategy. Front Environ Sci Eng 17:120. https://doi.org/10.1007/s11783-023-1720-2
83. Zhang Q, Huang R, Jiang L et al (2021) Enhancing nitrogen removal and reducing aeration energy for wastewater treatment with intermittent modified Ludzack-Ettinger process: a field demonstration. J Water Process Eng 43:102303. https://doi.org/10.1016/j.jwpe.2021.102303
84. Zhao J, He Q, Chen N et al (2020) Denitrification behavior in a woodchip-packed bioreactor with gradient filling for nitrate-contaminated water treatment. Biochem Eng J 154:107454. https://doi.org/10.1016/j.bej.2019.107454

An Overview Towards Mitigating the Dissemination of Antibiotic Resistance Genes Through Implementing Advanced Oxidation Processes (AOPs) for Wastewater Treatment

Achinta Bordoloi and Gangadhar Andaluri

Abstract Antibiotic resistance genes (ARGs) as emerging contaminants (ECs) of concern, has raised alarming health concerns globally across geographical locations. Municipal wastewater treatment plants (WWTPs) have emerged as a significant cause for these ARGs and associated antibiotic-resistant bacteria (ARBs). WWTPs receives an influx of antibiotics from a plethora of anthropogenic sources ranging from municipal wastes, livestock wastes, hospitals, and pharmaceutical manufacturing industry. Conventional treatment methods have shown limited capabilities in effectively removing these emerging contaminants ranging from antibiotic residues, active pharmaceutical ingredients (APIs) and ARGs. It puts selection pressure on the proliferating microbes to develop antimicrobial resistance (AMR) and aid in their distribution through the system into the environment. Various disinfection methods have been implemented to remove ARGs in WWTPs such as chlorination and UV to eliminate microbial loads in wastewater, but ARGs have been shown to be persistent and survive these conventional treatments. In recent times, advanced oxidation processes (e.g. Ozone, UV/H_2O_2, Fenton) have gained attention due to their increased efficacies in ARG abatement, albeit showing variable results. However, adequate data on the effect of influencing factors on the occurrence, dissemination, and fate of ARGs is lacking in the literature. Therefore, the key objective of this review chapter is to provide a comprehensive overview of the fate of ARGs in WWTP systems and discuss the influence of operational factors and the efficacy of treatment technologies regarding its mitigating potential. In addition, current challenges pertaining to AOPs are discussed, and key knowledge gaps are suggested for future research which could

A. Bordoloi (✉) · G. Andaluri
Department of Civil and Environmental Engineering, Temple University, Philadelphia, PA 19122, USA
e-mail: achintab@gmail.com; achinta.bordoloi@temple.edu

G. Andaluri
e-mail: gangadhar@temple.edu

S. Yadav et al. (eds.), *Emerging Technologies in Biological and Hybrid Wastewater Treatment: Lessons from Developed to Enhancing Practices in Developing Countries*, Earth and Environmental Sciences Library,
https://doi.org/10.1007/978-3-031-74515-7_17

aid in better understanding the impact of influencing factors on the ARGs fate during implementation of various AOPs.

Keywords Antibiotic resistance genes (ARGs) · Antibiotic resistant bacteria (ARBs) · Anti-microbial resistance (AMR) · Advanced oxidation processes (AOPs) · Wastewater treatment

1 Introduction

The rapid spread of antimicrobial resistance (AMR) is an emerging cause of global health concerns. AMR has been linked as a significant cause for 1.2 million deaths in 2019 [48]. The problem is further intensified by the rise of superbugs, often harboring resistance against multiple critical drugs of clinical importance [6]. This can lead to serious consequences as a potential 10 million annual deaths from these resistant bugs is projected by 2050 unless adequate measures are employed to curb the spread of AMR [35]. Wastewater treatment plants (WWTPs) have emerged as a key hotspot harboring and facilitating distribution of these ARGs and ARBs [15, 38, 56]. In addition to the spread of ARGs through effluents, WWTPs divert the significant quantity of generated sludge through anaerobic digestion (AD) processes which has been shown to hold substantial ARG load that disseminate throughout the treatment process [54].

While WWTPs are pivotal for control and degradation of these ARGs, they were not traditionally designed to treat these kinds of emerging contaminants of concerns. The conventional biological treatment processes have limited capabilities in removing these ARGs [37]. In addition, conventional disinfection methods such as UV and chlorination have shown limited, albeit variable effectiveness in the removal of these ARGs. These ARGs are persistent and routinely found in numerous wastewater effluents and other aquatic bodies [36, 76]. Recently, various AOP based treatment methods have shown promising results in eliminating the ARBs and the more persistent ARGs which could help curb the spread of AMR in the environment [13, 39, 41, 55]. Therefore, it is essential to improve the elimination efficacies by incorporating AOPs for water treatment to effectively eliminate and impede the dissemination of AMR through the effluent discharges.

This section reviews various emerging water treatment technologies and compares their efficacies to some conventional methods. It also analyses potential of sequential combination treatments in terms of enhanced abetment and degradation of ARGs. This review chapter further discusses the potential of these advanced technologies, identified knowledge gaps and laid down a framework for adequate implementation of these technologies for curbing the proliferation of AMR.

2 Occurrence and Dissemination of ARGs

The rapid use of antibiotics for healthcare has caused its increased abundance in the environment as detected by their presence in soil, surface water, groundwater and WWTPs. Oftentimes, WWTPs serves as a central accumulation point for these residual antibiotics from various anthropogenic sources [51]. A significant portion of administered antibiotics (10–70%) for humans and livestock are released into the treatment-influent streams in unmetabolized forms [62]. The persistence of sublethal residual antibiotics concentrations, nutrients and abundance of microbial loads present an ideal microcosm for developing resistance and their subsequent proliferation in WWTPs. It puts selection pressure on the microbes to acquire resistance and leading to further proliferation of AMR in the environment [59]. Furthermore, presence of other abiotic stressors such as biocides and heavy metals also influences the *in-situ* transfer of resistance in the natural environment. Bacteria develops resistance via various intrinsic mechanisms as well as by acquiring resistance genes from plasmids and associated mobile genetic elements [11]. Usually, resistance is mediated through chromosomal mutations and proliferated to the progeny known as vertical gene transfer. The other *in-situ* mechanism which leads to rapid distribution of ARGs is through horizontal gene transfer (HGT) where competent microbes take up ARG harboring free genetic elements from the environment [68]. HGT is primarily mediated through conjugation, transduction and transformation mechanisms [5]. Detection of resistance genes for commonly used antibiotics such as tetracycline, sulfonamide and other class such as quinolones across various water bodies is a cause for concern [30, 45]. Figure 1 outlines the different mode of action and potential resistance mechanisms deployed by bacteria against different class of antibiotics and their eventual dissemination into various water bodies.

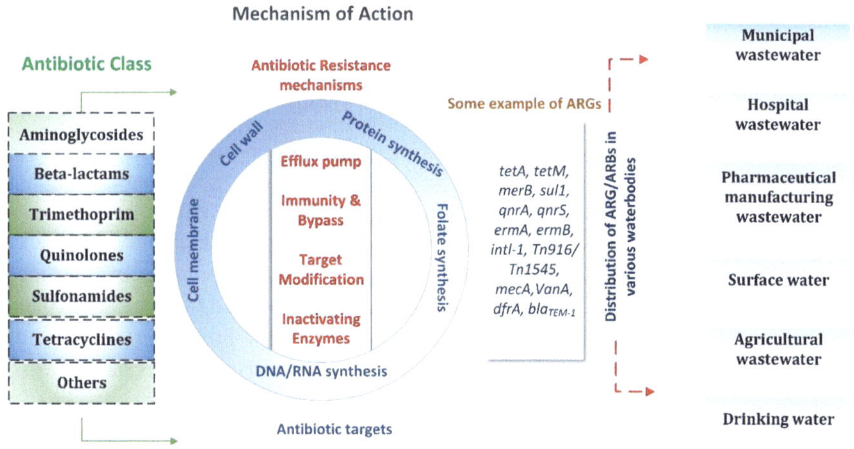

Fig. 1 Illustration of the flow of antibiotic resistance acquiring mechanisms to eventual spread of ARGs through aquatic bodies

3 Wastewater Treatment Technologies

3.1 Ultraviolet Light (UV)

UV treatment has been used as a promising tertiary disinfection method in WWTPs for effectively eliminating pathogens, including ARBs albeit showing variable results for complete removal of ARGs [44]. UV lights penetrate the cells and induce DNA damage leading to cell lysis and death with no disinfection by-products (DBPs) release [18]. UV mediated DNA damage occurs primarily via two pathways—generation of the cyclobutene-pyrimidine dimers (CPDs) and (6-4) photoproduct formation [9, 44, 72]. It is most potent at a wavelength of 254 nm leading to maximum absorption of UV radiation by the DNA [7]. Adjacent TT dimer in the amplicon sequence is the most active site for UV induced lesion and has been strongly correlated to ARG degradation by UV treatment [9, 50]. Elimination of both ARGs and ARBs from WW effluents via UV disinfection has been reported [7, 16, 27, 31]. UV treatment showed between 1 and 4 log reduction of ARGs from wastewater and required much higher UV dosage for degrading ARGs compared to that of ARBs [12, 49, 71]. However bacteria have shown repair capabilities of damages caused during UV treatments via photoreactivation and dark repair mechanisms which can decrease the effectiveness of UV disinfection for mitigating ARBs [26]. In addition, the persistence of residual ARGs after UV treatment makes it susceptible to transfer these viable residual ARG fragments to competent microbes in the environment, thus posing further AMR dissemination risks [12]. Wastewater matrix components such as the presence of humic acids, and effluent organic matter can significantly impact the UV induced elimination efficiency of genes and microbes [49]. More studies integrating UV mediated AOPs such as UV/H_2O_2 and UV-photo Fenton needs to be evaluated for enhanced removal of resistant genes and bacteria from spreading in the environment.

3.2 Chlorination

Chlorine has been used as a disinfect in water treatment for a long time. Freely available chlorine is present as HOCl with good oxidizing potential which directly attacks DNA and associated intracellular proteins and cell wall components in microbes [18]. While it is effective as a disinfectant for microbes, limited efficiency has been shown for deactivation of ARGs by chlorination. Around 40–80% of erythromycin-resistant genes (*ereA, ereB, ermA, ermB*) and tetracycline-resistant genes (*tet-A, tet-B, tet-M, tet-O*) persisted in wastewater after chlorination showing ineffectiveness in degrading these ARGs [74]. Some studies have reported a relative i-ARGs concentration increase in the water matrix after chlorine treatment and suggested that chlorine treatment can help select resistant bacteria and promote conjugative transfer of ARGs

[25, 64]. Additionally, usage of higher chlorine dosages may lead to the creation of toxic disinfection by-products (DBPs) like halo-acetic acids, tri-halomethanes [42].

3.3 *Advanced Oxidation Processes (AOPs) Used for Wastewater Treatments*

AOPs are regarded as viable and promising advanced treatment technologies for treating and eliminating contaminants of emerging concerns such as ARGs and ARBs as opposed to limited efficacies shown by traditionally implemented methods such as chlorination and UV treatment [29]. They act via generation of strong radicals (e.g. $\cdot OH$, SO_4^-,) which non-selectively attack and effectively removes a range of organic contaminants [30]. Of late, they are gaining increasing traction for the removal of emerging biobased contaminants of concerns such as ARGs and ARBs much more effectively compared to conventional methods [22, 58]. This section outlines and discusses about some currently used AOP treatments for the abatement of ARGs and ARBs from water sources.

3.3.1 Ozonation

Ozone has a strong oxidation potential and instantly reacts with a range of compounds and microbes including intracellular genes and cellular components [43]. Ozonation works via two pathways, through direct oxidation reaction mediated by O_3 and generation of radicals such as $\cdot OH$, superoxide ($\cdot O_2^-$) and hydroperoxyl ($HO_2\cdot$) mediated by the pH of the matrix [21, 52]. $\cdot OH$ is also a strong oxidant (reduction potential of 2.8 V) compared to O_3 (2.07 V) and reacts non-selectively with genes and causes strand fragmentation [21]. Ozone reacts with a range of inorganic compounds by oxygen transfer mechanisms and organic compounds by attacking double bonds and functional groups [18, 47]. Ozone at higher concentration can eliminate the ARBs and can diffuse into the cells to degrade the intracellular ARGs (iARGs) as well as the extracellular ARGs (eARGs) [13]. It degrades cell walls leading to leaking of cytoplasmic constituents causing cell lysis/death as well as structural damage to DNA inside the cells [66]. Some dissolved organic matter or floc can impede the efficiency of ozone treatment by reducing ozone stability and protect microorganisms from its direct impact [10]. However, ozone treatment has shown good results in degrading ARGs from a variety of water matrices. Table 1 compiles various literature studies investigating ozone treatment for ARB and ARG removal from water matrices. Up to 1–5 log removal of different ARB species such as *E. coli*, *P. aeruginosa* and *Staphylococus* have been reported [4, 66, 67]. Yoon et al [73] reported 1–4 log removal of plasmid encoded ampicillin resistant genes (amp^R). Furthermore, up to 5 log removal from different ARG types, forms and amplicon lengths were reported [13, 28]. Choi et al [9] reported > 6 log reduction of chromosomal encoded *mecA*

ARG in *Staphylococcus aureus* with varying degradation rates for intracellular and extracellular ARGs. Although, a few studies have also reported limited removal potential for ARGs via ozonation [10, 77]. While variable, albeit effective results for the degradation of ARGs have been reported attributing it to be dependent on factors like different ARG forms, types and the nature of the water matrix [13, 76]. Although, ozonation at higher dosages does produce DBPs, including some with carcinogenic traits [53]. Thus, further investigation would be needed in the future with actual wastewater samples to adequately assess implementation and optimize process parameters.

3.3.2 Fenton

Fenton oxidation process is mediated via (Fe^{2+}/H_2O_2) by generating ·OH radicals that can non-selectively destroy organic contaminants [29]. Fenton process has shown effectiveness in inactivating ARBs and degradation of ARGs in wastewater. The process efficacy is dependent on a range of operating parameters ranging from Fe^{2+}/H_2O_2 molar ratio, H_2O_2 concentration, pH and exposure time [75]. Due to operation in the acidic pH range, treated effluents often need a neutralization step before discharge. In the photo-Fenton process, ·OH radical generation is accelerated via irradiation with UV light which has shown complete inactivation of various ARBs [61]. The UV light mediated irradiation requirement can be supplemented by a more cost-effective solar-Fenton process which has also shown significant ARB and ARG removal potential [47]. Table 2 compiles various literature studies investigating Fenton oxidation treatment for ARB and ARG removal from water matrices. Some limitations of homogenous Fenton processes are generation of iron rich sludge and low pH operation requirements which increases operation costs. Heterogenous Fenton process has gained attention which can alleviate these limitations via use of zerovalent iron (nZVI), iron oxide and iron sulfide enabling operation at neutral pH [2]. Ahmed et al [2] reported 6–7 log reduction of *E. coli* and associated extracellular ARGs (*tetA*, $bla_{TEM\text{-}1}$) in 7.5 min, although intracellular ARG degradation was comparatively lower at 3.9–4.2 log reduction taking much longer contact time of 2 h. Overall, Fenton processes offers an effective approach using AOPs to eliminate ARGs and ARBs to curb the dissemination of AMR in the environment.

3.3.3 UV/H_2O_2

UV based AOPs such as UV/H_2O_2 works via generation of ·OH radicals upon irradiation by UV light wavelength primarily < 300 $_{nm}$ [58]. Extent of DNA degradation from direct UV treatment is caused by producing lesions which is dependent on the nature and composition of the base sequence of the genes [18]. TT dimers are the most active sites for UV induced damage [28]. While additional ·OH radicals can often enhance degradation rates due to lack of selectivity in attacking genes, inflicting DNA damage via oxidation of base sequences and causing strand breakage

Table 1 Application of ozone treatment for removal of ARBs and ARGs

Types and forms of ARBs/ ARGs	Operating conditions	Removal of ARG and ARB	Reference
E. coli and plasmid encoded *sul1* gene	Batch operation, Ozone dose: 0.02–0.2 mgO_3L^{-1}, contact time > 30 min	• 4 log reduction of *sul1* and *16 s rRNA* • Inactivation of *E. coli* and loss of membrane integrity	[10]
Antibiotic resistant *enterococci* and *pseudomonas* *ARGs: vanA, ampC, ecfX, mecA, ermB, blaVIM*	Ozone dose: 0.9 g Ozone. $gDOC^{-1}$,	• Degradation varied across different types of ARGs, some more robust • ARBs were reduced by 2 logs with different coping mechanisms observed for ARBs	[4]
Cultivable antibiotic resistant fungal and bacterial spp., ARGs: *vanA, blaTEM, sul1, qnrS and intl1*	Bench scale, Ozone dose: 50 g N m^{-3}	• 4 log reduction of ARBs • ARG removal ranged from 2 log to below LOD	[66]
E. coli and *enterococcus faecium*, ARGs: *sul1, sul2, ermB, ampC, tetA*,	Batch set-up, Ozone dose: 0–13 mgL^{-1}, exposure time: 5 min	• 5 log removal of ARBs • 4.3–4.6 log removal of ARGs	[67]
Antibiotic-resistant *E. coli*, ARGs: *intl1,aadA1, dfrA1,qacE∇1, sul1*	Continuous ozonation, HRT-10- 60 min, Ozone dose: 0.25–0.75 gO_3gDOC^{-1},	• Higher ozone dose required to degrade ARGs compared to antibiotic compounds • ARBs were totally inactivated • 2–6 log reduction of ARGs with increasing dose	[32]
Bacillus subtilis 1A189, (10^6 CFU mL^{-1}) ARG: *blt*	Ozone dose: 1 mgL^{-1}	• ARG degradation rate were correlated to gene sequence • Up to 5 log reduction of e-ARGs observed • i-ARG degradation lagged ARB inactivation	[28]

(continued)

Table 1 (continued)

Types and forms of ARBs/ ARGs	Operating conditions	Removal of ARG and ARB	Reference
pUC19 plasmid encoded ARG *(amp^R), 10^{14} copies/mL*	Bench scale, Ozone dose: (0–0.15 mgO_3 x min) at pH 7	• Longer amplicon showed 4 log reduction as opposed to 1.2 log for smaller amplicon size • Deactivation of ARGs evaluated through transformation assays varied (1–4 log reduction), depending on type of recipient *E. coli* strain used	[73]
Methicillin-resistant *Staphylococcus aureus* (MRSA) with *mecA* ARG	Batch reactor, Ozone dose: 2.6–10.4 μM for 45 s	• 5–6 log reduction of ARGs • Difference in degradation rates observed for eARG and iARG	[9]
Antibiotic-resistant *E. coli,* Chromosomal and plasmid encoded ARGs: *intl1, tetA, gyrA, qnrS*	Semi batch bubble column reactor, Ozone dose: 2 mgO_3 L^{-1}, Time 0–15 min, PBS buffer matrix, DI and WW matrix	• 3.6–5.2 log reduction of eARGs • Plasmid encoded ARGs showed 4–5 log removal • Electrical energy per order (E_{EO}) ranged between 0.68–1.1 $kWhm^{-3}$	[13]

[17, 65]. UV/H_2O_2 based process has been reported to effectively remove ARBs and ARGs to varying degrees often influenced by the nature of genes investigated and water matrix composition [12, 49, 72]. Table 3 outlines literature reports of UV/H_2O_2 treatment for eliminating resistant microbes and genes. Yoon et al. [71] reported up to 4 log removal of e-ARGs (*amp^R, kan^R*) while degradation was comparatively slower for i-ARGs due to ·OH radical scavenging by intracellular components. Das et al. [12] reported similar trends of slower removal rates (1.3–2x) for chromosomal encoded i-ARGs in comparison to e-ARGs while effectively inactivating ARBs, although showing persistence of free residual ARGs. Table 3 compiles various literature studies investigating UV/H_2O_2 treatment for ARB and ARG removal from water matrices. Overall, UV based AOPs allows for relatively easy integration into the existing UV systems but requires more investigation for adequate application in real wastewater systems to discern inhibitory effects of water matrix impeding process efficacy,

Table 2 Application of Fenton oxidation for removing ARBs and ARGs from water matrices

Types and forms of ARBs/ARGs	Operating conditions	Removal of ARG and ARB	References
sulI, tetX, tetG, intlI, 16 s rRNA	Fe^{2+}/ H_2O_2 molar ratio 0.1, H_2O_2 conc 0.01 molL^{-1}, pH 3 and 7, reaction time 2 h	2.58–3.79 log removal	[75]
E. coli (10^6–10^7 CFU mL^{-1})	Photo-Fenton:1mgL^{-1} Fe^{2+} and 10 mgL^{-1} H_2O_2, 300 s, UV 254 nm	Complete inactivation, bacteria re-growth not observed	[61]
E. coli K12 (10^9 CFU mL^{-1})	Photo-Fenton: 1mgL^{-1} Fe^{2+} and 10 mgL^{-1} H_2O_2, 120 min, solar light (620 Wm^{-2})	4-log removal	[60]
Enterococcus resistant to clarithromycin and sulfamethoxazole (10^6 CFU mL^{-1})	Photo-Fenton: 5mgL^{-1} Fe^{2+} and 50 mgL^{-1} H_2O_2, 250 min, solar light (30 Wm^{-2}), pH4, pilot-scale	5-log removal	[33]
E. coli (10^6 CFU mL^{-1}), ARG: *tetA, bla$_{TEM-1}$*	Photo-Fenton: 0.5 mM Fe^{2+} (2.8 mgL^{-1}), and 10 mM H_2O_2 (340.2 mgL^{-1}), 1:20 ratio, 60 min, solar light intensity (2 Wm^{-2}), pH 7.4	• 6 log removal of ARB within 30 min, no re-growth • 6.7- 8.6 log removal of e-ARGs • i-ARG degradation was slower and required higher Fenton reagent	[1]
E. coli (10^6 CFU mL^{-1}), ARG: *tetA, bla$_{TEM-1}$*	Modified Photo-Fenton: a) Fe(III): EDDS:H_2O_2 (0.1:0.2:0.3 mM), 30–60 min b) Solar simulated light 130 mW cm^{-2}	• 6 log removal of ARB by 30 min, no re-growth • 6 log removal of ARGs after 10 min of treatment	[3]
S. aureus, E. coli, K.pneumoniae (10^6 CFU mL^{-1}) ARGs: *bla$_{CTX-M-1}$, bla$_{CTX-M-9}$*	Photo-Fenton: 1mgL^{-1} and 10 mgL^{-1} H_2O_2, 240 min, solar light (750 Wm^{-2})	• Complete inactivation of ARB, no re-growth • Qualitative PCR amplification data showed variable ARG damage with exposure time	[22]

(continued)

Table 2 (continued)

Types and forms of ARBs/ARGs	Operating conditions	Removal of ARG and ARB	References
E. coli, P. aeruginosa and Klebsiella spp. harbouring ARGs:*ermB, sul1, mecA, ampC, Enc, ecfX*	Photo-Fenton: 5mgL^{-1} and 50 mgL^{-1} H_2O_2, 240 min, Natural solar light	• ARBs were efficiently removed • Total DNA concentration was removed by 97% • Sul1 and *ermB* ARGS were not completely removed	[34]
Plasmid pSB1A2 and pSB1K3 harbouring ARGs resistant to ampicillin and kanamycin	Photo-Fenton: 30 mgL^{-1} and 50 mgL^{-1} H_2O_2, 60 min, solar simulated light (268 Wm^{-2}) ni	• Plasmid were removed from water samples • Transformation capabilities were eliminated	[70]
E.coil containing ARGs *(tetA, bla$_{TEM-1}$)*, 10^6 CFU mL^{-1}	FeS_2@GO based Heterogenous Photo-Fenton: 0.25 gL^{-1} catalyst, 1 mM H_2O_2 (34.2 mgL^{-1}), 8:2 ration of FeS_2, 30 min, solar simulated light (130 mWcm^{-2}), pH 7.4	• 6 log removal of ARBs after 30 min • 7 log reduction of ARGs after 7 min • No re-growth observed	[2]
Pseudomonas putida MX-2, 7 ARGs encoded in chromosome *(acrB, tetA-02, sul1,mexF)* and plasmids *(tetA-01,aac6-ib, strA)*	Zero-valent iron (ZVI) activated Fenton-like process: ZVI/H_2O_2 and ZVI/PDS	• Pre-acidification was effective in inactivating ARB and promoting e-ARG degradation • In wastewater, ARB inactivation was 5.7–5.9 log while ARG degradation ranged from 2.5–2.9 logs for long amplicons • Plasmid encoded were more resistant to degradation	[46]

Table 3 Application of UV/H_2O_2 treatment of ARGs and ARBs from wastewater matrices

Types and forms of ARBs/ARGs	Operating conditions	Removal of ARG and ARB	References
Coliforms, E. coli and antibiotic resistant *E. coli* ARGs: bla_{TEM}, *qnrS, tetW*	UV dose: 0–2.5 × 10^5 μW s cm^{-2}, H_2O_2: 20 mgL^{-1} Experiments done with WW samples	• ARBs were inactivated • ARG removal was not very effective	[19]
Antibiotic resistant *E. coli* ARGs: *blaTEM, qnrS, tetW,*	UV dose: 0–2.5 × 10^4 μW s cm^{-2}, time 0–300 min, H_2O_2: 20 mgL^{-1}	• *E. coil* was inactivated • ARG removal was not very effective after 90 min	[20]
ARGs: *sul1, sul2, qnrS*, bla_{TEM}, $bla_{OXA\text{-}A}$, *intI1*	WWTP tertiary treatment: UVC dosage- (42 -170 JL^{-1}, contact time (4-18 s), Reactor vol:140 L	• Direct UV induced photolysis of DNA was still effective than radical induced damage	[58]
K. pneumoniae ARG: *blaKPC-3*	UV dose: 0–2.5 × 10^5 μW cm^{-2}, H_2O_2: 1 mM, deionized water (DI) and WW samples	• ARB was inactivated, WW sample needed longer exposure that DI water • ARGs requires higher dose than inactivating ARB	[63]
E. coli DH5α with plasmid pUC4k (10^9 CFU mL^{-1}), ARGs: amp^R and kan^R	Bench scale collimated beam, UV dose: 0.3 mW cm^{-2}, H_2O_2: 10 mgL^{-1}, PBS buffer matrix	• 4 log removal of e-ARG • i-ARG removal slower due to ·OH radical scavenging by cellular matrix	[71]
Staphylococcus aureus and Psuedomonas aeruginosa ARG: *mecA, ampC*	UV fluence: 12 mJ cm^{-2}, H_2O_2: 0–100 mmol L^{-1}, PBS matrix	• Up to 4 log removal of ARBs • > 5 log reduction of ARGs, *mecA* degraded at a higher rate than *ampC*	[23]
E. coli DH5α with pUC19 plasmid encoding amp^R ARG (10^{11} copies per mL)	Bench scale collimated beam, UV dose: 0.3 mW cm^{-2}, H_2O_2: 10 mgL^{-1}, PBS buffer matrix	• 4 log removal of amp^R • amp^R transforming activity was also reduced by 4 logs • UV dose higher than 150 mJ cm^{-2} is needed for 4 log removal	[72]

(continued)

Table 3 (continued)

Types and forms of ARBs/ARGs	Operating conditions	Removal of ARG and ARB	References
ARGs: *intl1, qnrS, tetA, sul2, and 16 s rRNA*	UV dose: 4.9 mW cm^{-2}, H_2O_2: 50 mgL^{-1}, WW sample	• Post treatment resulted in higher abundance of ARGs	[14]
Chromosomal and plasmid encoded ARGs: *intl1, qnrS, tetA, gyrA, and 16 s rRNA*	Bench scale collimated beam, UV dose: 1 mW cm^{-2}, time: 0–14 min, H_2O_2: 10 mgL^{-1}, PBS buffer matrix, DI and WW matrix	• e-ARG degradation (4 log) was significantly enhanced by radicals • WW matrix did not impact degradation, potentially generated other radical species • ARBs were inactivated but free-residual ARGs persisted	[12]

3.3.4 Photocatalysis

The use of photocatalysis for generation of free radicals for oxidative elimination of ARBs and ARGs is gaining traction due to their cost effectiveness and reuse potential of catalysts [40]. TiO_2 based photocatalytic degradation has been widely used for water treatment of organics, while also showing promising effects on the elimination of these emerging contaminants [47]. Numerous studies have also focused on doping TiO_2 with supplemental metals (CO, Ag, etc.) to improve efficacy and widen the spectrum of usable light wavelength to increase photocatalytic efficiency [69]. Table 4 compiles various literature studies investigating photocatalytic treatment for ARB and ARG removal from water matrices. Guo and Tain [24] showed reduced ARB viability with the use of TiO_2/Ag/GO composite which impacted ARBs resistance against tetracycline. They, however, observed an increased propensity towards horizontal gene transfer by 1–2 folds. UV/TiO_2 combination resulted in 4.7–5.8 log removal of *ampC* and *mecA* genes [23]. Some studies, in contrast, have reported lower removal efficiency and persistence of some ARGs [24, 56, 69]. While photocatalysis has shown promising results, future investigations tailored towards developing efficient material composites, evaluating the impact of wastewater matrix composition and increasing the elimination potential of ARGs/ARBs is critical for the broader application of this emerging treatment technology.

Table 4 Application of photocatalysis for treatment of ARGs and ARBs from WW matrices

Types and forms of ARBs/ARGs	Operating conditions	Removal of ARG and ARB	References
E. coli	Nitrogen-doped TiO_2 (0.025–0.5 gL^{-1}), simulated solar radiation, natural light	• Complete ARB deactivation was achieved by 60 min • Resistance to tetracycline and vancomycin was not significantly affected	[57]
Staphylococcus aureus and Psuedomonas aeruginosa ARG: *mecA, ampC*	UV/TiO_2/H_2O_2 (120 mJ cm^{-2} UV_{254}), H_2O_2: 0–100 mmol L^{-1}	• 5.8–4.7 log reduction of ARGs	[23]
K. pneumoniae ARGs: ampC, bla_{TEM}, sul1, tetA, tetM	Mn and Co-doped TiO_2, (0.1 wt % Mn:TiO_2, 0.1 wt% CO:TiO_2, 0.04 wt% Mn/Co:TiO_2) Solar irradiation (12.7–13.4 Wm^2),	• ARB is quickly inactivated • Treatment changed resistance profile in the ARBs, some ARGs persisted (up to 2 log reduction)	[69]
Antibiotic resistant *E. coli* HB101 spp. (108 CFU mL^{-1})	TiO_2/ Ag/GO, Time: 10 min, stimulated sunlight	• ARB was inactivated as a function of exposure time (0.8–0.9 log) • Gene transfer can be promoted under investigated conditions	[24]
Antibiotic resistant *E. coli* with *bla_{NDM-1}* ARG (10^6 CFU mL^{-1})	UV-A/TiO_2, Time: 0–240 min, TiO_2 concentration: 0–1000 mgL^{-1}, UVA intensity (0–8.7mW cm^{-2})	• ARB inactivation was up to 5 log reduction • ARG degradation was between 0.7 and 1.5 logs • Residual ARG had transformation potential	[8]

4 Recommendation/Future Perspectives

There is an urgent need to eliminate and curb the spread of AMR in the environment given its major health concern for both humans and animals. While showing promise in the lab scale, many of these AOP based technologies still needs to demonstrate scalability and cost effectiveness for practical applications. The bench scale results also need to be implemented to actual practical application at scale in WWTPs to discern matrix effect of complex wastewater samples. Focus should be also on the implementation and integration of theses treatment methods to the existing infrastructures, ideally without the need for major capital investments. In addition, establishment of a clear framework for proper assessment of removal efficacy across the board is essential for meeting the objectives of curbing and eliminating these emerging contaminants is paramount. A key factor going forward would be the need for implementing regulatory discharge limits from the governing bodies for ARGs as well as antibiotics/APIs at sub-inhibitory levels which can mitigate the spread of AMR. This would entail in targeted research approach while generating all-encompassing dataset to cover minimum inhibitory concentration for antimicrobials and levels at which selection is happening in-situ. Future research should also integrate the wet lab techniques of ARG detection via PCR with novel high throughput metagenomics to better access the totality of the ARG spread in the samples. This will facilitate more robust interpretation of the nature and potential transfer mechanisms in the environment and address the current knowledge gaps. More in-depth study of the microbial community and its correlation with harboring specific ARGs can be implemented to generate predictive models to identify potential host. In addition, more focus should be placed on understanding the in-situ gene transfer mechanism and the role of mobile genetic elements and environmental stressors contributing to the spread of AMR.

5 Conclusion

AOPs are a promising treatment option that has gained significant consideration due to improved removal potential over conventional disinfection treatments through free radical generations. AOP based processes with improved efficacy, cost-effectiveness, sustainability and lack of toxic by-products formation are highly desirable aspects for new technologies to expand its footprints. They can play a vital role in adequately eliminating ARBs and ARGs to stop the spread in the environment and mitigate critical health concerns for human and livestock associated with these contaminants. The AOP processes can be influenced by the nature and type of ARBs/ARGs, dosage, contact time and water matrices, which need robust evaluations.

Some major shortcomings are high capital associated with the consumption of chemicals and energy and translating lab-scale results to real application at full-scale WWTPs. In addition, process parameters need to be optimized (dosage, exposure time, effect of water matrix) for the complete removal of genes as well as evaluating the viability of residual ARGs. Also, the re-growth potential of treated ARBs should be completely eliminated, and the treatment should enable selection of resistant species. The fate of residual ARGs and their potential to transfer resistance to competent microbes in the environment requires proper investigation and needs to be adequately neutralized to control the risk of spreading. Overall, AOPs have been identified as a viable alternative for implementation as advanced treatment technologies for proper abatement of ECs such as ARBs and ARGs to hinder the dissemination of AMR within the environment.

References

1. Ahmed Y, Lu J, Yuan Z, Bond PL, Guo J (2020) Efficient inactivation of antibiotic resistant bacteria and antibiotic resistance genes by photo-Fenton process under visible LED light and neutral pH. Water Res 179:115878. https://doi.org/10.1016/j.watres.2020.115878
2. Ahmed Y, Zhong J, Wang Z, Wang L, Yuan Z, Guo J (2022) Simultaneous removal of antibiotic resistant bacteria, antibiotic resistance genes, and micropollutants by FeS_2@GO-based heterogeneous photo-Fenton process. Environ Sci Technol 56(21):15156–15166. https://doi.org/10.1021/acs.est.2c03334
3. Ahmed Y, Zhong J, Yuan Z, Guo J (2021) Simultaneous removal of antibiotic resistant bacteria, antibiotic resistance genes, and micropollutants by a modified photo-Fenton process. Water Res 197:117075. https://doi.org/10.1016/j.watres.2021.117075
4. Alexander J, Knopp G, Dötsch A, Wieland A, Schwartz T (2016). Ozone treatment of conditioned wastewater selects antibiotic resistance genes, opportunistic bacteria, and induce strong population shifts. Sci Total Environ 559:103–112. https://doi.org/10.1016/j.scitotenv.2016.03.154
5. Barancheshme F, Munir M (2018a) Strategies to combat antibiotic resistance in the wastewater treatment plants. In: Frontiers in microbiology (vol 8). Frontiers Media S.A.
6. Barancheshme F, Munir M (2018b) Strategies to combat antibiotic resistance in the wastewater treatment plants. Front Microbiol 8(JAN). https://doi.org/10.3389/fmicb.2017.02603
7. Chang PH, Juhrend B, Olson TM, Marrs CF, Wigginton KR (2017) Degradation of extracellular antibiotic resistance genes with UV254 treatment. Environ Sci Technol 51(11):6185–6192. https://doi.org/10.1021/acs.est.7b01120
8. Chen X, Han W, Patel M, Wang Q, Li Q, Zhao S, Jia W (2022) Inactivation of a pathogenic NDM-1-positive *Escherichia coli* strain and the resistance gene blaNDM-1 by TiO2/UVA photocatalysis. Sci Total Environ 846:157369. https://doi.org/10.1016/j.scitotenv.2022.157369
9. Choi Y, He H, Dodd MC, Lee Y (2021) Degradation kinetics of antibiotic resistance gene mecA of methicillin-resistant staphylococcus aureus (MRSA) during water disinfection with chlorine, ozone, and ultraviolet light. Environ Sci Technol. https://doi.org/10.1021/acs.est.0c05274
10. Czekalski N, Imminger S, Salhi E, Veljkovic M, Kleffel K, Drissner D, Hammes F, Bürgmann H, von Gunten U (2016) Inactivation of antibiotic resistant bacteria and resistance genes by ozone: from laboratory experiments to full-scale wastewater treatment. Environ Sci Technol 50(21):11862–11871. https://doi.org/10.1021/acs.est.6b02640
11. Darby EM, Trampari E, Siasat P, Gaya MS, Alav I, Webber MA, Blair JMA (2023) Molecular mechanisms of antibiotic resistance revisited. Nat Rev Microbiol 21(5):280–295. https://doi.org/10.1038/s41579-022-00820-y
12. Das D, Bordoloi A, Achary MP, Caldwell DJ, Suri RPS (2022) Degradation and inactivation of chromosomal and plasmid encoded resistance genes/ARBs and the impact of different matrices on UV and UV/H_2O_2 based advanced oxidation process. Sci Total Environ 833:155205. https://doi.org/10.1016/j.scitotenv.2022.155205
13. Das D, Bordoloi A, Yuan H, Caldwell DJ, Suri RPS (2024) Degradation of chromosomal and plasmid encoded extracellular and intracellular resistance genes across different types and sizes in water matrices during ozone-based oxidation process. J Water Process Eng 57:104560. https://doi.org/10.1016/j.jwpe.2023.104560
14. Di Cesare A, De Carluccio M, Eckert EM, Fontaneto D, Fiorentino A, Corno G, Prete P, Cucciniello R, Proto A, Rizzo L (2020) Combination of flow cytometry and molecular analysis to monitor the effect of UVC/H_2O_2 vs UVC/H_2O_2/Cu-IDS processes on pathogens and antibiotic resistant genes in secondary wastewater effluents. Water Res 184:116194. https://doi.org/10.1016/j.watres.2020.116194
15. Di Cesare A, Eckert EM, D'Urso S, Bertoni R, Gillan DC, Wattiez R, Corno G (2016) Co-occurrence of integrase 1, antibiotic and heavy metal resistance genes in municipal wastewater treatment plants. Water Res 94:208–214. https://doi.org/10.1016/j.watres.2016.02.049

16. Dias MF, Leroy-Freitas D, Machado EC, da Silva Santos L, Leal CD, da Rocha Fernandes G, de Araújo JC (2022) Effects of activated sludge and UV disinfection processes on the bacterial community and antibiotic resistance profile in a municipal wastewater treatment plant. Environ Sci Pollut Res. https://doi.org/10.1007/s11356-022-18749-3
17. Dizdaroglu M, Jaruga P (2012) Mechanisms of free radical-induced damage to DNA. Free Radical Res 46(4):382–419. https://doi.org/10.3109/10715762.2011.653969
18. Dodd MC (2012) Potential impacts of disinfection processes on elimination and deactivation of antibiotic resistance genes during water and wastewater treatment. J Environ Monit 14(7):1754–1771. https://doi.org/10.1039/c2em00006g
19. Ferro G, Guarino F, Castiglione S, Rizzo L (2016) Antibiotic resistance spread potential in urban wastewater effluents disinfected by UV/H2O2 process. Sci Total Environ 560–561:29–35. https://doi.org/10.1016/j.scitotenv.2016.04.047
20. Ferro G, Guarino F, Cicatelli A, Rizzo L (2017) β-lactams resistance gene quantification in an antibiotic resistant *Escherichia coli* water suspension treated by advanced oxidation with UV/H_2O_2. J Hazard Mater 323:426–433. https://doi.org/10.1016/j.jhazmat.2016.03.014
21. Foroughi M, Khiadani M, Kakhki S, Kholghi V, Naderi K, Yektay S (2022) Effect of ozonation-based disinfection methods on the removal of antibiotic resistant bacteria and resistance genes (ARB/ARGs) in water and wastewater treatment: a systematic review. Sci Total Environ 811:151404. https://doi.org/10.1016/j.scitotenv.2021.151404
22. Giannakis S, Le T-TM, Entenza JM, Pulgarin C (2018) Solar photo-Fenton disinfection of 11 antibiotic-resistant bacteria (ARB) and elimination of representative AR genes. Evidence that antibiotic resistance does not imply resistance to oxidative treatment. Water Res 143:334–345. https://doi.org/10.1016/j.watres.2018.06.062
23. Guo C, Wang K, Hou S, Wan L, Lv J, Zhang Y, Qu X, Chen S, Xu J (2017) H_2O_2 and/or TiO_2 photocatalysis under UV irradiation for the removal of antibiotic resistant bacteria and their antibiotic resistance genes. J Hazard Mater 323:710–718. https://doi.org/10.1016/j.jhazmat.2016.10.041
24. Guo M-T, Tian X-B (2019) Impacts on antibiotic-resistant bacteria and their horizontal gene transfer by graphene-based TiO_2&Ag composite photocatalysts under solar irradiation. J Hazard Mater 380:120877. https://doi.org/10.1016/j.jhazmat.2019.120877
25. Guo M-T, Yuan Q-B, Yang J (2015) Distinguishing effects of ultraviolet exposure and chlorination on the horizontal transfer of antibiotic resistance genes in municipal wastewater. Environ Sci Technol 49(9):5771–5778. https://doi.org/10.1021/acs.est.5b00644
26. Guo MT, Kong C (2019) Antibiotic resistant bacteria survived from UV disinfection: safety concerns on genes dissemination. Chemosphere 224:827–832. https://doi.org/10.1016/j.chemosphere.2019.03.004
27. Guo MT, Yuan QB, Yang J (2013) Ultraviolet reduction of erythromycin and tetracycline resistant heterotrophic bacteria and their resistance genes in municipal wastewater. Chemosphere 93(11):2864–2868. https://doi.org/10.1016/j.chemosphere.2013.08.068
28. He H, Zhou P, Shimabuku KK, Fang X, Li S, Lee Y, Dodd MC (2019) Degradation and deactivation of bacterial antibiotic resistance genes during exposure to free chlorine, monochloramine, chlorine dioxide, ozone, ultraviolet light, and hydroxyl radical. Environ Sci Technol 53(4):2013–2026. https://doi.org/10.1021/acs.est.8b04393
29. Herraiz-Carboné M, Cotillas S, Lacasa E, Sainz de Baranda C, Riquelme E, Cañizares P, Rodrigo MA, Sáez C (2021) A review on disinfection technologies for controlling the antibiotic resistance spread. Sci Total Environ 797:149150. https://doi.org/10.1016/j.scitotenv.2021.149150
30. Hiller CX, Hübner U, Fajnorova S, Schwartz T, Drewes JE (2019) Antibiotic microbial resistance (AMR) removal efficiencies by conventional and advanced wastewater treatment processes: a review. Sci Total Environ 685:596–608. https://doi.org/10.1016/j.scitotenv.2019.05.315
31. Huang JJ, Xi J, Hu HY, Li Y, Lu SQ, Tang F, Pang YC (2016) UV light tolerance and reactivation potential of tetracycline-resistant bacteria from secondary effluents of a wastewater treatment plant. J Environ Sci (China) 41:146–153. https://doi.org/10.1016/j.jes.2015.04.034

32. Iakovides I, Michael-Kordatou I, Moreira N, Ribeiro A, Fernandes T, Pereira MFR, Nunes OC, Manaia C, Silva A, Fatta-Kassinos D (2019) Continuous ozonation of urban wastewater: removal of antibiotics, antibiotic-resistant *Escherichia coli* and antibiotic resistance genes and phytotoxicity. Water Res 159:333–347
33. Karaolia P, Michael I, García-Fernández I, Agüera A, Malato S, Fernández-Ibáñez P, Fatta-Kassinos D (2014) Reduction of clarithromycin and sulfamethoxazole-resistant Enterococcus by pilot-scale solar-driven Fenton oxidation. Sci Total Environ 468–469:19–27. https://doi.org/10.1016/j.scitotenv.2013.08.027
34. Karaolia P, Michael-Kordatou I, Hapeshi E, Alexander J, Schwartz T, Fatta-Kassinos D (2017) Investigation of the potential of a Membrane BioReactor followed by solar Fenton oxidation to remove antibiotic-related microcontaminants. Chem Eng J 310:491–502. https://doi.org/10.1016/j.cej.2016.04.113
35. Kent AG, Vill AC, Shi Q, Satlin MJ, Brito IL (2020) Widespread transfer of mobile antibiotic resistance genes within individual gut microbiomes revealed through bacterial Hi-C. Nat Commun 11(1):4379. https://doi.org/10.1038/s41467-020-18164-7
36. Krzeminski P, Tomei MC, Karaolia P, Langenhoff A, Almeida CMR, Felis E, Gritten F, Andersen HR, Fernandes T, Manaia CM, Rizzo L, Fatta-Kassinos D (2019) Performance of secondary wastewater treatment methods for the removal of contaminants of emerging concern implicated in crop uptake and antibiotic resistance spread: a review. Sci Total Environ 648:1052–1081. https://doi.org/10.1016/j.scitotenv.2018.08.130
37. Li B, Zhang T (2010) Biodegradation and adsorption of antibiotics in the activated sludge process. Environ Sci Technol 44(9):3468–3473. https://doi.org/10.1021/es903490h
38. Li F, Liu K, Bao Y, Li Y, Zhao Z, Wang P, Zhan S (2024) Molecular level removal of antibiotic resistant bacteria and genes: a review of interfacial chemical in advanced oxidation processes. Water Res 254:121373. https://doi.org/10.1016/j.watres.2024.121373
39. Li H, Dechesne A, He Z, Jensen MM, Song HL, Smets BF (2023) Electrochemical disinfection may increase the spread of antibiotic resistance genes by promoting conjugal plasmid transfer. Sci Total Environ 858:159846. https://doi.org/10.1016/j.scitotenv.2022.159846
40. Li S, Zhang C, Li F, Hua T, Zhou Q, Ho S-H (2021) Technologies towards antibiotic resistance genes (ARGs) removal from aquatic environment: a critical review. J Hazard Mater 411:125148. https://doi.org/10.1016/j.jhazmat.2021.125148
41. Li Y, Li R, Hou J, Sun X, Wang Y, Li L, Yang F, Yao Y, An Y (2024) Mobile genetic elements affect the dissemination of antibiotic resistance genes (ARGs) of clinical importance in the environment. Environ Res 243:117801. https://doi.org/10.1016/j.envres.2023.117801
42. Liu X, Hu Y (2020) Effect of DNA sizes and reactive oxygen species on degradation of sulphonamide resistance sul1 genes by combined UV/free chlorine processes. J Hazard Mater 392:122283–122283. https://doi.org/10.1016/j.jhazmat.2020.122283
43. Lüddeke F, Heß S, Gallert C, Winter J, Güde H, Löffler H (2015) Removal of total and antibiotic resistant bacteria in advanced wastewater treatment by ozonation in combination with different filtering techniques. Water Res 69:243–251. https://doi.org/10.1016/j.watres.2014.11.018
44. McKinney CW, Pruden A (2012) Ultraviolet disinfection of antibiotic resistant bacteria and their antibiotic resistance genes in water and wastewater. Environ Sci Technol 46(24):13393–13400. https://doi.org/10.1021/es303652q
45. Meng L-X, Sun Y-J, Zhu L, Lin Z-J, Shuai X-Y, Zhou Z-C, Chen H (2022) Mechanism and potential risk of antibiotic resistant bacteria carrying last resort antibiotic resistance genes under electrochemical treatment. Sci Total Environ 821:153367. https://doi.org/10.1016/j.scitotenv.2022.153367
46. Meng X, Li H, Yan S, Ning Z, Yuan Q, Liu C, Zhou L, Zheng G (2023) Bacterial cell lysis induced by pre-acidification greatly promoting the degradation of carried antibiotic resistance genes in zero-valent iron activated Fenton-like processes. Chem Eng J 452:139413. https://doi.org/10.1016/j.cej.2022.139413
47. Michael-Kordatou I, Karaolia P, Fatta-Kassinos D (2018) The role of operating parameters and oxidative damage mechanisms of advanced chemical oxidation processes in the combat against antibiotic-resistant bacteria and resistance genes present in urban wastewater. Water Res 129:208–230. https://doi.org/10.1016/j.watres.2017.10.007

48. Murray CJL, Ikuta KS, Sharara F, Swetschinski L, Robles Aguilar G, Gray A, Han C, Bisignano C, Rao P, Wool E, Johnson SC, Browne AJ, Chipeta MG, Fell F, Hackett S, Haines-Woodhouse G, Kashef Hamadani BH, Kumaran EAP, McManigal B, Lopez AD, Andy S, Naghavi M (2022) Global burden of bacterial antimicrobial resistance in 2019: a systematic analysis. The Lancet 399(10325):629–655. https://doi.org/10.1016/S0140-6736(21)02724-0
49. Nihemaiti M, Yoon Y, He H, Dodd MC, Croué J-P, Lee Y (2020a) Degradation and deactivation of a plasmid-encoded extracellular antibiotic resistance gene during separate and combined exposures to UV254 and radicals. Water Res 115921. https://doi.org/10.1016/j.watres.2020.115921
50. Nihemaiti M, Yoon Y, He H, Dodd MC, Croué JP, Lee Y (2020) Degradation and deactivation of a plasmid-encoded extracellular antibiotic resistance gene during separate and combined exposures to UV254 and radicals. Water Res 182:115921–115921. https://doi.org/10.1016/j.watres.2020.115921
51. Oberoi AS, Jia Y, Zhang H, Khanal SK, Lu H (2019) Insights into the fate and removal of antibiotics in engineered biological treatment systems: a critical review. Environ Sci Technol 53(13):7234–7264. https://doi.org/10.1021/acs.est.9b01131
52. Oh J, Salcedo DE, Medriano CA, Kim S (2014) Comparison of different disinfection processes in the effective removal of antibiotic-resistant bacteria and genes. J Environ Sci (China) 26(6):1238–1242. https://doi.org/10.1016/S1001-0742(13)60594-X
53. Park K-Y, Choi S-Y, Lee S-H, Kweon J-H, Song J-H (2016) Comparison of formation of disinfection by-products by chlorination and ozonation of wastewater effluents and their toxicity to Daphnia magna. Environ Pollut 215:314–321. https://doi.org/10.1016/j.envpol.2016.04.001
54. Pourrostami Niavol K, Bordoloi A, Suri R (2024) An overview of the occurrence, impact of process parameters, and the fate of antibiotic resistance genes during anaerobic digestion processes. Environ Sci Pollut Res 31(29):41745–41774. https://doi.org/10.1007/s11356-024-33844-3
55. Rathinavelu S, Shanmugam MK, Gummadi SN, Nambi IM (2023) Inactivation of antibiotic resistant bacteria and elimination of transforming ability of plasmid carrying single and dual drug resistance genes by electro-oxidation using Ti/Sb-SnO_2/PbO_2 anode. Chem Eng J 461:141807. https://doi.org/10.1016/j.cej.2023.141807
56. Rizzo L, Manaia C, Merlin C, Schwartz T, Dagot C, Ploy MC, Michael I, Fatta-Kassinos D (2013) Urban wastewater treatment plants as hotspots for antibiotic resistant bacteria and genes spread into the environment: a review. Sci Total Environ 447:345–360. https://doi.org/10.1016/j.scitotenv.2013.01.032
57. Rizzo L, Sannino D, Vaiano V, Sacco O, Scarpa A, Pietrogiacomi D (2014) Effect of solar simulated N-doped TiO_2 photocatalysis on the inactivation and antibiotic resistance of an *E. coli* strain in biologically treated urban wastewater. Appl Catal B: Environ 144:369–378. https://doi.org/10.1016/j.apcatb.2013.07.033
58. Rodríguez-Chueca J, della Giustina SV, Rocha J, Fernandes T, Pablos C, Encinas Á, Barceló D, Rodríguez-Mozaz S, Manaia CM, Marugán J (2019) Assessment of full-scale tertiary wastewater treatment by UV-C based-AOPs: removal or persistence of antibiotics and antibiotic resistance genes? Sci Total Environ 652:1051–1061
59. Rodriguez-Mozaz S, Chamorro S, Marti E, Huerta B, Gros M, Sànchez-Melsió A, Borrego CM, Barceló D, Balcázar JL (2015) Occurrence of antibiotics and antibiotic resistance genes in hospital and urban wastewaters and their impact on the receiving river. Water Res 69:234–242. https://doi.org/10.1016/j.watres.2014.11.021
60. Rommozzi E, Giannakis S, Giovannetti R, Vione D, Pulgarin C (2020) Detrimental vs. beneficial influence of ions during solar (SODIS) and photo-Fenton disinfection of *E. coli* in water: (Bi)carbonate, chloride, nitrate and nitrite effects. Appl Catal B: Environ 270:118877. https://doi.org/10.1016/j.apcatb.2020.118877
61. Rubio D, Nebot E, Casanueva JF, Pulgarin C (2013) Comparative effect of simulated solar light, UV, UV/H_2O_2 and photo-Fenton treatment (UV–Vis/H_2O_2/$Fe^{2+,3+}$) in the *Escherichia coli* inactivation in artificial seawater. Water Res 47(16):6367–6379. https://doi.org/10.1016/j.watres.2013.08.006

62. Selvam A, Xu D, Zhao Z, Wong JWC (2012) Fate of tetracycline, sulfonamide and fluoroquinolone resistance genes and the changes in bacterial diversity during composting of swine manure. Biores Technol 126:383–390. https://doi.org/10.1016/J.BIORTECH.2012.03.045
63. Serna-Galvis EA, Salazar-Ospina L, Jiménez JN, Pino NJ, Torres-Palma RA (2018) Elimination of carbapenem resistant Klebsiella pneumoniae in water by UV-C, UV-C/persulfate and UV-C/ H_2O_2. Evaluation of response to antibiotic, residual effect of the processes and removal of resistance gene. J Environ Chem Eng 102196
64. Shi P, Jia S, Zhang X-X, Zhang T, Cheng S, Li A (2013) Metagenomic insights into chlorination effects on microbial antibiotic resistance in drinking water. Water Res 47(1):111–120. https://doi.org/10.1016/j.watres.2012.09.046
65. Sonntag V (2006) Free-radical-induced DNA damage and its repair: a chemical perspective. Springer
66. Sousa JM, Macedo G, Pedrosa M, Becerra-Castro C, Castro-Silva S, Pereira MFR, Silva AMT, Nunes OC, Manaia CM (2017) Ozonation and UV254nm radiation for the removal of microorganisms and antibiotic resistance genes from urban wastewater. J Hazard Mater 323:434–441. https://doi.org/10.1016/j.jhazmat.2016.03.096
67. Stange C, Sidhu J, Toze S, Tiehm A (2019) Comparative removal of antibiotic resistance genes during chlorination, ozonation, and UV treatment. Int J Hyg Environ Health 222(3):541–548
68. Van Hoek AHAM, Mevius D, Guerra B, Mullany P, Roberts AP, Aarts HJM (2011) Acquired antibiotic resistance genes: an overview. In: Frontiers in microbiology (vol 2). Frontiers Research Foundation
69. Venieri D, Gounaki I, Bikouvaraki M, Binas V, Zachopoulos A, Kiriakidis G, Mantzavinos D (2017) Solar photocatalysis as disinfection technique: Inactivation of Klebsiella pneumoniae in sewage and investigation of changes in antibiotic resistance profile. J Environ Manage 195:140–147
70. Vilela PB, Martins AS, Starling MCVM, de Souza FAR, Pires GFF, Aguilar AP, Pinto MEA, Mendes TAO, de Amorim CC (2021) Solar photon-Fenton process eliminates free plasmid DNA harboring antimicrobial resistance genes from wastewater. J Environ Manage 285:112204. https://doi.org/10.1016/j.jenvman.2021.112204
71. Yoon Y, Chung HJ, Wen Di DY, Dodd MC, Hur HG, Lee Y (2017) Inactivation efficiency of plasmid-encoded antibiotic resistance genes during water treatment with chlorine, UV, and UV/H_2O_2. Water Res 123:783–793 https://doi.org/10.1016/j.watres.2017.06.056
72. Yoon Y, Dodd MC, Lee Y (2018) Elimination of transforming activity and gene degradation during UV and UV/H_2O_2 treatment of plasmid-encoded antibiotic resistance genes. Environ Sci: Water Res Technol 4(9):1239–1251. https://doi.org/10.1039/c8ew00200b
73. Yoon Y, He H, Dodd MC, Lee Y (2021) Degradation and deactivation of plasmid-encoded antibiotic resistance genes during exposure to ozone and chlorine. Water Res 202:117408. https://doi.org/10.1016/j.watres.2021.117408
74. Yuan Q-B, Guo M-T, Yang J (2015) Fate of antibiotic resistant bacteria and genes during wastewater chlorination: implication for antibiotic resistance control. PLoS ONE 10(3):e0119403. https://doi.org/10.1371/journal.pone.0119403
75. Zhang Y, Zhuang Y, Geng J, Ren H, Xu K, Ding L (2016) Reduction of antibiotic resistance genes in municipal wastewater effluent by advanced oxidation processes. Sci Total Environ 550:184–191. https://doi.org/10.1016/j.scitotenv.2016.01.078
76. Zheng Q, Zhang Y, Qianxin Z, Wang Y, Yu G (2023) Removal of antibiotic resistant bacteria and plasmid-encoded antibiotic resistance genes in water by ozonation and electro-peroxone process. Chemosphere 319:138039. https://doi.org/10.1016/j.chemosphere.2023.138039
77. Zhuang Y, Ren H, Geng J, Zhang Y, Zhang Y, Ding L, Xu K (2015) Inactivation of antibiotic resistance genes in municipal wastewater by chlorination, ultraviolet, and ozonation disinfection. Environ Sci Pollut Res 22(9):7037–7044. https://doi.org/10.1007/s11356-014-3919-z

Life Cycle Assessment as an Alternate Tool to Evaluate Wastewater Treatment Technologies

Bikash R. Tiwari, Dhananjaya Panda, Koteswara Rao Peta, and Satinder K. Brar

Abstract The efficacy of wastewater treatment technologies is primarily evaluated based on their ability to remove the desired contaminants. Once, the performance parameters are established, attention is paid to the economic feasibility of the field scale application of the technology. However, in the present era of awareness, a third evaluation criteria alternate to the technical feasibility and economic suitability of the treatment options, is the consideration of the environmental sustainability. The impacts of a technology application are not only limited to receiving water systems in concern to removal of contaminants present in wastewater. With a wide vision, the impacts encompass the effect of the technology on the flora and fauna, human health, and the exploitation of natural resources. This is an upcoming parameter of consideration for evaluation of new technologies as well as upgradation of existing options for wastewater treatment. One such tool is the life cycle assessment (LCA), which evaluates the impacts associated with a process/technology inclusive of the material and energy consumption and the overall emissions. Various studies have carried out LCA for lab-scale or field scale treatment technologies. LCA studies can estimate significant environmental parameters such as global warming potential, eutrophication, ozone layer depletion, water consumption, land utilization, etc. This book chapter will provide views and discussion on the potential of LCA as a tool for evaluation of various wastewater treatment technologies.

B. R. Tiwari
Centre Eau Terre Environnement, Institut national de La recherche scientifique, Université du Québec, Québec City, Canada

D. Panda · K. R. Peta
Department of Electronic Science, University of Delhi South Campus, Benito Juarez Road, New Delhi 110021, India

S. K. Brar (✉)
Department of Civil Engineering, Lassonde School of Engineering, York University, North York, Toronto, Canada
e-mail: satinder.brar@lassonde.yorku.ca

S. Yadav et al. (eds.), *Emerging Technologies in Biological and Hybrid Wastewater Treatment: Lessons from Developed to Enhancing Practices in Developing Countries*, Earth and Environmental Sciences Library,
https://doi.org/10.1007/978-3-031-74515-7_18

Keywords Life cycle assessment · Wastewater treatment · SimaPro · ReCiPe · Climate change · Eutrophication

1 Introduction

The United Nations (UN) sustainable development goal 6 "Clean Water and Sanitation" is an important criterion to achieve multifaceted development that encompasses social, economic, and environmental dimensions [13]. The target time period to achieve this goal is set to 2030 which needs the utilization of various wastewater treatment technologies. Apart from conventional technologies such as activated sludge process (ASP), primary and secondary sedimentation, sand filtration, many advanced treatment technologies such as membrane bioreactor, moving bed bioreactor, sequencing batch reactors have been implemented in WWTPs to facilitate the reclamation of water resources in various public-owned entities, companies, and organizations, regardless of their size. Hence, wastewater treatment technologies are targeted at mitigating the adverse impacts on the environment and promoting water conservation. The WWTPs treat the waste and untreated water originating from various sources such as residences, factories, medical facilities, and laboratories. The resulting effluents retrieved from this process can potentially be utilized for irrigation applications. Typically, the aforementioned intervention consists of a two-step procedure. During the initial stage, the procedure involves the elimination of floating debris, grit, greases, and other microparticles present in the accumulated wastewater within the collection tank. During the secondary treatment phase, the primary treated effluent undergoes a degradation of organic matter in the aeration tank. This degradation is facilitated by microorganisms, which aid in the removal of phosphorus and nitrogen from the sewage. Subsequently, the sludge and water are separated through the clarifier operations. The partially treated water undergoes additional treatment through the disinfection process within the chlorination tank. Ultimately, the dual filter media is employed for the purpose of treating the disinfected water, after which the completely treated wastewater effluent is afterwards utilized for various applications. It is imperative to properly cleanse effluents and sewage prior to their disposal. Each process utilizes various material and energy inputs such as chemicals, electrical energy, fuel, construction material, etc. [43]. Considering the aforementioned point, it is imperative on a worldwide scale to evaluate and scrutinize treatment methodologies with respect to their efficiency to achieve environmental regulatory standards, and cost of implementation. In addition, it is equally important to consider other aspects of their implementation such as regional and global environmental impacts [35]. Life cycle assessment (LCA) is a methodological framework to estimate the environmental consequences associated with a process or technology by taking into consideration the set of inputs and outputs throughout the life cycle of the process or technology [40, 43]. This assessment encompasses both direct factors such as wastewater management and sludge, as well as indirect factors including gas emissions, soil quality, and effluent discharge [39]. LCA is aimed at identifying the factors responsible for accumulating environmental burdens during the life cycle of a technology and then efforts can be made to eliminate or alleviate or improve the contribution of these factors [12]. LCA can also be used to compare alternate scenarios and technologies which can be more environmentally beneficial while serving the same purpose. In the long run it would be beneficial to incorporate the environmental indicators such as global warming potential, eutrophication potential, resource depletion etc. from the LCA to review the wastewater treatment technologies [44].

2 LCA for Wastewater Treatment

The strategic environmental assessment (SEA), net environmental benefit analysis (NEBA), material flow analysis (MFA), environmental impact assessment (EIA), and LCA are among the different environmental analysis tools utilized to examine the environmental implications associated with wastewater treatment plants (WWTPs) [10, 16]. During a comparison study it was found that LCA emerges as the most comprehensive environmental analysis technique, as supported by the data collected from the system [30]. Within this framework, LCA stands out as the most effective tool for determining the environmental impact associated with the aforementioned entities [16, 32]. The LCA research comprises four distinct phases, first with the establishment of the process's aim and scope. This is then followed by the life cycle inventory (LCI), the life cycle impact assessment (LCIA) phase, and finally, the interpretation of the resulting data (Fig. 1). In this chapter, the case studies have been discussed with respect to the different steps of LCA to allow the readers a better understanding and co-relation between the concept of LCA and application of these topics to wastewater treatment technologies.

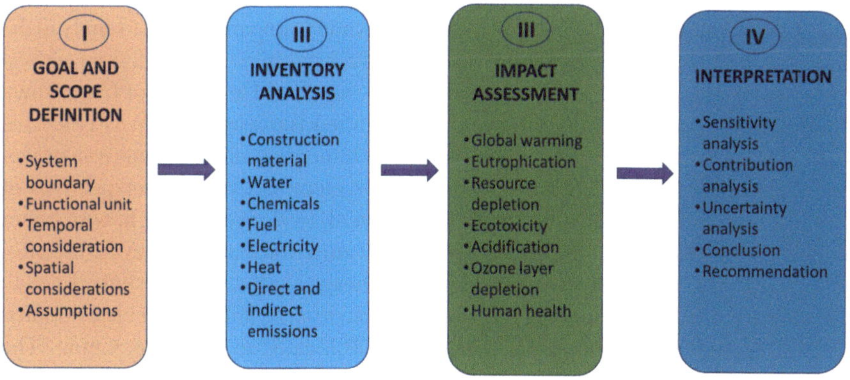

Fig. 1 Four steps of life cycle assessment

Various academics have conducted LCAs to evaluate the environmental impact of WWTPs around the world. Niero et al. [31] conducted a comparative LCA study to evaluate several wastewater treatment methods, including anaerobic versus aerobic treatment, as well as chemical treatment versus combined chemical-biological treatment. According to his findings, the recycling of phosphorus does not have any discernible impact on agricultural land when compared to the process of sludge incineration. Houillon and Joilet [19] conducted a comparative LCA in Switzerland and France, examining six WWTPs scenarios in terms of emissions and energy contribution. This study demonstrates the relationship between global warming and energy use, specifically in relation to emissions and sludge incineration. A separate investigation conducted in Galicia, a region in Northwestern Spain, conducted an LCA on 13 WWTPs. The findings of this study indicated that the aeration treatment method had a greater environmental impact compared to the aerobic-anoxic treatment approach [14].

Use of renewable energy such as wind has higher energy generation, lower gas and pollutants emission and less fossil fuel diminution compared to the other energy sector. Hence, the wind energy has lower environmental impacts as illustrated by the LCA study of WWTPs in China [1]. Finding the characterization factors for assessing pharmaceutical and personal care products (PPCPs) in wastewater allowed for a more nuanced assessment of LCA on a WWTP [1]. LCA for a Spanish Mediterranean wastewater treatment plant compared various options for reclaiming and reusing municipal wastewater [36]. They found that substituting desalinized water resulted in significant energy savings, whereas saving 1.1 m^3 of freshwater was the result of replacing potable water. Another study revealed a novel approach to identify the best procedure for WWTP that intended to integrate environmental considerations with LCA methodology and economic parameters [29]. The environmental impacts of two plants of different sizes and locations were compared in another study. According to the findings, the two biggest factors in the adverse environmental effects of wastewater treatment are the use of energy and the disposal of sludge in agricultural sector [27]. By applying LCA to a WWTP in Xi'an, China, Zhang et al. [50] showed that the benefits of reusing treated wastewater far outweighed the energy required to do so. The LCA technique was also utilized by Ontiveros and Campanella [33] to analyze the environmental performance of three different advanced biological nutrient removal processes in Argentina. These processes are the modified University of Cape Town process, the five stages Bardenpho, and the modified Bardenpho for WWTPs. This study can be used to assist the selection of the biological nutrient removal procedure that will be appropriate for the context of Argentina from the perspectives of both technology and the environment. Yoshida et al. [49] investigated how to better account for onsite GHG emissions and long-term emission data in life cycle inventories and methodologies for land application of sewage sludge. In their study, Li et al. [24] employed LCA methodology to assess the environmental consequences related to the wastewater treatment process in a WWTP located in Kunshan, China. The aim of their study was to evaluate the environmental advantages and disadvantages of this WWTP in relation to other WWTPs that employ various advanced treatment technologies.

3 Four Steps of LCA

3.1 Goal and Scope Definition

3.1.1 Goal and Scope

During the aim and scope definition phase of an ISO-standardized LCA, the primary objective of the assessment is determined, and deliberations are conducted regarding the specific characteristics of the product system under investigation [40]. The objective and extent of the study are established prior to the collection of any data. The significance of the initial phase, as outlined in ISO 14040 and 14044, is frequently undervalued due to its comprehensive nature that extends beyond a mere introduction to the LCA process. The determination of the precise technique to be followed is a critical element within the framework of LCA methodology. Nevertheless, it is important to note that both the objective and the extent of the study can be subject to adjustments throughout the research process, particularly as data is gathered and novel insights are obtained. For instance, it is plausible that the suggested method for allocating co-products proves to be ineffective or that there is insufficient data to construct a comprehensive life cycle inventory. Transparent descriptions of such alterations must be provided in the data spreadsheets and final report, as they are necessary in certain instances. The standards purposely provide flexible guidelines to accommodate a wide range of alternatives in defining the goal and scope. The goal and scope definition phase encompasses the identification and clarification of various elements, such as the functional unit, system boundaries, exclusion of certain life cycle stages or inputs, data granularity (which can reflect either process-specific or generic data), and the selection of impact indicators and characterization factors (Fig. 2).

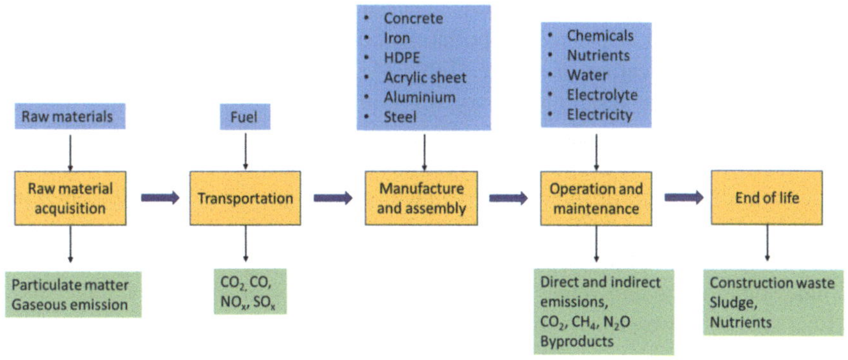

Fig. 2 Life cycle stages and corresponding inventory for wastewater treatment technologies

3.1.2 Functional Unit

The functional unit is necessary to build and model a process/product system in LCA. It relates to the quantitative description of the function of a product i.e., services, performance, aesthetics, costs, etc. This further serves as the reference basis to all the calculations in concern to impact assessment (Table 1). The volume based functional unit is most common for wastewater treatment which is related to the amount of influent wastewater or the treated effluent [6, 7]. Either a unitary functional unit such as 1 m^3 wastewater can be considered or the capacity of the treatment unit such as 2 MGD of wastewater for 30 years can also be chosen [3, 28]. However, this functional unit does not take into consideration the pollutant load or the treatment efficiency of the system. Hence, some studies have also selected pollutant based functional unit based on the specific treatment technology used such as 1 kg COD removed [21] and kg PO_4 eq. removed [51]. The pollutant based functional unit is more representative of the treatment system as compared to the volumetric functional unit. Hence, certain studies have also selected population equivalent functional units to identify with the number of individuals in the locality who are being served by the wastewater treatment plant. This is usually selected in case of large domestic or municipal treatment plants.

3.1.3 System Boundary

In LCA, it is imperative to establish clear and defined system boundaries across various dimensions. These dimensions include demarcating the boundaries between the system and the natural environment, defining the area and time horizon under consideration, distinguishing between the production of capital goods, and delineating the boundaries between the life cycle of the product being studied and the interconnected life cycles of other products. In case of wastewater treatment system, the system boundary could include raw material acquisition, wastewater treatment, sludge treatment and disposal, transportation, construction and operation and recycling and land application of sludge (Table 1). Based on the extend of life cycle stages included in the system boundary, it is commonly classified as cradle to gate, cradle to grave, gate to gate, gate to grave [48]. The cradle to grave is the most comprehensive study where all stages of the life cycle starting from the raw material acquisition till the end of life of the treatment system is evaluated for the environmental burdens. However, it is possible that the scope of the study encompasses a specific stage/stages of the treatment processes. A cradle to gate boundary would not include some of the end stages such as fate of sludge or material recycling. Similarly, a gate to grave process may choose to exclude one or many initial stages of raw material acquisition, transportation and construction. If the LCA study is particularly based on a specific stage say only wastewater treatment or sludge disposal or construction, this can be termed as a gate to gate LCA. The discourse examines the principles governing the selection of system limits, with particular emphasis on the final dimension. LCA research typically involves the analysis of two distinct types

Table 1 LCA of wastewater treatment technologies

Technology	System boundary	Functional unit	Methodology	References
Coagulation/filtration	Wastewater treatment	1 ton COD eq. removal	CML	[46]
Activated sludge process	Wastewater treatment, sludge treatment and disposal	2.56×10^4 population equivalent (p.e)/d	CML	[46]
Wastewater treatment plants	Wastewater treatment, sludge treatment, disposal, and reuse	kg PO_4 eq. removed	CML	[51]
Municipal wastewater treatment plants	Wastewater treatment, sludge treatment and disposal	1 p.e. per year	CML	[22]
Upflow anaerobic sludge blanket, anaerobic filter, subsurface flow constructed wetlands, and photoreactors	Wastewater treatment and reuse	333.9 tons of treated wastewater over 10 years	ReCiPe	[25]
Anaerobic membrane bioreactor	Wastewater treatment, sludge disposal and reuse	1 kg of COD removal	ReCiPe	[21]
Rotating biological contactor and anaerobic baffled reactor	Wastewater treatment, sludge disposal	2 MGD of domestic wastewater over 30 years	IPCC 2017, ReCiPe, IMPACT 2002 +	[28]
Upflow anaerobic sludge blanket reactor + integrated fixed film activated sludge (IFAS); high rate activated sludge + IFAS, and rotating belt filter + chemical enhanced primary treatment + IFAS	Wastewater treatment, sludge disposal	1 m^3 of treated wastewater and 1 kWh of energy produced	Recipe and CML	[3]

of systems: foreground systems and background systems. In the case of foreground systems, data that is specific to the primary site is typically gathered, while background systems rely on secondary data obtained from databases, public references, or estimated data derived from input-output life cycle assessment models. The facilitation of product LCA can be enhanced by decomposing the sequential operations related to a product or material into a main system and a set of subsystems. The primary system encompasses a set of activities that directly contribute to the major life cycle stages of a product, including its creation, utilisation, and recycling. The secondary tier encompasses supplementary items or processes that directly interact with the primary activity sequence. Multiple levels of supplementary materials or procedures have the potential to progressively expand beyond the primary sequence. The discrete auxiliary materials or processes are characterized as subsystems that constitute components of the established manufacturing system.

3.2 Life Cycle Inventory

LCIs are utilized to monitor and analyze the utilization of materials, energy, co-product creation, wastes, and emissions over a series of sequential processing stages. These stages encompass the extraction of raw materials, their further refinement and processing, the manufacturing of the final product, its utilisation by consumers, and finally, the recycling or disposal of the product. Firstly, it entails gathering essential data to define all unit processes that will be encompassed within the system boundaries. Secondly, it involves conducting allocation steps for multifunctional processes to establish the relationship between inputs (such as energy and material resources) and outputs (including end-products, GHG emissions and wastes) with the corresponding products and by-products, if applicable. Lastly, LCI necessitates performing final calculations, validating the collected data, and applying allocation procedures. In case of wastewater treatment technologies, the holistic inventory would include data pertaining to construction materials, transportation fuel, operation electricity, direct and indirect emissions, any byproduct recovery and recycling component.

3.3 Life Cycle Impact Assessment

The life cycle impact assessment is the third phase of the LCA process, which converts the input data in the form of material, energy and emissions into the corresponding environmental consequences [42]. The LCIA phase consists of two mandatory steps: (1) classifying the product categories into several distinct impact categories, and (2) characterizing the potential impacts by assessing and exploiting scientific factors. For example, the impact categories determined by the LCI include human toxicity, climate change, acidification, eutrophication, ozone depletion, fossil depletion, ecotoxicity, photochemical oxidation, particulate matter formation, and global warming [20] which are discussed elaborately with excerpts from LCA studies.

3.3.1 Global Warming Potential

The concept of global warming potential (GWP) provides a straightforward representation of the relative radiative forcing that results from the emission of a unit mass of a greenhouse gas (GHG), when compared to CO_2, a typical reference compound [47]. The indicator signifies the uptick in global average temperatures ensuing from the release of GHGs, with the predominant source being the combustion of fossil fuels such as coal, oil, and natural gas. The collective warming potential of all GHG emissions is measured in kg CO_2 equivalent, wherein each GHG is evaluated relative to the warming potential of 1 kg of CO_2 [45]. This concept is exercised to balance the ecosystem exchanges of multiple GHGs, including CO_2, CH_4, and NO_2. These gases trap heat by absorbing energy thus reducing the rate at which the trapped energy escapes the atmosphere. For example, CH_4 has a higher GWP, as it traps more heat than CO_2 per molecule [47]. Each gas has a specific GWP, which allows the comparison of the amount of energy that the emissions of one ton of a gas will absorb over a given time (usually a 20-year or 100-year), relative to the emissions of one ton of CO_2.

The study conducted by Kalbar et al. [22] applied the LCA method to evaluate the environmental consequences of sequencing batch reactors (SBRs), which are fundamental components of numerous WWTPs. It was observed that the SBR exhibits the highest energy consumption among all the wastewater treatment technologies. Additionally, the SBR has other detrimental effects, as it releases CO_2 which leads to the increase in GWP. According to Guinee's baseline method, the GWP of ASP and SBR was found to be ~ 18 and ~ 31 kg of CO_2 eq., respectively [15]. It was found that the direct and indirect emissions of CO_2 and NO_2 from the electricity use contribute to maximizing the adverse impacts [4]. As a result, the electricity consumption, in addition to sludge treatment and disposal, are reported to be significant contributors to global warming.

Cornejo et al. [5] found that methane emissions significantly impact GWP when the input value was increased by 20% for two treatments in Bolivia, which utilized an up-flow anaerobic sludge blanket reactor or a facultative pond, followed by two maturation ponds in sequence. Another study contrasted the biogas production from a constructed wetland (CH_4 recovery) and an ASP (biogas produced through sludge digestion) [23]. The investigation determined that in comparison to ASP, constructed wetlands generated a lower GWP as opposed to ASP, particularly when the replaced energy primarily originates from fossil fuels, particularly coal, as the latter generate less biogas, thereby only partially replacing energy.

An investigation delved into the environmental impacts of an anaerobic baffled reactor (ABR) both independently and in conjunction with various configurations [41]. Their findings demonstrated that capturing dissolved methane as gaseous methane and converting it into heat and power yields increased benefits for the end point impacts of resources, ecosystem quality, and human health across all ABR options studied. In a separate study, the environmental burdens were compared for different scenarios involving the simultaneous utilization of the ABR system and the rotating biological contactor (RBC) [28]. The ABR-RBC system reported that climate change impact is reduced by 87% in comparison to the ABR system owing to the fact that 80% of the dissolved methane undergoes oxidation in the RBC.

3.3.2 Toxicity

Human toxicity pertains to the consequences of harmful materials on the human environment. It does not cover the hazards of exposure in workplaces. The Uniform System for the Evaluation of Substances adapted for LCA (USES-LCA) is used to calculate the characterization factors, which are the human toxicity potentials (HTP) [9]. HTP assesses the impact of chemicals on human health, particularly emphasizing the repercussions of chemical release into the environment, often stemming from electricity generation from fossil fuels. These factors describe the fate, exposure, and impact of toxic substances for an infinite time period [9]. The HTPs are expressed in units of 1,4-dichlorobenzene equivalents per kg of emission [20]. The scope of this indicator can vary depending on the geographic area, ranging from a local to a global scale. Ecotoxicity addresses potential toxic impacts on ecosystems, posing risks to both individual species and overall ecosystem health [20]. Certain substances have a propensity to accumulate within living organisms, heightening concerns. Measurement is conducted using comparative toxic units for ecosystems. Freshwater Aquatic Ecotoxicity Potential (FAETP) gauges the toxic effects of chemicals specifically within freshwater environments, potentially leading to biodiversity loss and species extinction. Characterization factors are expressed as ETP in kg 1.4-dichlorobenzene equivalent.

The EDIP97 method can be used to evaluate potential human toxicity and environmental ecotoxicity [30]. An LCA was carried out to compare the environmental performance of vertical flow and horizontal flow constructed wetland [11]. The authors exploited both the Eco-Indicator 99 and the CML2 Baseline 2000 methods to evaluate the environmental impact of nitrogen and carbon GHG emissions. It was mentioned that gaseous emissions from wastewater treatment were not considered for calculations which affected the results. It is important to consider all relevant factors in order to obtain accurate results. Munoz et al. [30] indicated that treated wastewater exhibited lower environmental impact compared to effluents. Furthermore, the authors identified PPCPs as significant contributors to ecotoxicity. Specifically, nicotine, ciprofloxacin, and fluoxetine within PPCPs raised concerns regarding wastewater toxicity, while hexachlorobenzene and nickel emerged as major wastewater contaminants.

3.3.3 Acidification

Acidification potential pertains to compounds acting as precursors to acid rain, including sulfur dioxide, nitrogen oxides, nitrogen monoxide, nitrogen dioxide, and various other substances [8]. Typically, acidification potential is measured in terms of

SO_2-equivalence. These acidifying gases predominantly originate from fuel combustion processes. Conversely, modern coal-fired power plants employ desulfurization techniques to mitigate SO_2 emissions into the environment. Acidification occurs through substances with varying degrees of acid-forming potential.

Hong et al. [18] accentuated the importance of employing LCA to scrutinize sewage sludge treatment methodologies, particularly by gauging the environmental and economic implications of commonly utilized alternatives in Japan. These encompass dewatering, incineration, dewatered sludge melting, composting, acidification, and drying. This study revealed that the drying process exerts the most pronounced effects on GWP. However, the incinerated ash melting along with acidification exhibit the greatest impact on human health. Conversely, methods such as dewatering, composting, and incineration demonstrate the least adverse effects on human health. Another study utilized the Eco-Indicator 99 method to evaluate environmental impacts stemming from a pumping station wastewater treatment plant [38]. Their aim was to compare the outcomes of this method with those generated by the CML2 method. Specifically, the CML2 method was applied to refine assessments of environmental impacts related to eutrophication and acidification, ensuring congruence with results obtained from the Eco-Indicator 99 method, which proved consistent. Consequently, the CML2 method garnered considerable attention and was recommended for addressing high-impact categories.

3.3.4 Eutrophication

Eutrophication potential encompasses the environmental impacts arising from nutrient overloading, particularly focusing on nitrogen and phosphorus, which are key substances in this context [35]. This nutrient overload can lead to oxygen depletion or the presence of harmful toxins, creating conditions detrimental to aquatic life. Typically, either phosphorus or nitrogen acts as the limiting nutrient for algal plankton and blue-green algae, depending on the environmental conditions. Consequently, oxygen depletion and the accumulation of toxins can result in the formation of "dead zones," where both flora and fauna struggle to survive.

Rahman et al. [37] demonstrated that by utilizing reduced amounts of total nitrogen (TN) and total phosphorus (TP)—specifically, TN of 3 mg/L and TP of 0.1 mg/L as opposed to TN of 8 mg/L and TP of 1 mg/L—the average potential for eutrophication could be decreased by up to 70%. Pan et al. [34] conducted an analysis to estimate the emissions of three greenhouse gases (CH_4, NO_2, and CO_2) from a vertical subsurface flow constructed wetland and two conventional WWTPs in Changzhou, China, utilizing LCA. Their findings revealed that for every kilogram of biological oxygen demand (BOD) removed, conventional WWTPs emit 7.3 kg CO_2 equivalent, while the vertical subsurface flow system emits only 3.18 kg CO_2-eq, half the amount emitted by WWTPs. This suggests that its implementation could significantly reduce GHG emission in the wastewater sector, both in conventional centralized systems and on-site greywater treatment. Additionally, their study indicated that the dry composting toilet system yielded the lowest eutrophication potential, whereas nutrients in wastewater effluent were found to have the greatest contribution to eutrophication potential. The authors underscored eco-digestion along with energy recovery as the most influential parameters in determining GWP results. Amini et al. [2] employed the TRACI 2.1 method to evaluate impact categories such as smog formation, human health, ozone depletion, acidification, global warming, and ecotoxicity. The utilization of this method for assessing the environmental impacts of wastewater treatment yielded positive results, as it considered eutrophication not only in the receiving water body but also in downstream water bodies.

3.3.5 Fossil Depletion

Abiotic resources encompass nonliving natural substances like copper ore, crude oil, and energy sources such as wind energy. One of the key impact categories under scrutiny is abiotic depletion potential (ADP), which evaluates the utilization of nonrenewable sources for energy production. This predominantly concerns the extraction of non-renewable raw materials. The unit of measurement for ADP is typically kg Sb eq. The primary concern within this category is the impact on human and ecosystem health stemming from mineral and fossil fuel extraction [8]. For each extraction event, an abiotic depletion factor is determined.

Niero et al. [31] conducted a comparative analysis using LCA tools alongside sensitivity and uncertainty analyses to evaluate the performance of four different sizes of WWTPs employing diverse treatment technologies. This comparison involved contrasting aerobic versus anaerobic treatments, as well as chemical versus combined chemical and biological treatments. Employing the ReCiPe method, the authors assessed the potential impact of various sludge disposal methods, including combustion with energy production versus agricultural application. The findings highlighted recycling phosphorus into agricultural soils as the most sustainable alternative for mitigating climate change and fossil depletion impacts. However, regarding eutrophication potential and toxicity impact categories, the uncertainty and sensitivity analyses revealed inconclusive results.

3.3.6 Photochemical Oxidation

The key photochemical oxidants in the atmosphere are NO_2, ozone, and peroxyacetyl nitrate, often categorized as secondary pollutants, stemming from the interaction of sunlight with byproducts of fuel combustion. Ozone may descend from the stratosphere through vertical winds or be formed during electrical storms. Notably, it is generated when sunlight reacts with hydrocarbons and nitrogen oxides, often originating from burning waste and combustion of fossil fuels, notably exhaust gases from internal combustion engines. Ozone is renowned for its crucial role in shielding life against ultraviolet radiation by forming the ozone layer. However, ground-level ozone concentrations are rising due to increased emissions of ozone precursors, particularly in polluted areas.

Traditionally, indirect emissions primarily the electricity consumption have contributed to the primary source of impacts such as global warming, acidification, or photochemical oxidation, particularly in intensive WWT technologies [26]. In addition, a small percentage is also contributed by other sources such as the production and consumption of chemicals also contribute to these impacts. In addressing this issue, Hernández-Padilla et al. [17] conducted a study to compare extended aeration and pond system technologies. They investigated the consequences of specific country electricity mixes for 22 Latin American and Caribbean countries, as opposed to an average mix for the entire region. It was observed that country-specific electricity mixes are important as it led to a variation in the selection of the best treatment technology for human health damage for 16 out of 22 countries (73%).

3.4 Interpretation

This stage of the life cycle entails the process of interpreting the gathered information. This involves doing a thorough and critical evaluation of the collected data, assessing the level of sensitivity of the data, and presenting the results in a clear and concise manner. The impacts estimated in the previous steps are corelated with the future actions which could alleviate the adverse impacts. Different scenarios are proposed such as an environmentally benign replacement of a chemical or a construction material utilized for wastewater treatment which could eventually reduce the overall impacts of a system. Another well know case scenario for LCA where transportation fuel contributes to a significant impact is the reduction in the distance of raw material transport. Similarly, an electrical mix composed of more renewable is expected to generate lower environmental burdens as compared to a fossil dependent electricity mix. Apart from scenario analysis, sensitivity analysis and uncertainty analysis are also carried out to take into consideration the variation in results owing to the variation in the input inventory.

4 Conclusion

Conventionally, wastewater treatment systems are evaluated using performance parameters such as organic matter removal, solids removal, nutrient removal, metal removal, etc. However, in the recent times the associated environmental impacts have been taken into consideration and LCA has emerged as an important tool to estimate the impacts associated with the treatment systems. Different studies have defined different system boundaries and functional units for evaluating various environmental parameters such as GWP, eutrophication potential, fossil depletion, ecotoxicity, ozone depletion potential, etc. This difference in system boundary and functional unit hinders the comparison of various LCA studies for a holistic review. Similarly, the use of different models for impact characterization such as ReCiPe,

Ecoindicator 99, USLCI, ILCD 2011, TRACI 2.1 is also a reason for not ensuring a comparative analysis. Although spatial boundaries are always defined, some inventory parameters are not available from all locations. For example, electricity mix is an important parameter which affects the impacts. However, in absence of a location specific electricity mix, an alternative electricity mix which may not be representative is utilized to perform the LCA. This will result in increasing the uncertainties for various impact categories. Hence, the lack of country specific databases can influence the environmental assessment. Moreover, it is possible that the inventory data used are also the source of some variation which are indicated in the life cycle impacts. Hence, it would be beneficial if some universal guidelines and standards are drafted specifically for wastewater treatment systems in addition to development of national and regional databases and inclusion of construction phase in the system boundary to achieve maximum inferences from LCA studies. Finally, the study of environmental impacts could be extended to social variables in social LCA and economic consideration in techno-economic study to arrive at a complete criteria-based assessment of wastewater treatment systems.

References

1. Alfonsín C, Hospido A, Omil F et al (2014) PPCPs in wastewater—update and calculation of characterization factors for their inclusion in LCA studies. J Clean Prod 83:245–255. https://doi.org/10.1016/j.jclepro.2014.07.024
2. Amini A, Kim Y, Zhang J et al (2015) Environmental and economic sustainability of ion exchange drinking water treatment for organics removal. J Clean Prod 104:413–421. https://doi.org/10.1016/j.jclepro.2015.05.056
3. Arias A, Feijoo G, Moreira MT (2020) Linking organic matter removal and biogas yield in the environmental profile of innovative wastewater treatment technologies. J Clean Prod 276:124292. https://doi.org/10.1016/j.jclepro.2020.124292
4. Chai C, Zhang D, Yu Y et al (2015) Carbon footprint analyses of mainstream wastewater treatment technologies under different sludge treatment scenarios in China. Water 7:918–938
5. Cornejo PK, Zhang Q, Mihelcic JR (2013) Quantifying benefits of resource recovery from sanitation provision in a developing world setting. J Environ Manage 131:7–15. https://doi.org/10.1016/j.jenvman.2013.09.043
6. Corominas L, Byrne DM, Guest JS et al (2020) The application of life cycle assessment (LCA) to wastewater treatment: a best practice guide and critical review. Water Res 184:. https://doi.org/10.1016/j.watres.2020.116058
7. Corominas L, Larsen HF, Flores-Alsina X, Vanrolleghem PA (2013) Including life cycle assessment for decision-making in controlling wastewater nutrient removal systems. J Environ Manage 128:759–767
8. Dincer I, Abu-Rayash A (2019) Energy sustainability. Academic Press
9. Durante S, Comoglio M, Ridgway N (2015) Life cycle assessment in nanotechnology, materials and manufacturing A2-Qin, Yi, chap 34. Micromanuf Eng Technol
10. Finnveden G, Moberg Å (2005) Environmental systems analysis tools—an overview. J Clean Prod 13:1165–1173. https://doi.org/10.1016/j.jclepro.2004.06.004

11. Fuchs VJ, Mihelcic JR, Gierke JS (2011) Life cycle assessment of vertical and horizontal flow constructed wetlands for wastewater treatment considering nitrogen and carbon greenhouse gas emissions. Water Res 45:2073–2081. https://doi.org/10.1016/j.watres.2010.12.021
12. Furness M, Bello-Mendoza R, Dassonvalle J, Chamy-Maggi R (2021) Building the 'bio-factory': a bibliometric analysis of circular economies and life cycle sustainability assessment in wastewater treatment. J Clean Prod 323:129127. https://doi.org/10.1016/j.jclepro.2021.129127
13. Gallego-Schmid A, Tarpani RRZ (2019) Life cycle assessment of wastewater treatment in developing countries: a review. Water Res 153:63–79. https://doi.org/10.1016/j.watres.2019.01.010
14. Gallego A, Hospido A, Moreira MT, Feijoo G (2008) Environmental performance of wastewater treatment plants for small populations. Resour Conserv Recycl 52:931–940. https://doi.org/10.1016/j.resconrec.2008.02.001
15. Guinee JB (2002) Handbook on life cycle assessment operational guide to the ISO standards. Int J Life Cycle Assess 7:311–313. https://doi.org/10.1007/BF02978897
16. Harun SN, Hanafiah MM, Noor NM (2022) Rice straw utilisation for bioenergy production: a brief overview. Energies 15
17. Hernández-Padilla F, Margni M, Noyola A et al (2017) Assessing wastewater treatment in Latin America and the Caribbean: enhancing life cycle assessment interpretation by regionalization and impact assessment sensibility. J Clean Prod 142:2140–2153. https://doi.org/10.1016/j.jclepro.2016.11.068
18. Hong J, Hong J, Otaki M, Jolliet O (2009) Environmental and economic life cycle assessment for sewage sludge treatment processes in Japan. Waste Manag 29:696–703. https://doi.org/10.1016/j.wasman.2008.03.026
19. Houillon G, Jolliet O (2005) Life cycle assessment of processes for the treatment of wastewater urban sludge: energy and global warming analysis. J Clean Prod 13:287–299. https://doi.org/10.1016/j.jclepro.2004.02.022
20. Huijbregts MAJ, Steinmann ZJN, Elshout PMF et al (2017) ReCiPe2016: a harmonised life cycle impact assessment method at midpoint and endpoint level. Int J Life Cycle Assess 22:138–147. https://doi.org/10.1007/s11367-016-1246-y
21. Jiménez-Benítez A, Vázquez JR, Seco A et al (2023) Life cycle assessment of AnMBR technology for urban wastewater treatment: a case study based on a demo-scale AnMBR system. J Environ Chem Eng 11:111141. https://doi.org/10.1016/j.jece.2023.111141
22. Kalbar PP, Karmakar S, Asolekar SR (2013) Assessment of wastewater treatment technologies: life cycle approach. Water Environ J 27:261–268. https://doi.org/10.1111/wej.12006
23. Laitinen J, Moliis K, Surakka M (2017) Resource efficient wastewater treatment in a developing area—climate change impacts and economic feasibility. Ecol Eng 103:217–225. https://doi.org/10.1016/j.ecoleng.2017.04.017
24. Li Y, Luo X, Huang X et al (2013) Life cycle assessment of a municipal wastewater treatment plant: a case study in Suzhou, China. J Clean Prod 57:221–227. https://doi.org/10.1016/j.jclepro.2013.05.035
25. Lutterbeck CA, Kist LT, Lopez DR et al (2017) Life cycle assessment of integrated wastewater treatment systems with constructed wetlands in rural areas. J Clean Prod 148:527–536. https://doi.org/10.1016/j.jclepro.2017.02.024
26. Machado AP, Urbano L, Brito AG et al (2007) Life cycle assessment of wastewater treatment options for small and decentralized communities. Water Sci Technol 56:15–22
27. McNamara G, Fitzsimons L, Horrigan M et al (2016) Life cycle assessment of wastewater treatment plants in Ireland. J Sustain Dev Energy, Water Environ Syst 4:216–233
28. Mohammed M, Sills DL (2022) Coupling a rotating biological contactor with an anaerobic baffled reactor for sustainable energy recovery from domestic wastewater. Environ Sci Water Res Technol 8:1822–1835. https://doi.org/10.1039/D2EW00260D
29. Morera S, Comas J, Poch M, Corominas L (2015) Connection of neighboring wastewater treatment plants: economic and environmental assessment. J Clean Prod 90:34–42. https://doi.org/10.1016/j.jclepro.2014.11.047

30. Muñoz I (2006) Life cycle assessment as a tool for green chemistry: application to different advanced oxidation processes for wastewater treatment
31. Niero M, Pizzol M, Bruun HG, Thomsen M (2014) Comparative life cycle assessment of wastewater treatment in Denmark including sensitivity and uncertainty analysis. J Clean Prod 68:25–35. https://doi.org/10.1016/j.jclepro.2013.12.051
32. Nizam NUM, Hanafiah MM, Mahmoudi E et al (2021) The removal of anionic and cationic dyes from an aqueous solution using biomass-based activated carbon. Sci Rep 11:8623. https://doi.org/10.1038/s41598-021-88084-z
33. Ontiveros GA, Campanella EA (2013) Environmental performance of biological nutrient removal processes from a life cycle perspective. Bioresour Technol 150:506–512. https://doi.org/10.1016/j.biortech.2013.08.059
34. Pan T, Zhu X-D, Ye Y-P (2011) Estimate of life-cycle greenhouse gas emissions from a vertical subsurface flow constructed wetland and conventional wastewater treatment plants: a case study in China. Ecol Eng 37:248–254. https://doi.org/10.1016/j.ecoleng.2010.11.014
35. Panigrahi S, Sharma HB, Tiwari BR et al (2021) Insight into understanding the performance of electrochemical pretreatment on improving anaerobic biodegradability of yard waste. Renew Energy. https://doi.org/10.1016/j.renene.2021.08.123
36. Pasqualino JC, Meneses M, Castells F (2011) Life cycle assessment of urban wastewater reclamation and reuse alternatives. J Ind Ecol 15:49–63. https://doi.org/10.1111/j.1530-9290.2010.00293.x
37. Rahman SM, Eckelman MJ, Onnis-Hayden A, Gu AZ (2016) Life-cycle assessment of advanced nutrient removal technologies for wastewater treatment. Environ Sci Technol. https://doi.org/10.1021/acs.est.5b05070
38. Renzoni R, Germain A (2007) Life cycle assessment of water: from the pumping station to the wastewater treatment plant (9 pp). Int J Life Cycle Assess 12:118–126. https://doi.org/10.1065/lca2005.12.243
39. Sabeen AH, Noor ZZ, Ngadi N et al (2018) Quantification of environmental impacts of domestic wastewater treatment using life cycle assessment: a review. J Clean Prod 190:221–233. https://doi.org/10.1016/j.jclepro.2018.04.053
40. Saini R, Tiwari BR, Brancoli P et al (2024) Environmental assessment of rhodosporidium toruloides-1588 based oil production using wood hydrolysate and crude glycerol. Bioresour Technol 393:130102. https://doi.org/10.1016/j.biortech.2023.130102
41. Sills DL, Wade VL, DiStefano TD (2016) Comparative life cycle and technoeconomic assessment for energy recovery from dilute wastewater. Environ Eng Sci 33:861–872. https://doi.org/10.1089/ees.2016.0153
42. Tiwari BR, Bhar R, Dubey BK et al (2023) Life cycle assessment of microbial 2,3-butanediol production from brewer's spent grain modeled on pinch technology. ACS Sustain Chem Eng. https://doi.org/10.1021/acssuschemeng.3c00616
43. Tiwari BR, Brar SK (2024) A Life cycle assessment and economic perspective of biohydrogen production. In: Biohydrogen-advances and processes. Springer, pp 547–564
44. Tiwari BR, Brar SK (2022) a life cycle assessment perspective to conventional and modular wastewater treatment, chap 11. In: Kaur Brar S, Kumar P, Cuprys ABT-MTA for DW and W (eds). Elsevier, pp 187–205
45. Vallero DA (2019) Air pollution biogeochemistry. Air Pollut Calc 175–206
46. Wang D, Guo F, Wu Y et al (2018) Technical, economic and environmental assessment of coagulation/filtration tertiary treatment processes in full-scale wastewater treatment plants. J Clean Prod 170:1185–1194. https://doi.org/10.1016/j.jclepro.2017.09.231
47. Yang WH, Hall SJ, McNicol G (2021) Global gases. In: Principles and applications of soil microbiology. Elsevier, pp 557–579
48. Yilmaz M, Guven H, Ozgun H et al (2024) The application of Life cycle assessment (LCA) to anaerobic technologies for the treatment of municipal wastewater: a review. Process Saf Environ Prot 182:357–370. https://doi.org/10.1016/j.psep.2023.11.078
49. Yoshida H, Clavreul J, Scheutz C, Christensen TH (2014) Influence of data collection schemes on the life cycle assessment of a municipal wastewater treatment plant. Water Res 56:292–303. https://doi.org/10.1016/j.watres.2014.03.014

50. Zhang QH, Wang XC, Xiong JQ et al (2010) Application of life cycle assessment for an evaluation of wastewater treatment and reuse project—case study of Xi'an, China. Bioresour Technol 101:1421–1425. https://doi.org/10.1016/j.biortech.2009.05.071
51. Zhu L, Liu B, Wang F, Bi J (2013) Raising discharge standards leads to environmental problem shifting in China. Water Sci Technol 68:2605–2612

SPRINGER NATURE

GPSR Compliance

The European Union's (EU) General Product Safety Regulation (GPSR) is a set of rules that requires consumer products to be safe and our obligations to ensure this.

If you have any concerns about our products, you can contact us on ProductSafety@springernature.com

In case Publisher is established outside the EU, the EU authorized representative is:

Springer Nature Customer Service Center GmbH
Europaplatz 3
69115 Heidelberg, Germany

The manufacturer's authorised representative in the EU is Springer Nature Customer Service Centre GmbH, Europaplatz 3, 69115 Heidelberg, Germany. If you have any concerns regarding our products, please contact ProductSafety@springernature.com

Printed and bound by CPI Group (UK) Ltd, Croydon, CR0 4YY
01/10/2026
02225966-0015